The Nutrition of the Rabbit

Edited by

C. de Blas
Departamento de Producción Animal
Universidad Politécnica
Madrid, Spain

and

J. Wiseman
Division of Agriculture and Horticulture
University of Nottingham
Nottingham, UK

CABI *Publishing*

CABI is a trading name of CAB International

CABI Head Office
Nosworthy Way
Wallingford
Oxfordshire OX10 8DE
UK

Tel: +44 (0)1491 832111
Fax: +44 (0)1491 833508
Email: cabi@cabi.org
Web site: www.cabi.org

CABI North American Office
875 Massachusetts Avenue
7th Floor
Cambridge, MA 02139
USA

Tel: +1 617 395 4056
Fax: +1 617 354 6875
Email: cabi-nao@cabi.org

A catalogue record for this book is available from the British Library,
London, UK.

Library of Congress Cataloging-in-Publication Data
The nutrition of the rabbit / edited by C. de Blas and J. Wiseman.
    p.   cm
    Includes bibliographical references and index.
    ISBN 0-85199-279-X (alk. paper)
    1.   Rabbits- - Feeding and feeds. I. Blas, C. de (Carlos de)
II. Wiseman J. (Julian)
SF454.N88 1998
636.932′2- -dc21   98-7573
                                                                                     CIP

ISBN-13: 978-0-85199-279-2
ISBN-10: 0- 85199-279-X

First published 1998
Reprinted 2003
Transferred to print on demand 2006

Printed and bound by CPI Antony Rowe, Eastbourne

Contents

Contributors

D. Allain, *INRA Centre de Toulouse, Station d'Amélioration Génétique des Animaux, BP 27, 31326 Castanet-Tolosan, France.*

E. Blas, *Departamento de Ciencia Animal, Universidad Politécnica de Valencia, Camino de Vera, Apdo 22012, 46071 Valencia, Spain.*

R. Carabaño, *Departamento de Producción Animal, ETS Ingenieros Agronomos, Universidad Politécnica de Madrid, Ciudad Universitaria, 28040 Madrid, Spain.*

C. Cervera, *Departamento de Ciencia Animal, Universidad Politécnica, Camino de Vera, Apdo 22012, 46071 Valencia, Spain.*

C. de Blas, *Departamento de Producción Animal, ETS Ingenieros Agronomos, Universidad Politécnica de Madrid, Ciudad Universitaria, 28040 Madrid, Spain.*

J. Fernández Carmona, *Departamento Ciencia Animal, Universidad Politécnica de Valencia, Camino de Vera, Apdo 22012, 46071 Valencia, Spain.*

M.J. Fraga, *Departamento de Producción Animal, ETS Ingenieros Agrónomos, Universidad Politécnica de Madrid, Ciudad Universitaria, 28040 Madrid, Spain.*

J. García, *Departamento de Producción Animal, ETS Ingenieros Agronomos, Universidad Politécnica de Madrid, Ciudad Universitaria, 28040 Madrid, Spain.*

T. Gidenne, *Station de Recherches Cunicoles, INRA Centre de Toulouse, BP 27, 31326 Castanet-Tolosan Cedex, France.*

F. Lebas, *Station de Recherches Cunicoles, INRA Centre de Toulouse, BP 27, 31326 Castanet-Tolosan Cedex, France.*

D. Licois, *INRA Station de Pathologie Aviai et de Parasitologie, Laboratoire de Pathologie du Lapin, 37380 Nouzilly, France.*

J.A. Lowe, *Gilbertson & Page Ltd, PO Box 321, Welwyn Garden City, Hertfordshire AL7 1LF, UK.*

L. Maertens, *Agricultural Research Centre–Ghent, Rijksstation voor Kleinveeteelt, Burg. Van Gansberhelaan 92, 98290 Merelbeke, Belgium.*

G.G. Mateos, *Departamento de Producción Animal, ETS Ingenieros Agronomos, Universidad Politécnica de Madrid, Ciudad Universitaria, 28040 Madrid, Spain.*

J. Mendez, *Cooperativas Orensanas Sociedad Cooperativa Ltda, Juan XXIII 33, 32003 Orense, Spain.*

J. Ouhayoun, *Station de Recherches Cunicoles, INRA Centre de Toulouse, BP 27, 31326, Castanet-Tolosan Cedex, France.*

R. Parigi Bini, *Dipartimento di Scienze Zootecniche, University of Padua, Agripolis, 35020 Legnaro (Padova), Italy.*

J.M. Perez, *Station de Recherches Cunicoles, INRA Centre de Toulouse, BP 27, 31326, Castanet-Tolosan Cedex, France.*

J. Piquer, *Pfizer Salud Anima, C/Principe Vergara 108, 28002 Madrid, Spain.*

E. Rial, *Cooperativas Orensanas Sociedad Cooperativa Ltda, Juan XXIII 33, 32003 Orense, Spain.*

G. Santomá, *Agrovic, C/Mejia Lequerica 22-24, 08028 Barcelona, Spain.*

R.G. Thébault, *INRA Centre Poitou-Charentes, U.E. Génétique Animale Phanères, Domaine du Magneraud, BP 52, 17700 Surgères, France.*

M.J. Villamide, *Departamento de Producción Animal, ETS Ingenieros Agrónomos, Universidad Politécnica de Madrid, Ciudad Universitaria, 28040 Madrid, Spain.*

G. Xiccato, *Dipartimento di Scienze Zootecniche, University of Padua, Agripolis, 35020 Legnaro (Padova), Italy.*

Preface

In the last 20 years, rabbit production has become an increasingly intensive system, such that productivity is now equivalent to that obtained in other intensively farmed species.

The importance of nutrition has increased significantly as feed costs, pathological conditions associated with energy and nutrient deficiencies, and considerations of product quality have become limiting factors to economic output from a unit.

The rabbit is unique. It requires a high daily nutrient and energy intake but, because it is a herbivore, it also needs a diet with a high concentration of fibre to ensure optimum performance and, in addition, to minimize the incidence of digestive disorders.

Diets of rabbits are closer to those of dairy cows than to other intensive meat producers such as pigs or poultry. This means use of a wider range of raw materials (forages, but also those with high concentration of energy and nutrients) and greater complexity in both formulation of optimum diets and the overall feed manufacturing process.

Furthermore, the unusual digestive physiology includes several characteristics such as the mechanism of particle separation at the ileo-caecal junction and the recycling of soft faeces through caecotrophy, both of which have specific nutritional and pathological implications.

The objective of this book has been to update the wealth of scientific information on rabbit feeding and nutrition. The chapters have been written by distinguished research workers from around the world who are recognized specialists in their field. The contents cover the physiological basis of nutrition, nutrient requirements, feeding value and management, feed manufacturing, interaction of nutrition with environment, pathology and carcass quality. The final two chapters have been devoted to Angora and pet rabbits.

Abbreviations

ADL	acid detergent lignin	IDF	indigestible dietary fibre
AFB_1	aflatoxin B_1	INRA	Institut Nationale de la Recherche Agronomique
ANF	antinutritive factors		
ASESCU	Asociacion Española Cunicultura Cientifica	LA	linoleic acid
		LCT	lower critical temperature
CCW	caecal contents weight	LW	live weight
CF	crude fibre	ME	metabolizable energy
CP	crude protein	MEI	metabolizable energy intake
CPD	crude protein digestibility	MEn	metabolizable energy corrected to N equilibrium
CT	computerized tomography		
CV	coefficient of variation	MRT	mean retention time
DDP	dietary digestible protein	N-ADF	N bound to acid detergent fibre
DE	digestible energy	NDF	neutral detergent fibre
DF	dietary fibre	NE	net energy
DM	dry matter	NEFA	non-esterified fatty acids
d.p.	degree of polymerization	NMR	nuclear magnetic resonance
DP	digestible protein	NSP	non-starch polysaccharide
DWG	daily weight gain	PCW	plant cell walls
EAA	essential amino acids	PTH	parathyroid hormone
EBG	empty body gain	PUFA	polyunsaturated fatty acids
EE	ether extract	RE	retained energy
EEd	ether extract digestibility	SAA	sulphur amino acids
EFA	essential fatty acids	SFA	saturated fatty acids
EGRAN	European Group on Rabbit Nutrition	THI	temperature–humidity index
		TNZ	thermoneutral zone
FA	fatty acids	Tobec	total body electrical conductivity
FCR	feed conversion ratio		
FE	faecal energy	TT	transit time
GasE	intestinal fermentation energy associated with gas production	UCT	upper critical temperature
		UE	urine energy
		UFA	unsaturated fatty acids
GE	gross energy	UN	urinary N
HE	heat energy	VFA	volatile fatty acids
HI	heat increment	VFI	voluntary feed intake
ICPD	ileal digestibility coefficient of crude protein		

1. The Digestive System of the Rabbit

R. Carabaño[1] and J. Piquer[2]

[1]*Departamento de Producción Animal, Universidad Politécnica de Madrid,
ETS Ingenieros Agronomos, Ciudad Universitaria, 28040 Madrid, Spain;
and* [2]*Pfizer Salud Anima, C/Principe Vergara 108, 28002 Madrid, Spain*

Introduction

The digestive system of the rabbit is characterized by the relative importance of the caecum and colon when compared with other species (Portsmouth, 1977). As a consequence, the microbial activity of the caecum is of great importance for the processes of digestion and nutrient utilization. Furthermore, caecotrophy, the behaviour of ingestion of soft faeces of caecal origin, makes microbial digestion in the caecum more important for the overall utilization of nutrients by the rabbit. Additionally, the rabbit has developed a strategy of high feed intake (65–80 g kg^{-1} body weight (BW)) and a rapid transit of feed through the digestive system to meet nutritional requirements.

To reach its full functional capacity, the digestive system of the growing rabbit must go through a period of adaptation from a milk-base feeding to the sole dependence on solid feed without milk or its by-products. It is intended in this chapter: (i) to give a general and brief description of the morphological and functional characteristics of the digestive system of the rabbit that may be important for understanding the digestive processes explained in the following chapters; and (ii) to explain how these characteristics change from the time of weaning until attainment of maturity.

The digestive system of the rabbit

The first important compartment of the digestive system of the rabbit is the stomach, which has a very weak muscular layer and is always partially filled. After caecotrophy the fundic region of the stomach acts as a storage cavity for caecotrophs. Thus, the stomach is continuously secreting and the pH is acid. The stomach pH ranges from 1 to 5, depending on site of determination (fundus vs. cardiac–pyloric region), the presence or absence of soft faeces (Griffiths and Davies, 1963), the time from the feed intake (Alexander and

Chowdhury, 1958) and the age of the rabbit (Grobner, 1982). The lowest figures (from 1 to 2.5) are determined in the cardiac region, in the absence of soft faeces, after 4 h of diet ingestion, and rabbits older than 5 weeks. The capacity of the stomach is about 0.34 of the total capacity of the digestive system (Portsmouth, 1977). The stomach is linked with a coiled caecum by a small intestine approximately 3 m long where the secretion of bile, digestive enzymes and buffers occurs. The pH of the small intestine is close to 7 (Vernay and Raynaud, 1975). The caecum is characterized by having a weak muscular layer and contents with a dry matter of 200 g kg^{-1}. The pH of the caecal contents is slightly acid (5.6–6.2) (Candau et al., 1986; Carabaño et al., 1988). The capacity of the caecum is approximately 0.49 of the total capacity of the digestive tract (Portsmouth, 1977). The colon can be divided in two portions, the proximal colon (approximately 35 cm long) and the distal colon (80–100 cm long). The proximal colon can be further divided into three segments: the first segment possesses three taeniae with the formation of haustra between them, while the second segment has a single taenia covering half of the circumference of the digestive tube, and the third segment or fusus coli has no taeniae or haustra but is densely enervated. Thus, it acts as a pacemaker for the colon during the phase of hard faeces formation (Snipes et al., 1982).

Age-related changes in the morphology and function of the digestive system of the rabbit

The different segments of the digestive system of the rabbit grow at different rates until reaching maturity. The capacity for milk intake increases threefold from the time of birth until the peak of milk production (12–35 g milk day^{-1}). Caecum and colon develop faster than the rest of the body from 3 to 7 weeks of age whereas the relative size of intestine and stomach decreases from 3 to 11 weeks of age (Fig. 1.1; Lebas and Laplace, 1972). The fast growth of the caecum during this period is more evident if the caecal contents are included. Caecum and caecal contents reach a peak of about 0.06 of total body weight at 7–9 weeks of age. The pH of the caecum is also affected by age and decreases from 6.8 at 15 days of age to 5.6 at 50 days of age (Padilha et al., 1995).

Very marked changes also occur in the activity of the different digestive enzymes. In the 4-week-old rabbit, the activity of gastric lipase represents most of the lipolytic activity of the whole digestive tract, whereas this activity is not detectable in the 3-month-old rabbit (Marounek et al., 1995). As the activity of gastric lipase decreases, pancreatic lipase activity increases, both when expressed as specific activity (μmol of substrate degraded per unit of time and mg of protein) or as total activity (μmol of substrate degraded per unit of time for the whole organ) after 14 days of age. Prior to this age, the specific activity is constant or increases slightly (Lebas et al., 1971; Corring et al., 1972).

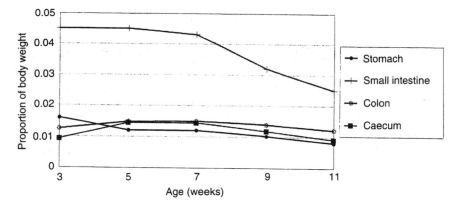

Fig. 1.1. Development of different segments of the digestive system of the rabbit from 3 to 11 weeks (Lebas and Laplace, 1972).

The main proteolytic activity is also localized in the stomach of the young rabbit and its importance decreases with age as proteolytic activity in the caecum, colon and pancreas increases (Marounek *et al.*, 1995). In the case of the pancreatic enzymes trypsin and chymotrypsin, their total activity increases markedly after 32 and 21 days of age, respectively (Corring *et al.*, 1972). However, the specific activities of these two enzymes decrease during the period 1–43 days of age (Lebas *et al.*, 1971).

The other main enzyme activity at the pancreatic level is amylase. The age-related changes in the activity of this enzyme are similar to those of the pancreatic lipase, with marked increases both in specific and total activities after 14 days of age and a slight decrease in specific activity from 1 to 14 days of age (Lebas *et al.*, 1971; Corring *et al.*, 1972). The carbohydrase activity of the pancreas is complemented by the activities of disaccharidases located mainly in the small intestine. Lactase activity decreases with age whereas that of invertase and maltase increases (Marounek *et al.*, 1995). Other enzyme activities that increase markedly with the age of the rabbit are those due to the presence of microorganisms that will determine the ability of the rabbit to utilize fibre sources. Cellulase, pectinase, xylanase and urease are some of the main activities provided by the intestinal microflora.

Role of the intestinal flora in the digestion and absorption of nutrients by the rabbit

The presence of the microbial population in the caecum, together with caecotrophy, permits the rabbit to obtain additional energy, amino acids and vitamins. The main genus of the microbial population in the caecum of the adult rabbit is *Bacteroides* (Gouet and Fonty, 1973). The *Bacteroides* population comprises 10^9–10^{10} bacteria g^{-1} and other genera such as *Bifidobacterium, Clostridium, Streptococcus* and *Enterobacter* complete this population to give a bacterial load of 10^{10}–10^{12} bacteria g^{-1} (Bonnafous and Raynaud, 1970; Gouet and Fonty, 1979; Forsythe and Parker, 1985; Penney *et al.*, 1986; Cortez *et al.*, 1992).

The presence of cellulolytic bacteria in the caecum of the rabbit has already been indicated by Hall (1952) and Davies (1965). Later, Emaldi *et al.* (1979) studied the enzymatic activities of this microflora and indicated that the main activities were, in decreasing order, ammonia-use, ureolytic, proteolytic and cellulolytic. The great importance of other activities, i.e. xylanolytic and pectinolytic, has been indicated in studies conducted by Forsythe and Parker (1985) and Marounek *et al.* (1995). Forsythe and Parker (1985) estimate populations of 10^8 and 10^9 xylanolytic and pectinolytic bacteria, respectively. The composition of the microflora does not remain constant throughout the life of the rabbit and is strongly influenced by the time of weaning (Padilha *et al.*, 1996). During the first week of age, the digestive system of the rabbit is colonized by strict anaerobes, predominantly *Bacteroides*. At 15 days of age, the numbers of amylolytic bacteria seem to be stabilized, whereas those of colibacilli decrease as the numbers of cellulolytic bacteria increase (Padilha *et al.*, 1995). However, milk intake may delay the colonization by cellulolytic flora but does not seem to affect the evolution of the population of colibacilli (Padilha *et al.*, 1996). As a consequence of the age-related changes in the microbial population, production of volatile fatty acids (VFA) increases with age (Bellier *et al.*, 1995; Padilha *et al.*, 1995). Moreover, as caecotrophy is initiated, the presence of bacteria of caecal origin can be detected. Smith (1965) and Gouet and Fonty (1979) were able to detect precaecal microbial flora after only 16 and 17 days of age, respectively. The presence of these precaecal microbes is dependent on caecotrophy, with high counts after caecotrophy and no viable cells after 5–6 h (Jilge and Meyer, 1975). The composition of the microflora does not remain constant during the life of the rabbit.

As a result of the fermentative activity of the microflora, VFA are produced in the proportion of 60–80 moles of acetate, 8–20 moles of butyrate, and 3–10 moles of propionate per 100 moles of VFA (Gidenne, 1996). However, this proportion changes with the time of the day, as described in the caecotrophy section of this chapter, and with the developmental stage of the rabbit, with increases in the acetate concentration from 15 to 25 days of age and a reversal of the propionate to butyrate ratio from 15 to 29 days of age (Padilha *et al.*, 1995). The potential of modification of VFA production by dietary changes will be described in the following chapters of this book. According to Marty and Vernay (1984), VFA can be metabolized in the hindgut tissues, with butyrate being the preferred substance for the colonocytes. The liver is the main organ metabolizing absorbed propionate and butyrate. However, acetate is available for extrahepatic tissue metabolism. It is estimated that the rabbit obtains up to 0.40 of its maintenance energy requirement from VFA produced by fermentation in the hindgut (Parker, 1976; Marty and Vernay, 1984).

Caecotrophy

Patterns of daily feed intake and soft faeces excretion

Soft faeces are excreted according to a circadian rhythm which is the opposite of that of feed intake and hard faeces excretion. Caecotrophy occurs mainly during the light period, whereas feed intake and hard faeces excretion occur during darkness (Lebas and Laplace, 1974, 1975; Fioramonti and Ruckebush, 1976; Ruckebush and Hörnicke, 1977; Battaglini and Grandi, 1988; Merino, 1994; Bellier *et al.*, 1995; Bellier and Gidenne, 1996; El-Adawy, 1996). Figure 1.2 shows the pattern of faeces excretion and feed intake for adult rabbits under a schedule of 12 h light/12 h dark and *ad libitum* access to feed (Carabaño and Merino, 1996). Most of the rabbits showed monophasic patterns of soft faeces excretion from 08.00 to 17.00 h, with a maximum at 12.00 h. However, 0.25 of them showed a diphasic pattern, with a second period of excretion during the night. The occurrence of diphasic patterns is more frequent when the length of the light period is reduced. Under continuous light conditions (24 h) caecotrophy runs freely and mono-phasically (Jilge, 1982). During the caecotrophy period, lasting from 7 to 9 h, there is an absence of hard faeces excretion and the feed intake is low.

Feed intake and hard faeces excretion occur along the complementary period, showing two phases (Fig. 1.2). Feed intake increases from 15.00 to 18.00 h and then remains high until 24.00 h. After this period, rabbits reduce feed intake until 02.00 h and then a new phase starts, with a maximum at 06.00 h. The second phase finishes at 08.00 h. Hard faeces excretion (from 18.00 to 08.00 h) shows a similar pattern, with two maxima at 24.00 h and 06.00 h.

The age of the rabbits, their physiological status or restricted access to feed can alter this pattern. Bellier *et al.* (1995) observed that weaned rabbits (6 weeks old) show a greater incidence of diphasic patterns and a longer caecotrophy period than adults (14 weeks old) from 04.00 to 12.00 h and from 22.00 to 24.00 h vs. from 08.00 to 14.00 h, respectively. Lactating does show a different pattern of excretion from that described previously for non-lactating adult rabbits. During the lactation period, does exhibit an alternated rhythm of soft and hard faeces excretion. Caecotrophy occurs during two periods from 02.00 to 09.00 h (0.40 of total excretion) and from 13.00 to 17.00 h (0.60 of total excretion), with a lack of excretion from 09.00 to 13.00 h (Lorente *et al.*, 1988). This pattern could be mainly related to maternal behaviour of does through the morning rather than to physiological status. All the experiments described above were carried out with *ad libitum*

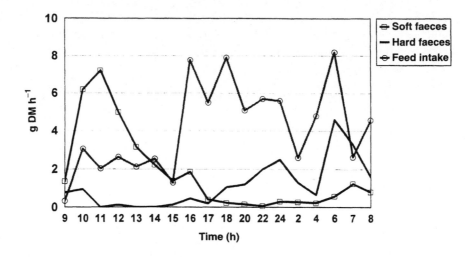

Fig. 1.2. Soft and hard faeces excretion and dry matter intake throughout the day (Carabaño and Merino, 1996).

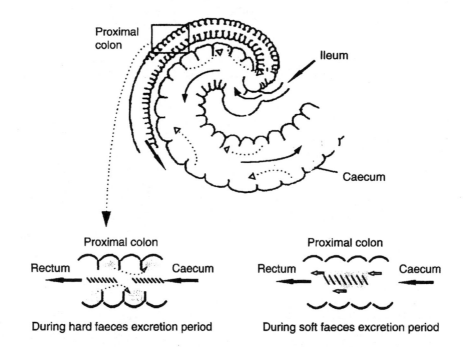

Fig. 1.3. Movement of digesta in the rabbit ileo-caeco-colic segment. Courtesy of T. Gidenne. ▨ liquids and fine particles; \\\\\\ large particles (>300 µm); → peristaltic digesta movement; ⇒ antiperistaltic digesta movement.

access to feed. When the feeding regime is changed from *ad libitum* to a restricted access to feed the rhythm of excretion is profoundly altered, whatever the length of the light period. In these situations, the time for soft faeces excretion depends on the time of feed distribution (Fioramonti and Ruckebush, 1976). Disruption of the internal cycle may have important practical implications. Lebas and Laplace (1975) recommend distributing the feed once per day late in the afternoon. In other situations (one meal at 09.30 h, or two meals at 09.30 and 16.30 h), changes in faecal excretion patterns and a lower growth rate should be expected.

Determination of soft faeces excretion and consumption

Several authors have tried to explain the physiological mechanisms that determine the differentiation and recognition of the two types of faeces in rabbits, according to the circadian patterns described above. The results obtained allow a partial understanding of the complex regulation of this behaviour.

Differentiation between soft and hard faeces begins during the transit of digesta through the caecum and proximal colon. From the results obtained by Björnhag (1972) and Pickard and Stevens (1972) it can be assumed that the formation of hard faeces is not by resorption of some components of caecal contents in the colon, but by mechanical separation of the different components of digesta. As is shown in Fig. 1.3, during hard faeces excretion, water-soluble substances and fine particles (smaller than 0.3 mm diameter) (including microorganisms) are brought back to the caecum by means of antiperistaltic movements and retrograde flow. Coarse particles (larger than 0.3 mm diameter) pass to the distal part of the colon. In contrast, the motility of both the caecal base and proximal colon decreases during the formation of soft faeces (Ruckebush and Hörnicke, 1977). Endogenous prostaglandins play an important role in the motor function involved in soft faeces formation. The infusion of both PGE_2 and $PGF_{2\alpha}$ produces an inhibition of proximal colon movements, a stimulation of the distal colon and is followed by soft faeces production (Pairet *et al.*, 1986). Changes in VFA concentrations and caecal pH occurring after a meal have been proposed as primary signals leading to a period of soft faeces excretion. Ruckebush and Hörnicke (1977) observed soft faeces excretion after an intracaecal infusion of VFA in rabbits with restricted access to feed. However, postprandial VFA variations are not so evident in rabbits fed *ad libitum*, and therefore factors other than those mentioned above could also be implicated. Structures typically involved in feed intake regulation, such as lateral hypothalamus and the hypothalamic ventromedian nodes do not seem to have the same roles as those described in other non-ruminant species. Damage to these structures does not imply changes in feed intake behaviour in rabbits (Gallouin, 1984).

During soft faeces excretion, caecal contents are covered by a mucous envelope secreted at the proximal colon according to the described circadian rhythms. Therefore, the soft faeces consist of small pellets of 5 mm diameter that rabbits can recognize. Soft faeces are taken directly from the anus, swallowed without mastication, and stored intact in the fundus of the stomach for 3–6 h (Gidenne and Poncet, 1985). The mechanisms of recognition are unclear. The special smell of soft faeces compared with that of hard faeces or the existence of mechanoreceptors in the rectum have been proposed as factors involved in reingestion of soft faeces. However, results obtained from rabbits deprived of olfactory bulbs and supporting an artificial anus which bypasses the rectum show that rabbits are still able to recognize and reingest soft faeces (Gallouin, 1984).

Nutritional implications

Caecotrophy in rabbits does not occur as a response to a nutritional imbalance, but represents a specialized digestive strategy. Caecotrophy in rabbits begins at 3 weeks of age, when rabbits begin to consume solid food. In postweaned rabbits (4 weeks old), soft faeces production linearly increases with age, showing a maximum at 63–77 days old (25 g dry matter (DM) day^{-1}). This period corresponds to the maximum growth requirements and to the greatest increment in feed intake. From 77 to 133 days old (2.5 vs. 3.9 kg, respectively) growth rate decreases, feed intake slightly increases and soft faeces excretion is stabilized (20 g DM day^{-1}) (Gidenne and Lebas, 1987). Similar figures (21.8 g DM day^{-1}) have been reported for adult females during pregnancy. However, lactating does showed greater soft faeces production (34 g DM day^{-1}) related to the higher feed intake (Lorente et al., 1988). In these situations caecotrophy represents from 0.09 to 0.15 of total DM intake (feed intake + soft faeces). The importance of caecotrophy also varies with the nutritive characteristics of diet as will be discussed in the following chapters.

As a consequence of the mechanical separation of digesta at the caecum and proximal colon, the chemical composition of soft faeces is similar to that of the caecal contents but quite different from that of hard faeces (Table 1.1). Soft faeces contain greater proportions of protein, minerals and vitamins than hard faeces. In contrast with this, hard faeces are enriched in fibrous components compared with soft faeces. As far as nutrient supply through soft faeces is concerned, protein represents from 0.15 to 0.22 of the total daily protein intake, either in growing rabbits or lactating does. Protein of soft faeces is high in essential amino acids such as lysine, sulphur amino acids or threonine (Proto, 1976; Spreadbury, 1978; C. de Blas, personal communication) which represents from 0.10 to 0.23 of total intake. The importance of these amino acids depends on the efficiency of microbial protein synthesis. The proportion of microbial protein with respect to total protein of soft faeces varies with the diet from 0.30 to 0.60 (Spreadbury, 1978; García et al., 1995). Microbial activity is also responsible for the high content of K and B vitamins in soft faeces.

Table 1.1. Average chemical composition of caecal contents and soft and hard faeces.

	Caecum	Soft faeces	Hard faeces	Ref.
Dry matter (g kg⁻¹)	200	340	470	3,4,5,6,7
Crude protein (g kg⁻¹ DM)	280	300	170	3,4,5,6,7
Crude fibre (g kg⁻¹ DM)	170	180	300	3,4,5,6,7
MgO (g kg⁻¹ DM)		12.8	8.7	2
CaO (g kg⁻¹ DM)		13.5	18.0	2
Fe₂O₃ (g kg⁻¹ DM)		2.6	2.5	2
Inorganic P (g kg⁻¹ DM)		10.4	6.0	2
Organic P (g kg⁻¹ DM)		5.0	3.5	2
Cl⁻ (mmol kg⁻¹ DM)		55	33	2
Na⁺ (mmol kg⁻¹ DM)		105	38	2
K⁺ (mmol kg⁻¹ DM)		260	84	2
Bacteria (10¹⁰ g⁻¹ DM)		142	31	2
Nicotinic acid (mg kg⁻¹)		139	40	1
Riboflavin (mg kg⁻¹)		30	9	1
Panthotenic acid (mg kg⁻¹)		52	8	1
Cianocobalamine (mg kg⁻¹)		3	1	1

1, Kulwich *et al.* (1953); 2, adapted from Hörnicke and Björnhag (1980); 3, Carabaño *et al.* (1988); 4, Carabaño *et al.* (1989); 5, Fraga *et al.* (1991); 6, Motta-Ferreira *et al.* (1996); 7, Carabaño *et al.* (1997).

In conclusion, caecotrophy could overcome poor quality protein or low vitamin diets in traditional rearing conditions, but it is necessary to supply extra B vitamins, minerals and limiting amino acids in intensive rearing conditions.

Methodological implications of caecotrophy on physiological research work

The marked circadian rhythms of caecotrophy and feed intake imply changes both in the organ content weights and the chemical composition of their contents throughout the day. These circumstances make it necessary to take into account the sampling time in the experimental procedures to obtain reliable digestibility data. The lack of homogeneity in the sampling procedure between different studies leads to difficulties in making comparisons and considerable misunderstanding. The diurnal variations of the main physiological parameters will now be summarized.

Weight and chemical composition of the organ contents

The weight of the stomach and caecal contents reflects the diurnal rhythm of intake and soft faeces production. Stomach contents show greater weights during the morning than during the night. The opposite is found for the weight of caecal contents. Diurnal differences in the weight of caecal and stomach contents up to 20% and 30%, respectively, can be observed (Fraga et al., 1984; Gidenne and Lebas, 1987).

Differences in the origin of stomach contents are explained by the diurnal changes in chemical composition of stomach, duodenum, jejunum and ileum contents. Intact soft faeces in the stomach have been detected from 09.00 to 18.00 h (Gidenne and Poncet, 1985; Carabaño et al., 1988), representing about a half of the total weight of the stomach contents. During the complementary period, the stomach only contains food. Protein content of precaecal digesta is the chemical parameter more affected by sampling time, showing greater values (from 50 to 100%) during the soft faeces excretion period (Catala, 1976; Gidenne and Poncet, 1985; Merino, 1994). The same tendency has been observed for the chemical composition of colonic and rectal contents. However, the protein concentration of caecal contents remains stable throughout the day.

Ileal digestibility

The use of cannulated animals to determine ileal digestibility requires the use of markers to estimate the ileal flow of DM and the need to obtain an ileal sample which is representative of that present throughout the day. Merino (1994) observed in cannulated animals a diurnal variation in the crude protein (CP) content of ileal digesta, showing greater values during the soft faeces excretion period than during the hard faeces excretion period (180 vs. 120 g CP kg^{-1} DM). When caecotrophy was prevented, no variation was detected in the protein content of ileal digesta (average value 120 g kg^{-1} DM). These results suggest that it is essential to take samples throughout the day to estimate the average composition of ileal digesta. No diurnal changes were detected in the marker concentration or fibre content of ileal digesta.

Fermentation patterns

The results obtained by Fioramonti and Ruckebusch (1976) and Gidenne and Bellier (1992) in adult animals showed that VFA concentration in caecal contents depends on time of feeding, rising to a maximum 5 h after feeding. In weaned (4 weeks old) or growing rabbits (9 weeks old) fed ad libitum, diurnal differences in VFA concentrations and caecal pH of 50 and 10%, respectively, can be observed (Gidenne, 1986; Bellier et al., 1995; Bellier and Gidenne, 1996). Caecal VFA concentrations are greater during the hard faeces than during the soft faeces excretion period. According to Bellier et al. (1995), this increment could have two causes: (i) the greater flow of substrate to the caecum related to an increase in feed intake during this period, and (ii) enrichment of the microbial population as a consequence of antiperistaltic movements of the proximal colon. Caecal pH varies inversely to the increase in VFA concentration. Smaller values of caecal pH were observed during hard faeces excretion. In consequence, it is preferable to take the caecal samples during the hard faeces excretion period.

Transit time

Giving a marker as simple doses is the most frequent procedure used in transit time studies. The question is at what time must the doses be administered? According to Laplace and Lebas (1975), doses given before the caecotrophy period lead to a higher mean retention time (3–4 h) compared with doses given after caecotrophy. This effect can be explained by an increase in time before the first appearance of the marker in hard faeces. According to Jilge (1974), the time for the first appearance of the marker in faeces is the same (4 or 5 h) for doses given before or after the caecotrophy period. However, depending on the time of administration, the marker changes the site of its first appearance (soft or hard faeces) and, in consequence, the opportunity to detect it.

According to these results, and taking into account the fact that feed intake starts just after the caecotrophy period, the hard faeces excretion period is recommended as the best time for marker administration.

Rate of passage

The capacity of the rabbit to digest its feed depends not only on endogenous enzyme activities and digestion by the microbial population but also on the rate of passage of the feed. The passage of feed through the stomach of the rabbit and caecum is relatively slow and varies between 3–6 and 4–9 h, respectively, as measured by the technique of comparative slaughter (Gidenne and Poncet, 1985). However, transit is very fast in the small intestine. Estimated retention times in the jejunum and ileum are 10–20 and 30–60 min, respectively (Lebas, 1979). Taking into account the entire digestive tract, the mean retention time varies from 9 to 30 h, with an average of 19 h (Laplace and Lebas, 1975, 1977; Udén *et al.*, 1982; Fraga *et al.*, 1984; Ledin, 1984). More recently, with rabbits cannulated at the ileum, the mean retention times for the ileo-rectal and oro-ileal segments, and for the stomach, have been calculated as 7–24, 4–9 and 1–3 h, respectively (Gidenne and Ruckebush, 1989; Gidenne *et al.*, 1991; Gidenne and Pérez, 1993; Gidenne, 1994).

The wide variability in the results obtained might be related to factors such as the methodology used (type of marker, time and route of administration of the marker, mathematical calculations, etc.), the animal (age, physiological status, caecotrophy allowed or not), and feeding variables (feed intake, particle size and fibre concentration of the diet). It has been reported that the marker ytterbium was retained for 3 h longer than chromium (Gidenne and Ruckebush, 1989) and that liquid-phase markers are retained longer than solid-phase markers (Laplace and Lebas, 1975; Sakaguchi *et al.*, 1992). Preventing caecotrophy reduces the mean retention time by 0–7 h, depending on the type of diet fed (Fraga *et al.*, 1991; Sakaguchi *et al.*, 1992) whereas restricting feed intake to 0.50 and 0.60 of *ad libitum* levels increases mean retention time by 7 and 13 h, respectively (Ledin, 1984). Increasing the dietary fibre contents from 220 to 400 g kg^{-1} decreased the total mean retention time by 12 h (an 11 h reduction in the ileo-rectal mean retention time) (Gidenne, 1994). Particle size can also modify the rate of passage, with longer times being obtained using diets with smaller particle size (Laplace and Lebas, 1977; Auvergne *et al.*, 1987).

References

Alexander, F. and Chowdhury, A.K. (1958) Digestion in the rabbit's stomach. *British Journal of Nutrition* 12, 65–73.

Auvergne, A., Bouyssou, T., Pairet, M., Bouillier-Oudot, M., Ruckebush, Y. and Candau, M. (1987) Nature de l'aliment, finesse de mouture et données anatomo-fonctionnelles du tube digestif proximal du lapin. *Reproduction, Nutrition and Development* 27, 755–768.

Battaglini, M.B. and Grandi, A. (1988) Some observations of feeding behaviour of growing rabbits. In: *Proceedings of the 4th World Rabbit Congress, Budapest,* Vol. 3. Sandor Holdas, Hercegalom, pp. 79–87.

Bellier, R. and Gidenne, T. (1996) Consequences of reduced fibre intake on digestion, rate of passage and caecal microbial activity in the young rabbit. *British Journal of Nutrition* 75, 353–363.

Bellier, R., Gidenne, T. and M. Collin (1995) *In vivo* study of circadian variations of the caecal fermentation pattern in postweaned and adult rabbits. *Journal of Animal Science* 73, 128–135.

Björnhag, G. (1972) Separation and delay contents in the rabbit colon. *Swedish Journal of Agricultural Research* 2, 125–136.

Bonnafous, R. and Raynaud, P. (1970) Recherches sur le variations de la densité des microorganismes dans le colon du lapin domestique. *Experientia* 26, 52.

Candau, M., Auvergne, A., Comes, F. and Bouilllier-Oudot, M. (1986) Influence de la forme de présentation et de la finesse de mouture de l'aliment sur les performances zoootechniques et la fonction caecale chez le lapin en croissance. *Annales de Zootechnie* 35, 373–386.

Carabaño, R. and Merino, J.M. (1996) Effect of ileal cannulation on feed intake, soft and hard faeces excretion throughout the day in rabbits. In: Lebas, F. (ed.) *Proceedings of the 6th World Rabbit Congress.* Association Française de Cuniculture, Lempdes, Toulouse, pp. 121–126.

Carabaño, R., Fraga, M.J., Santomá, G. and De Blas, J.C. (1988) Effect of diet on composition of caecal contents and on excretion and composition of soft and hard faeces. *Journal of Animal Science* 66, 901–910.

Carabaño, R., Fraga, M.J. and De Blas, J.C. (1989) Effect of protein source in fibrous diets on performance and digestive parameters of fattening rabbits. *Journal of Applied Rabbit Research* 12, 201–204.

Carabaño, R., Motta-Ferreira, W., De Blas, J.C. and Fraga, M.J. (1997) Substitution of sugarbeet pulp for alfalfa hay in diets for growing rabbits. *Animal Feed Science and Technology* 65, 249–256.

Catala, J. (1976) Étude sur les répartitions hydriques, pondérals et azotées dans le material digestif chez le lapin, en relation avec la dualité de l'élimination fécale. In: Lebas, F. (ed.) *Proceedings of the 1st World Rabbit Congress.* Dijon. Communication no. 58, 6 pp.

Corring, T., Lebas, F. and Courtot, T. (1972) Contrôle de l'evolution de l'equipement enzymatique du pancréas exocrine du lapin de la naissance a six semaines. *Annales de Biologie Animale, Biochimie et Biophysique* 12, 221–231.

Cortez, S., Brandenburger, H., Greuele, E. and Sundrum, A. (1992) Untersuchungen zur Darmflora des Kaninchens in Abhängigkeit von der Fütterung und dem Gesund-heitsstatus. *Tierärztliche Umschau* 47, 544–549.

Davies, M.E. (1965) Cellulolytic bacteria in some ruminants and herbivores as shown by fluorescent antibodies. *Journal of General Microbiology* 39, 139–141.

El-Adawy, M.M. (1996) The influence of caecotomy on composition and excretion rate of soft and hard faeces, feed and water intake in rabbits. In: Lebas, F. (ed.) *Proceedings of the 6th World Rabbit Congress.* Association Française de Cuniculture, Lempdes, pp. 145–149.

Emaldi, O., Crociani, F., Matteuzi, D. and Proto, V. (1979) A note on the total viable counts and selective enumeration of anaerobic bacteria in the caecal content, soft, and hard faeces of rabbit. *Journal of Applied Bacteriology* 46, 169–172.

Fioramonti, J. and Ruckebush, Y. (1976) La motricité caecale chez le lapin. 3. Dualité de l'excretion fécale. *Annales de Recherches Vétérinaires* 7, 281–295.

Forsythe, S.J. and Parker, D.S. (1985) Nitrogen metabolism by the microbial flora of the rabbit caecum. *Journal of Applied Bacteriology* 58, 363–369.

Fraga, M.J., Barreno, C., Carabaño, R., Méndez, J. and De Blas, J.C. (1984) Efecto de los niveles de fibra y proteína del pienso sobre la velocidad de crecimiento y los parámetros digestivos de los conejos. *Anales del INIA. Serie Ganadera* 21, 91–110.

Fraga, M.J., Pérez de Ayala, P., Carabaño, R. and De Blas, J.C. (1991) Effect of type of fiber on the rate of passage and on the contribution of soft faeces to nutrient intake of fattening rabbits. *Journal of Animal Science* 69, 1566–1574.

Gallouin, F. (1984) Le comportement de caecotrophie chez le lapin. In: Finzi, A. (ed.) *Proceedings of the 3th World Rabbit Congress, Roma,* Vol. 2, pp. 363–408.

García, J., de Blas, J.C., Carabaño, R. and García, P. (1995) Effect of type of lucerne hay on caecal fermentation and nitrogen contribution through caecotrophy in rabbits. *Reproduction, Nutrition and Development* 35, 267–275.

Gidenne, T. (1986) Évolution nycthémérale des produits de la fermentation bactérienne dans le tube digestif du lapin en croissance. Relations avec la teneur en lignines de la ration. *Annales de Zootechnie* 35, 121–136.

Gidenne, T. (1994) Effets d'une réduction de la teneur en fibres alimentaires sur le transit digestif du lapin. Comparison et validation de modèles d'ajustement des cinétiques d'excretion fécale des marqueurs. *Reproduction, Nutrition and Development* 34, 295–306.

Gidenne, T. (1996) Nutritional and ontogenic factors affecting the rabbit caeco-colic digestive physiology. In: Lebas, F. (ed.) *Proceedings of the 6th World Rabbit Congress*, Vol. 1. Association Française de Cuniculture, Lempdes, pp. 13–28.

Gidenne, T. and Bellier, R. (1992). Étude *in vivo* de l'activité fermentaire caecale chez le lapin. Mise au point et validation d'une nouvelle technique de canulation caecale. *Reproduction, Nutrition and Development* 32, 365–376.

Gidenne, T. and Lebas, F. (1987) Estimation quantitative de la caecotrophie chez le lapin en croissance: variations en fonction de l'âge. *Annales de Zootechnie* 36, 225–236.

Gidenne, T. and Pérez, J.M. (1993) Effect of dietary starch origin on digestion in the rabbit. 2. Starch hydrolysis in the small intestine, cell wall degradation and rate of passage measurements. *Animal Feed Science and Technology* 42, 249–257.

Gidenne, T. and Poncet, C. (1985) Digestion chez le lapin en croissance, d'une ration à taux élevé de constituants pariétaux: étude méthodologique pour le calcul de digestibilité apparente, par segment digestif. *Annales de Zootechnie* 34, 429–446.

Gidenne, T. and Ruckebush, Y. (1989) Flow and passage rate studies at the ileal level in the rabbit. *Reproduction, Nutrition and Development* 29, 403–412.

Gidenne, T., Carré, B., Segura, M., Lapanouse, A. and Gómez, J. (1991) Fibre digestion and rate of passage in the rabbit: effect of particle size and level of lucerne meal. *Animal Feed Science and Technology* 32, 215–221.

Gouet, P. and Fonty, G. (1973) Evolution de la microflore digestive du lapin holoxénique de la naissance au sevrage. *Annales de Biologie Animale, Biochimie et Biophysique* 13, 733–735.

Gouet, P. and Fonty, G. (1979) Changes in the digestive microflora of holoxenic rabbits from birth until adulthood. *Annales de Biologie Animale, Biochimie et Biophysique* 19, 553–556.

Griffiths, M. and Davies, D. (1963) The role of soft pellets in the production of lactic acid in the rabbit stomach. *Journal of Nutriton* 80, 171–180.

Grobner, M.A. (1982) Diarrhea in the rabbit. A review. *Journal of Applied Rabbit Research* 5, 115–127.

Hall, E.R. (1952) Investigations on the microbiology of cellulose utilization in domestic rabbits. *Journal of General Microbiology* 7, 350–357.

Hörnicke, H. and Björnhag, G. (1980) Coprophagy and related strategies for digesta utilization. In: Ruckebusch, Y. and Thivend, P. (eds) *Digestive Physiology and Metabolism in Ruminants*. MTP Press, Lancaster, pp. 707–730.

Jilge, B. (1974) Soft faeces excretion and passage time in the laboratory rabbit. *Laboratory Animals* 8, 337–346.

Jilge, B. (1982) Monophasic and diphasic patterns of the circadian caecotrophy rhythm of rabbits. *Laboratory Animals* 16, 1–6.

Jilge, B. and Meyer, H. (1975) Coprophagy dependent changes of the anaerobic bacterial flora in the stomach and small intestine of the rabbit. *Zeitung Versuchstierkd* 17, 308–314.

Kulwich, R., Struglia, L. and Pearson, P.B. (1953) The effect of coprophagy in the excretion of B vitamins by the rabbit. *Journal of Nutrition* 49, 639–645.

Laplace, J.P. and Lebas, F. (1975) Le transit digetif chez le lapin. III. Influence de l'heure et du mode d'administration sur l'excrétion du cérium-141 chez le lapin alimenté ad libitum. *Annales de Zootechnie* 24, 255–265.

Laplace, J.P. and Lebas, F. (1977) Le transit digestif chez le lapin. VII. Influence de la finesse du broyage des constituents d'un aliment granulé. *Annales de Zootechnie* 26, 413–420.

Lebas, F. (1979) La nutrition du lapin: mouvement des digesta et transit. *Cuniculture* 6, 67–68.

Lebas, F., and Laplace, J.P. (1972) Mensurations viscérales chez le lapin. I. Croissance du foie, des reins et des divers segments intestinaux entre 3 et 11 semaines d'âge. *Annales de Zootechnie* 21, 337–347

Lebas, F. and Laplace, J.P. (1974) Note sur l'excretion fécale chez le lapins. *Annales de Zootechnie* 23, 577–581.

Lebas, F. and Laplace, J.P. (1975) Le transit digestif chez le lapin. 5. Evolution de l'excretion fécale en fonction de l'heure de distribution de l'aliment et du niveau de rationnement durant les 5 jours qui suivent l'application de ce denier. *Annales de Zootechnie* 24, 613–627.

Lebas, F., Corring, T. and Courtot, D. (1971) Equipement enzymatique du pancréas exocrine chez le lapin, mise en place en évolution de la naissance au sévrage. Relation avec la composition du régime alimentaire. *Annales de Biologie Animale, Biochimie et Biophysique* 11, 399–413.

Ledin, I. (1984) Effect of restricted feeding and realimentation on compensatory growth, carcass composition and organ growth in rabbit. *Annales de Zootechnie* 33, 33–50.

Lorente, M., Fraga, M.J., Carabaño, R. and de Blas, J.C. (1988) Coprophagy in lactating does fed different diets. *Journal of Applied Rabbit Research* 11, 11–15.

Marounek, M., Vovk, S.J. and Skřamová, V. (1995) Distribution of activity of hydrolytic enzymes in the digestive tract of rabbits. *British Journal of Nutrition* 73, 463–469.

Marty, J. and Vernay, M. (1984) Absorption and metabolism of the volatile fatty acids in the hindgut of the rabbit. *British Journal of Nutrition* 51, 265–277.

Merino, J.M. (1994) Puesta a punto de una técnica de canulación ileal en el conejo para el estudio del aprovechamiento de los nutrientes de la dieta. Doctoral Thesis. Universidad Complutense de Madrid.

Motta-Ferreira, W., Fraga, M.J. and Carabaño, R. (1996) Inclusion of grape pomace, in substitution for alfalfa hay, in diets for growing rabbits. *Animal Science* 63, 167–174.

Padilha, M.T.S., Licois, D., Gidenne, T., Carré, B. and Fonty, G. (1995) Relationships between microflora and caecal fermentation in rabbits before and after weaning. *Reproduction, Nutrition and Development* 35, 375–386.

Padilha, M.T.S., Licois, D., Gidenne, T., Carré, B., Coudert, P. and Lebas, F. (1996) Caecal microflora and fermentation pattern in exclusively milk-fed young rabbit. In: Lebas, F. (ed.) *Proceedings of the 6th World Rabbit Congress*, Vol. 1. Association Française de Cuniculture, Lempdes, pp. 247–251.

Pairet, M., Bouyssou, T. and Ruckebusch, Y. (1986) Colonic formation of soft faeces in rabbits: a role for endogenous prostaglandins. *American Journal Physiology* 250 (*Gastrointestinal. Liver Physiology* 13), G302–G308.

Parker, D.S. (1976) The measurement of production rates of volatile fatty acid production in rabbits. *British Journal of Nutrition* 36, 61–78.

Penney, R.L., Folk, G.E., Galask, R.P. and Petzold, C.R. (1986) The microflora of the alimentary tract of rabbits in relation to pH, diet and cold. *Journal of Applied Rabbit Research* 9, 152–156.

Pickard, D.W. and Stevens, C.E. (1972) Digesta flow through the rabbit large intestine. *American Journal of Physiology* 222, 1161–1166.

Portsmouth, J.I. (1977) The nutrition of rabbits. In: Haresign, W., Swan, H. and Lewis, D. (eds) *Nutrition and the Climatic Environment*. Butterworths, London, pp. 93–111.

Proto, V. (1976) Fisiologia della nutrizione del coniglio con particolare riguardo alla ciecotrofia. *Rivista di Conigliocoltura* 7, 15–33.

Ruckebush, Y. and Hörnicke, H. (1977) Motility of the rabbit's colon and caecotrophy. *Physiology and Behavior* 18, 871–878.

Sakaguchi, E., Kaizu, K. and Nakamichi, M. (1992) Fibre digestion and digesta retention from different physical forms of the feed in the rabbit. *Comparative Biochemistry and Physiology* 102A, 559–563.

Smith, H.W. (1965) The development of the flora of the alimentary tract in young animals. *Journal of Pathology and Bacteriology* 90, 495–513.

Snipes, R.L., Clauss, W., Weber, A. and Hörnicke, H. (1982) Structural and functional differences in various divisions of rabbit colon. *Cell and Tissue Research* 225, 331–346.

Spreadbury, D. (1978) A study of the protein and aminoacid requirements of the growing New Zealand White rabbit, with emphasis on lysine and the sulphur containing amino acids. *British Journal of Nutrition* 39, 601–613.

Udén, P., Rounsaville, T.R., Wiggans, G.R. and Van Soest, P.J. (1982) The measurement of liquid and solid digesta retention in ruminants, equines and rabbits given timothy (*Phleum pratense*) hay. *British Journal of Nutrition* 48, 329–339.

Vernay, M. and Raynaud, P. (1975) Répartition des gras volatils dans le tube digestif du lapin domestique. 1, Lapins alimentés en luzerne et avoine. *Annales des Recherches Vétérinaires* 6, 357–368.

2. Digestion of Starch and Sugars

E. Blas[1] and T. Gidenne[2]

[1]*Departamento de Ciencia Animal, Universidad Politécnica de Valencia, Camino de Vera, Apdo 22012, 46071 Valencia, Spain; and* [2]*Station de Recherches Cunicoles, INRA Centre de Toulouse, BP 27, 31326 Castanet-Tolosan Cedex, France*

It is possible to classify nutritionally into two groups the carbohydrate fractions of plants incorporated into animal feed: (i) those that are hydrolysable by the intestinal enzymes of the animal (these polysaccharides are located predominantly within the plant cell); and (ii) those that are hydrolysable only by enzymes produced by the microflora (these are principally cell wall polysaccharides and will be considered in more detail in Chapter 5). The former can be further separated into two main groups, first the simple sugars and the oligosaccharides (which are present at low levels in rabbit feeds, < 50 g kg^{-1}) and secondly the polysaccharides represented mainly by the starches (contributing 100–250 g kg^{-1}). The structure and digestion of starch will be described. Initially it is proposed to consider the digestion of simple sugars and oligosaccharides.

Simple sugars and oligosaccharides

These two types of carbohydrate are often simply classed under one general term 'the sugars'. Nevertheless, from a biochemical point of view, it is convenient to distinguish clearly between simple sugars and oligosaccharides because they are not digested by the same processes. For instance, the α-galactosides are only degraded by bacterial enzymes, whereas simple sugars are very rapidly absorbed in the small intestine.

Definition, structure and analysis

Sugars are generally of low concentration in animal feeds, although in some raw materials, such as molasses or beet, the level of sucrose could reach 500 g kg^{-1} (Table 2.1). Among the sugars found in common raw materials, glucose and fructose are the two major ones, found as monosaccharides or as sucrose (a disaccharide based on a combination of the two).

Compared with other mammals, lactose (glucose + galactose, $\alpha[1{\to}4]$) is

18 *E. Blas and T. Gidenne*

Table 2.1. Level of total sugars[a] in some raw materials (g kg⁻¹ air-dry) (INRA, 1989, 1992).

Beet molasses	Sugar beet	Sugar beet pulp	Citrus pulp	Apple pomace	Sweet lupin
490	160	50–70	100–200	140	110–130
Soybean meal	Sunflower meal	Brewer's grains	Wheat bran	Barley	Winter pea
80–110	70–90	50	50	25	20

[a]Analysed as total sugars soluble in ethanol (80%, v/v).

at a low level (31 g kg⁻¹ dry matter (DM)) in the milk of the rabbit female. Other disaccharides could also occur in the feed: maltose (two glucose units, $\alpha[1\rightarrow4]$) mainly originating from starch hydrolysis, and melibiose (galactose + glucose, $\alpha[1\rightarrow6]$, in some roots).

Oligosaccharides are defined as molecules with a low degree of polymerization (d.p.). Maltotriose corresponds to three units of glucose linked by $\alpha[1\rightarrow4]$ bonds, and originates from starch hydrolysis. The α-galactosides (d.p. 3–5) are a group of oligosaccharides (raffinose, stachyose, verbascose, ajugose) that are not digestible by the enzymes of the animal, but are rapidly degraded and fermented by the caecal microflora. They are found mainly in legume seeds or extracted legume meals: 50 g kg⁻¹ in pea seeds, 40 g kg⁻¹ in lupin seeds, 50 g kg⁻¹ in soybean meal, all on a DM basis (Thivend, 1981).

Simple sugars and oligosaccharides are solubilized in boiling ethanol (80%, v/v). Following acid hydrolysis (with hot concentrated sulphuric acid), the total sugars can be quantified (as monosaccharides) either through colorimetric or chromatographic methods. The choice of extraction process is important when analysing the sugars in a raw material or a feed. It is probable that the cold extraction (ethanol 40%, v/v, at ambient temperature) recommended in the EC method (AFNOR, 1985) is not adequate for extracting all sugars. This would explain the unexpected low values sometimes found for total sugars in soybean meals and legume seeds. Attention must be paid to the colorimetric determination of sugar in digesta and faeces because of interference by bile pigment (particularly for glucose determination in a starch analysis procedure). It is thus recommended to avoid the glucose-oxidase/peroxidase technique, and to use the hexokinase/glucose-6-P-dehydrogenase system (Kozlowski, 1994).

Digestion

Compared with starch, glucose and fructose are readily absorbed in the small intestine. However, fructose has been observed to be absorbed more slowly than glucose in pigs (Carré, 1992). In contrast with glucose, fructose is probably not absorbed through an energy-requiring mechanism. The level of sugars (soluble in ethanol 80%) in the ileal contents reached 25 g kg⁻¹ DM for adult rabbits fed a standard commercial diet (Gidenne and Ruckebusch, 1989), indicating that the flow of sugars entering the caecum is not negligible. Thus, further studies are necessary to quantify the digestion of sugars, especially in the young animal.

Complete digestion of α-galactosides has been observed in rats and pigs (Goodlad and Mathers, 1990, 1991), so it is assumed that in rabbits too they are totally digested by the caecal flora, although this has not yet been measured.

Starch

Definition, structure and analysis

Starch (α-glucan) is a major reserve polysaccharide of green plants and probably the second most abundant carbohydrate in nature next to cellulose. In some cases the reserve polysaccharides of the plant are α-fructans, such as inulin (linear α-fructan, d.p. 30) in the Jerusalem artichoke (*Helianthus tuberosus*), or levan (branched α-fructan, d.p. 100) in some grasses. Starch is found in nature as granules either in seeds, roots or tubers. The shape of the starch granule depends on the botanical source and many different sizes and forms are found: from tiny granules in oat or rice (5–6 μm) to the large ones in banana (38–50 μm). The interior of a granule is composed of alternating crystalline and amorphous regions. The disruption of this organization is the basis of gelatinization. The starch granule is modified by either chemical or physical treatment (heat, pressure, etc.). A prerequisite for digestion is that the enzymes are adsorbed on to the starch granule. Then hydrolysis may proceed through either surface erosion or penetration via pinholes. The physico-chemical and functional aspects of starch have been recently reviewed (Eliasson and Gudmundsson, 1996).

From a biochemical point of view, starch is a polysaccharide composed simply of D-glucose units. Starch basically consists of a mixture of two types of chains: amylose, a linear chain of glucose (α[1→4] links), and amylopectin, a branched chain with (α[1→4] + α[1→6] links). However, its polymeric structure is even more complicated, and its primary structure is not yet fully understood. Therefore starch is the subject of many investigations, because of its multiple uses in chemistry, fermentation processes, etc.

It is now recognized that some amylose molecules have several branches. In addition, the presence of materials intermediate between amylose and amylopectin has been suggested in amylomaize (maize rich in amylose) and wrinkled-pea starches (Hizukuri, 1996). The relative proportions of amylose and amylopectin could vary considerably according to the plant source. Cereal starch contains 200–250 g of amylose, in contrast with 250–650 g kg^{-1} in legume starch (Duprat *et al.*, 1980), and this could significantly affect its

E. Blas and T. Gidenne

digestion (Table 2.2). For instance, maize rich in amylose has a lower digestibility than standard maize. In addition, the starch granule is sometimes encapsulated in a protein matrix that reduces its accessibility to enzymes. Thus starch degradation is dependent on the biochemical and physical structure of the granule.

Numerous processes used currently in animal feed manufacturing are able to modify the starch granule, such as the use of temperature combined with hydration and pressure (extrusion process). The interaction between starch and water is well-known: the starch structure is strongly hydrophilic, and the starch–water ratio is inversely correlated to the gelatinization temperature (Champ and Colonna, 1993). In general, with an excess of water and temperatures over 55°C the granules swell and solubilize (disorganization/dispersion of the structure); this is the gelatinization step (in the case of pure starch, a viscous solution results). Following cooling, the chains of glucose could reassociate. This is termed retrogradation and can lead to forms of starch resistant to amylases. Complete gelatinization of the starch granule is essential for correct determination of the starch using the enzymatic procedure.

Starch is usually determined in animal feeds or raw materials through the Ewers EC (optical rotation determination) or enzymatic methods (hydrolysis followed by glucose determination). The two techniques provide, in general, very well correlated data, with slightly higher values for the Ewers EC method (+0.5 to +4%). The differences between the two methods are greater for legume than for cereal materials. The difference lies in the fact that the Ewers EC method may interfere with unextracted sugars or acid-labile polysaccharides. As an example, the recommended starch determination method for beet pulp is enzymatic hydrolysis. Specific procedures are also recommended for starch determination in the faeces or digesta (Kozlowski, 1994).

Table 2.2. Starch and amylose concentration in some raw materials, and respective faecal digestibility for the rat.

Source	Starch (g kg^{-1} DM)	Amylose (proportion of starch)	Faecal digestibility of starch in the rat[a]
Soft wheat	650–700	0.25–0.30	0.98–1.0
Maize	650–800	0.20–0.24	0.98–1.0
Maize rich in amylose	500–650	0.60–0.65	0.66–0.77
Smooth pea	430–480	0.31–0.35	0.99
Fava bean	300–430	0.31–0.34	0.99
Banana (green)	150–250	0.15–0.18	0.49
Cassava roots	800–850	0.17	0.95–0.97
Potato (uncooked)	600–650	0.20	0.27–0.28

[a]Champ and Colonna (1993).

Digestion of starch in the different parts of the gastro-intestinal tract

Starch is almost completely digested in the digestive tract of rabbits, as occurs in other livestock species. For this reason, faecal excretion of starch is generally minimal (less than 0.02 of intake), although in some cases it can reach 0.10–0.12 of intake, depending mainly on the age of the rabbit and the source of the starch, both of which will be discussed later.

It is acknowledged that starch digestion takes place mainly in the small intestine. Howewer, starch may also be degraded to some extent in other parts of the digestive tract, such as the stomach and the large intestine. It would be of particular interest to evaluate the degradation of starch (or of intermediate α-glucosides and glucose not absorbed in the small intestine) by the microflora in the large intestine and to review the factors affecting the ileal flow of starch and consequences on caeco-colic microbial activity. Therefore, in this section various aspects of the process of digestion of starch in the different parts of the gastro-intestinal tract of rabbits will be discussed.

Gastric digestion

There are no reliable measurements of the extent of starch hydrolysis in the stomach. It has been observed that starch concentration in the gastric digesta is clearly less than in the diet (Fraga *et al.*, 1984; Blas, 1986). Wolter *et al.* (1980) observed that, for restricted-fed rabbits slaughtered 4 h after feeding, 0.31 of the starch ingested had been hydrolysed in the stomach (recycling of the marker through the soft faeces was not taken into account). However, this is probably an overestimate because of possible faster passage through the stomach of the starch than of the rest of the diet, and by dilution of the diet with soft faeces recycled through caecotrophy.

Amylase in the stomach originates essentially from the soft faeces and saliva, at a constant level from the 4th week of life independently of starch intake (Blas, 1986). The gastric pH remains the main factor limiting its activity. Marounek *et al.* (1995) did not find amylase activity in the contents of the stomach of 4-week-old and 3-month-old rabbits, with enzyme–substrate incubations undertaken at pH 2.5.

In fact, the amylase activity of the stomach contents disappears completely if the pH is lower than 3.2 (Blas, 1986). The literature indicates that the gastric pH is normally around 2. However, the buffering capacity of the diet, soft faeces and saliva probably prevents immediate acidification. For instance, Blas (1986) found pH 4–4.5 in the stomach contents of growing rabbits 150 min after feeding following a 24-h fast; and Herrmann (1989) even reported a pH above 5 in certain areas of the stomach after a high feed intake. However, the rabbit is currently fed *ad libitum*, and has 20–30 'voluntary meals' a day. Thus gastric pH is normally less than 2.5. Nevertheless, during caecotrophy, the rabbit consumes only soft faeces and then the pH rises to 3.0 (Gidenne and Lebas, 1984).

Under these less acidic conditions, the amylase in the stomach contents, especially that of microbial origin from soft faeces, would maintain an appreciable activity (Alexander and Chowdhury, 1958; Griffiths and Davies, 1963; Hörnicke and Mackiewicz, 1976; Blas, 1986; Vernay, 1986). Table 2.3 illustrates this process of gastric fermentation (originating from starch, sugars and, perhaps, other carbohydrates), demonstrating that both the concentration of lactate in the gastric digesta (and not that in the other parts of the digestive tract) and the level of lactate in the blood fall significantly when caecotrophy is prevented.

Intestinal digestion

As stated above, it is acknowledged that starch digestion takes place mainly in the small intestine, and the most important enzyme involved is pancreatic amylase. Other enzymes of the epithelial cells of the intestinal brush border are also necessary (amyloglucosidase, maltase), resulting finally in the release of glucose, which in principle is absorbed *in situ*. It has been observed that both amylase activity in the pancreatic tissue and its pancreatic secretion increase rapidly between the 4th and 6th weeks of life, and that the amylase levels are higher when intake of starch increases (Corring *et al.*, 1972; Blas, 1986). This has also been observed in adult rabbits in both the pancreatic and the intestinal tissue, and also in the intestinal contents (Abbas *et al.*, 1991). The action of these enzymes is evidently helped by the close-to-neutral pH normally found in the contents of the small intestine.

Table 2.3. Effect of caecotrophy prevention on lactate concentration in the gastric contents and in the blood of rabbits (Vernay, 1986).

	Control	Caecotrophy prevention for 4 days
Feed intake (g day^{-1})	124	127
Gastric contents (mM)		
Fundus	4.7	1.9
Corpus	3.4	1.9
Antrum	2.4	1.6
Blood (mM in plasma)		
Venous		
Gastric	3.4	1.9
Ileal	3.8	2.1
Caecal	2.8	1.9
Portal	3.6	2.5
Arterial (abdominal aorta)	3.0	1.8

Caecal fermentation

Starch undigested in the small intestine is fermented by the microflora in the caeco-colic segment to lactate and volatile fatty acids, absorbed *in situ*. Different studies have demonstrated the presence of amylase activity in this part of the digestive tract (Yoshida *et al.*, 1968; Blas, 1986; Makkar and Singh, 1987; Marounek *et al.*, 1995). Some data suggest that amylase could be of microbial origin and also from the ileal digesta flow. For instance, amylase activity in the caecum and the colon is even greater in germ-free rabbits than in normal rabbits (Yoshida *et al.*, 1968), and it is more than double in the caecal contents of rabbits than in the contents of the rumen of steers given a maize concentrate supplement of 350 g kg^{-1} (Makkar and Singh, 1987). Blas (1986) observed that amylase activity in the caecal contents hardly varied with the age of 4- to 8-week-old rabbits, but was four times greater with a diet rich in starch than with a low-starch diet. It was also observed that caecal pH has little negative effect on this amylase activity.

Factors affecting starch digestion

Starch digestion is primarily affected by the age of the rabbit and by the dietary level and origin of starch. Other factors could also have some influence, such as the feed manufacturing process or the use of exogenous enzymes as diet supplements.

Age and starch in the diet

Adult rabbits. Table 2.4 shows starch digestibility data obtained in different studies of adult rabbits. It can be seen that faecal losses of starch were very low in all cases (less than 0.01 of intake). However, for diets rich in starch, these losses were significantly greater when the source of starch in the diet was maize compared with that from other sources (Gidenne and Perez, 1993b; Amber, 1997), although the differences may be considered to have little relevance. For adults, the faecal digestibility of starch also seems to be largely independent of the quantity of starch ingested. In fact, de Blas *et al.* (1995) found that there was no variation between lactating and non-lactating rabbit does, with the former consuming more than twice the amount of starch. As might be expected, almost all the starch was hydrolysed before reaching the caecum: its ileal digestibility was about 0.97 and seems to be largely independent of both the intake of starch and the source of starch used. In addition, the method employed by Amber (1997) to analyse the starch content evaluates starch, α-glucosides and glucose as a whole; thus the presence of a significant amount of intermediate α-glucosides and glucose not absorbed at ileal level could be excluded. Based on the above, the amount of starch fermented in the caeco-colic segment of adult rabbits is between 0.01 and 0.05 of starch intake.

Table 2.4. Ileal and faecal digestibility of starch in adult rabbits fitted with an ileal cannula.

Source of starch	Dietary level of starch (g kg^{-1} DM)	Feeding level	Digestibility of starch		Reference
			Ileal	Faecal	
Purified maize starch	160	*Ad libitum*	0.955	0.993	Gidenne (1992)
	353		0.947	0.996	
Purified maize starch	280	*Ad libitum*	0.992	0.998	Amber (1997)[a]
	351		0.991	0.999	
Maize	103		0.972	0.989	
	255		0.971	0.990	
Barley	102		0.970	0.990	
	251		0.984	0.994	
Barley	192–255	*Ad libitum*	0.975	0.994	Merino and Carabaño (1992)
Purified maize starch	256	Restricted		0.997	Gidenne and Perez (1993b)
Maize	292			0.990	
Barley	283			0.998	
Pea	280			0.996	

[a]Starch as starch + α-glucosides + glucose.

Young rabbits. Starch digestion in growing rabbits exhibits pronounced differences compared with adults, attributable to the relative importance of intestinal digestion and caecal fermentation on overall faecal digestibility.

Effects on faecal digestibility. The effects of age on faecal digestibility of starch and of the source of starch are summarized in Fig. 2.1 and Table 2.5.

The effect of age seems to be very limited for the majority of starch sources used (barley, wheat middlings or wheat). As for adult animals, faecal losses of starch just after weaning are generally less than 0.02 of intake, although in the case of peas these losses were in the 5th week almost triple those in the 8th and 11th weeks. In fact, Blas (1986) found that faecal digestibility of starch from barley or from wheat middlings was already between 0.98 and 0.99 when unweaned rabbit pups begin to consume feed (4th week), similar to that observed both in 2- to 4-week-old piglets fed pelleted feeds based on wheat or on purified wheat starch (Leibholz, 1982) and in 1- to 3-month-old infants given different cooked starches (de Vizia *et al.*, 1975).

However, faecal losses of starch greatly increase for maize and particularly for young rabbits. The resistance of maize starch to intestinal digestion is also found in pigs and ruminants, but not for poultry. The endosperm structure of maize seeds and also the resistance to grinding are considered as the main factors explaining this lower degradation (Rooney and Pflugfelder, 1986), which disappear in the process of manufacturing purified maize starch.

It must be stated that the differences between varieties of a particular

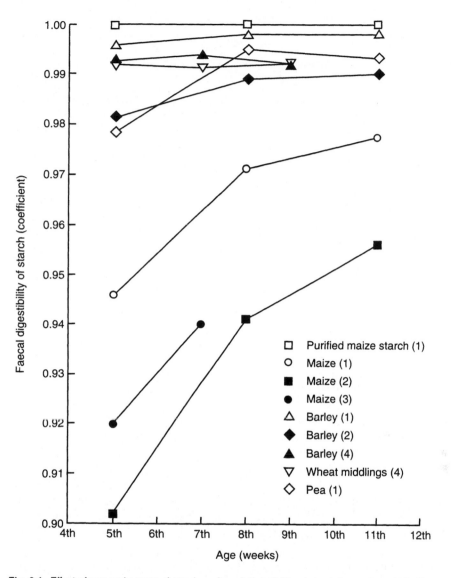

Fig. 2.1. Effect of age and source of starch on faecal digestibility of starch in growing rabbits (1, Gidenne and Perez (1993a); 2, Blas *et al.* (1990); 3, Maertens amd Luzi (1995); 4, Blas (1986)).

grain, with special reference to the amylose–amylopectin ratio, may affect the faecal losses of starch. This may help to explain both the abnormally low values of faecal digestibility of starch from barley and wheat middlings obtained by Parigi Bini *et al.* (1990) (see Table 2.5) and the differences in faecal digestibility of starch from maize ocurring in the various studies shown in Fig. 2.1.

Table 2.5. Faecal digestibility of starch from several studies on growing rabbits.

Age of animals (weeks)	Source of starch	Faecal digestibility of starch	Reference
5	Barley	0.997	Fernández *et al.* (1996)
5 and 7	Wheat middlings and wheat	0.985	Maertens and Luzi (1995)
8	Wheat and barley	0.989	Gidenne and Jehl (1996)
	Wheat middlings, wheat and barley	0.984	
8	Barley and wheat middlings	0.967	Parigi Bini *et al.* (1990)
	Wheat middlings and barley	0.945	
8–11	Barley and oats	0.986	Eggum *et al.* (1982)

In different studies with both growing and adult rabbits, comparing various dietary starch levels (Blas, 1986; Blas *et al.*, 1990; Parigi Bini *et al.*, 1990; Gidenne, 1992; de Blas *et al.*, 1995; Gidenne and Jehl, 1996; Amber, 1997), the faecal digestibility of starch tends to decrease systematically in diets with lower starch content in comparison with those having higher starch content (even with the same source of starch). Although statistically significant, this decrease remains small and may often be considered irrelevant. There is no clear explanation for these results. In fact, a lower dietary starch level corresponds frequently to a higher fibre level, and thus it could be hypothesized that it leads to a faster rate of passage and thus to a lower efficiency in starch degradation. A presence of endogenous α-linked glucose polymers (for example dextrans in the microbial reserves) being proportionally more important in diets lower in starch can also be hypothesized.

Effects on intestinal digestion and caecal fermentation. Unfortunately, no direct measurements of starch digestion in the intestine are presently available for growing rabbits. Lacking such measurements, some references provide data on the starch content in the digesta of the terminal ileum or the caecum of growing rabbits fed on different diets and slaughtered at different ages, which suggest that the amount of starch which reaches the caecum may vary widely, depending on both factors.

The starch content at the terminal ileum (Fig. 2.2) clearly decreases with age, and for lower starch intake. The results suggest that the amount of starch which reaches the caecum increases at earlier ages when the rabbit is fed on diets with high starch content, and this effect would be more pronounced with a maize-rich diet than with a barley-rich diet. This hypothesis may be supported by the lower availability of pancreatic amylase in the post-weaning period and by the lower accessibility of starch from maize, which have been mentioned above.

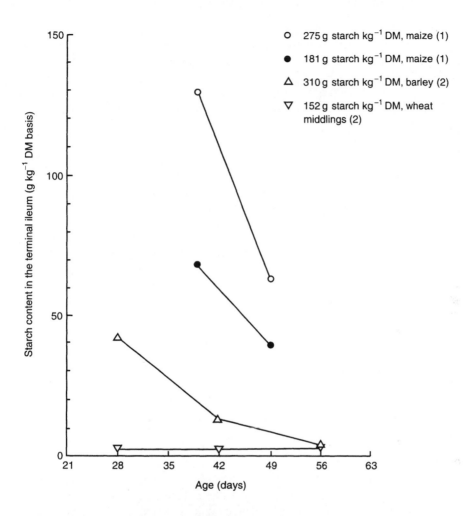

Fig. 2.2. Effect of age and dietary starch level or source on starch content of digesta in the terminal ileum of growing rabbits (1, spot-samples collected at 20 h from rabbits fed *ad libitum* (Blas *et al.*, 1994), starch as starch + α-glucosides + glucose; 2, spot-samples collected 150 min after feeding following a 24-h fast (Blas, 1986)).

Thus the amount of starch fermented in the caeco-colic segment could increase for younger animals. As shown in Fig. 2.2, Blas (1986) observed in 28-day-old and even in 49-day-old rabbits (fed on a barley-rich diet) a greater ileal starch level than in 56-day-old rabbits, whereas the faecal digestibility of starch was practically the same (0.99, Fig. 2.1). Furthermore, as mentioned above, the amylase activity in the caecal contents did not differ between 28 and 56 days of age, whereas the pancreatic amylase secretion clearly increased from 28 days of age. This may be interpreted as the result of an increment in the microbial contribution to the pool of amylase in the caecal contents of the youngest rabbits. On the other hand, the amylolytic microflora remains at a high level from 21 days of age (Padilha *et al.*, 1996).

Additionally, using single spot-samples of digesta, it has been found that, compared with ileal digesta, caecal digesta with a maize-rich diet contains 50% less starch (as starch + α-glucosides + glucose) in 38-day-old rabbits and 32% less in 49-day-old rabbits, while these figures are 24% and 3% respectively with a diet containing less maize (Blas *et al.*, 1994).

Therefore, there is a very clear need for more work on this subject, to estimate more precisely both the ileal flow of starch in recently weaned rabbits and its fermentation in their caeco-colic segment.

Feed manufacturing process

Figure 2.3 shows that oral administration of cooked purified maize starch in adult rabbits causes a clear post-prandial response of glucaemia in peripheral blood, similar to that produced by glucose, but somewhat later and more prolonged. However, uncooked purified maize starch scarcely affected basal glucaemia. This is due to a slower digestion leading to a prolonged increase of glucaemia in portal blood, but much less pronounced and therefore with little repercussion on glucaemia in the peripheral blood, after the extraction of glucose by the liver and tissues. It is unlikely that a slower digestion results in greater faecal losses of starch, but it could affect the amount of starch fermented in the large intestine.

In practice, the most interesting question would really be to clarify if, under normal conditions of feeding rabbits, the feed manufacturing process (usually involving heating, pressure and moisture) affects starch digestion, especially in young rabbits. Unfortunately, hardly any information on this matter is available.

All the results presented so far were obtained with pelleted diets, as pelleting is the usual process in rabbit feed manufacturing. Maertens and Luzi (1995) observed that extrusion of feed (which involves more intensive processing of the diet, at higher temperature, pressure and moisture) improved the solubility *in vitro* of dietary starch, but failed to reduce the faecal losses of starch in 5- or 7-week-old rabbits fed on a maize-rich diet. Surprisingly, these losses increased with the extruded diet, and the authors suggested that starch may be retrograded after cooling. Another alternative could be the inclusion of previously extruded maize in pelleted diets. Van der Poel *et al.* (1990) did not observe any change in the faecal digestibility of starch for 6-week-old piglets, and there was only a slight improvement in its ileal digestibility (0.997 vs. 0.981), although it led to a reduction in the amount of starch fermented in the large intestine to one-sixth. Similarly, in a recent study, Gidenne (1998) found that extruding maize clearly leads to a significant decrease (to approx. one-fifth) in the ileal level of starch in the growing rabbit, whatever the age (28 or 49 days).

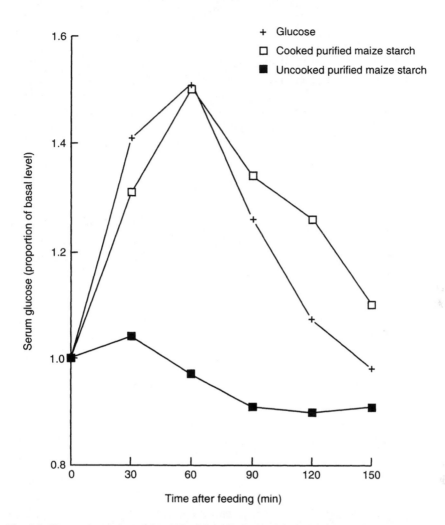

Fig. 2.3. Serum glucose in peripheral blood (marginal ear vein) of adult rabbits after oral administration of glucose or purified maize starch (0.5 g kg⁻¹) following a 12 h fast (Lee *et al.*, 1985).

Enzyme supplementation

Recent studies suggest that the inclusion of amylase, even thermostable, in rabbit diets has no effect either on starch digestion (Fernández *et al.*, 1996) or on mortality in the post-weaning period (Remois *et al.*, 1996). It is well established that the effectiveness of exogenous enzymes depends on their capacity to resist gastric pH and proteolytic attack by digestive enzymes, as well as to survive the rigours of the feed manufacturing process (Inborr, 1989; Bedford, 1995). In this context, Yu and Tsen (1993) observed that the incubation of thermostable amylase with rabbit intestinal contents at pH 7.5 did not greatly reduce its activity, while the activity fell to 0.2 in 10 min and reached negligible values in 30 min when the incubation was performed with the contents of the stomach at pH 2–3.2.

Finally, Mahagna *et al.* (1995) observed that the addition of amylase to a sorghum-based diet in meat-type chicks did not improve the faecal digestibility of starch and reduced the amount of amylase present in the intestinal contents, suggesting that the addition of an enzyme having activity similar to that of the pancreatic enzyme appears to have no benefit.

Consequences of starch digestion for the microbial ecosystem in the caeco-colic segment

Fermentation of starch not digested in the small intestine may affect the activity and stability of the microflora in the caeco-colic segment of the rabbit. This must be taken into account when considering the problem of digestive disorders in this species. Unfortunately, defining these effects and relating them to the factors which can affect starch digestion is a difficult task. The difficulty of quantifying starch fermentation, particularly in young rabbits, has already been mentioned.

On the other hand, variations in the level of starch intake are normally linked with substantial variations in fibre intake. The influence that both the level and the source of fibre has on digesta retention time and microbial activity in the caeco-colic segment is well established. The relationship between fibre deficiency and incidence of digestive disorders is presented in Chapter 12.

Extensive reviews of this matter (Gidenne, 1995, 1996) have emphasized these two problems: (i) the need for a correct assessment of the ileal flow of nutrients in order to evaluate the impact of diet on caeco-colic metabolism; and (ii) the difficulty of making a strict separation between the effects of fibre and those of starch on microbial activity and on the incidence of digestive disorders.

Adult rabbits

It has already been stated that, in adult rabbits, the starch fermented in the caeco-colic segment is a small proportion of starch intake. However, small variations in the amount of starch fermented may affect fibrolytic activity and even the frequency of digestive disorders. However, few results are available regarding possible interactions between starch and caecal microbial activity. Figure 2.4 shows that, when fibre and starch intake are constant, the faecal digestibility of NDF and hemicellulose for restricted-fed adult rabbits improves as the starch content of digesta in the terminal ileum increases as a consequence of varying the source of starch used. Amber (1997) found that both ileal flow and the amount of starch fermented in the caeco-colic segment of rabbits fed *ad libitum* are slightly greater with a maize-rich diet than with a barley-rich diet, but no significant differences in the faecal digestibility of the fibre were observed.

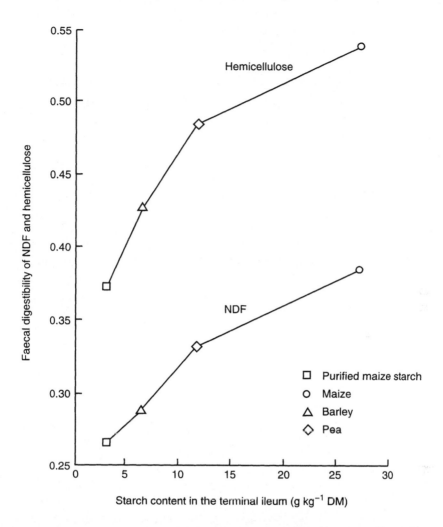

Fig. 2.4. Faecal digestibility of NDF and hemicellulose in adult rabbits with diets having different starch sources, resulting in varying starch content of digesta in the terminal ileum (spot-samples collected 2–4 h after feeding rabbits on restricted feeding (Gidenne and Perez, 1993b)).

The relationship of starch fermentation to the incidence of digestive disorders in adult rabbits is not at all clear, though in any case it seems to be very limited within the usual dietary starch levels. De Blas *et al.* (1995) have suggested a trend towards an increase in the replacement rate of rabbit does (associated with more diarrhoea and sudden deaths at parturition) as the starch content of the diet increased while the fibre content decreased. Lebas and Fortun-Lamothe (1996) found replacement rates of rabbit does very similar for feeds containing around 130 or 260 g starch kg^{-1} and, respectively,

190 or 140 g CF kg⁻¹ (DM basis), the starch intake being around 67% higher for the latter feed. Similarly, an increase in starch content from 124 to 198 g kg⁻¹ (DM basis) replacing vegetable or animal fat without varying the fibre, resulting in a 74% higher starch intake, did not affect replacement rate of rabbit does (J.J. Pascual and C. Cervera, personal communication).

Young rabbits

Given what has been described above, it is clearly of greater interest to address the matter of starch fermentation in the context of young rabbits, and particularly in those which are recently weaned. Herrmann (1989) studied caeco-colic composition in connection with microbial activity for the growing rabbit. Unfortunately, an increase in the dietary starch level from 220 to 380–510 g kg⁻¹ was accompanied by a reduction in the fibre level from 140 g CF to very low levels of 50–70 g CF kg⁻¹ (DM basis), which probably altered the retention time of digesta in the caeco-colic segment considerably and, as a result, the fermentation activity. Low-fibre/high-starch led to a rise in lactate concentration and a fall in the molar proportion of butyrate and propionate, as well as a drop in the lipopolysaccharide content. This suggests a change in the microflora, with stimulation of the Gram-positive bacteria to the detriment of the Gram-negative bacteria, though a small number of rabbits was used and the residual variabilities of these characteristics were considerable.

Jehl and Gidenne (1996) showed clearly that fibrolytic activity for the growing rabbit is greater when starch is replaced by digestible fibre (pectins and hemicellulose).

The results for faecal digestibility of fibre with diets having different starch content or source do not lend themselves to definitive conclusions, as they are normally accompanied by variations in fibre. If fibre intake is unchanged, and as mentioned previously for adults, the faecal digestibility of hemicellulose for growing rabbits appears greater for diets rich in maize than in other sources of starch (Blas et al., 1990; Gidenne and Perez, 1993a). The fermentation of starch may thus have a stimulating effect on some aspects of fibrolytic activity. If fibre intake decreases, it has been observed that the faecal digestibility of hemicellulose could decrease when the dietary starch content increases (Blas et al., 1990; Parigi Bini et al., 1990). Starch may thus be considered as a secondary factor able to affect caecal digestion, while fibre remains a primary factor.

Table 2.6 shows the results for rabbit mortality in the post-weaning period in four different large-scale trials, with diets of different starch content; fibre levels were kept above the generally recommended minima. In general, the mortality rate rose significantly as the starch intake rose, although in the work of Maître et al. (1990) the difference in the fibre sources could also be relevant. Finally, in the study of Lebas and Maître (1989) there was an interaction between location of trial and diet. This underlines the difficulty of measuring precisely the mortality rate.

Table 2.6. Effect of starch on mortality during post-weaning period in rabbits (5th–7th week of age).

Dietary level of starch (g kg⁻¹ DM)	Starch intake (g day⁻¹)	Dietary level of ADF (g kg⁻¹ DM)	Mortality (%)	No. of rabbits involved	Reference
181	15.7	236	4.7	1200	Blas *et al.* (1994)[a]
275	21.0	179	8.0		
120	9.9	192	1.2	491	Maertens and Luzi (1995)
216	17.7	175	5.7		
187	16.9	234	3.8	1520	Maître *et al.* (1990)
196	17.4	209	5.9		
200	17.1	191	7.8		
228	18.4	170	8.8		
Low[b]	10.6[c]	145[d]	4.8	2265[e]	Lebas and Maître (1989)
High[b]	16.2[c]	136[d]	11.8		
Low[b]	11.2[c]	149[d]	4.1	1106[f]	
High[b]	18.2[c]	136[d]	5.8		

[a]Starch as starch + α-glucosides + glucose.
[b]Around 150 g starch (low) or 250 g starch (high), air-dry basis; [c]approximate values.
[d]CF content, air-dry basis.
[e]In location A; [f]in location B.

It is noteworthy that the starch-rich diets in the studies of Blas *et al.* (1994) and Maertens and Luzi (1995) were based on maize (and no effect on mortality rate due to extrusion of the feed was observed in the latter). This might be related to less complete digestion of starch from maize in the small intestine.

In general, the results presented tend to support the early hypothesis suggested by Cheeke and Patton (1980) that an overload of rapidly fermentable carbohydrates in the large intestine increases the likelihood of digestive disorders, at least in the more susceptible recently weaned rabbits. *Clostridium spiriforme* and its glucose-dependent production of iota-toxin contribute to digestive disorders of growing rabbits. Peeters *et al.* (1993) reported that experimental infection with this bacterium caused clinical signs of iota-enterotoxaemia in early-weaned rabbits fed on a high-starch diet (based on maize) but not when fed on a low-starch diet, both diets having similar levels of fibre.

In any case, it seems clear that there is a need to limit the level of starch in the diet during the post-weaning period. A maximum of 135 g starch (air-dry basis) was proposed by Maertens (1992). It does not appear that slightly more severe restrictions have any clear effect on reducing the mortality rate during this period (Duperray, 1993; Mousset *et al.*, 1993).

Conclusions about the effect of dietary starch on mortality rate during the rest of the growing period are less firm. Among the papers reviewed, only that of Maître *et al.* (1990) presents specific data of relevance, which were obtained with a large number of rabbits, although differences between diets also involved the fibrous fraction to a significant extent. Based on these results, it cannot be ruled out that the effect observed in the post-weaning period persists during the rest of the growing period, although to a lesser extent, or that the effect observed during the rest of the growing period may depend on the earlier effect, as can also be deduced from the work of Lebas and Maître (1989).

On the other hand, according to the results presented in Table 2.7, it seems clear that dietary starch does not greatly affect the mortality rate of young rabbits from the time they begin to consume feed until weaning, although in the work of Lebas and Maître (1989) the above-mentioned interaction between location of trial and diet was also observed, but now with the opposite occurring. De Blas *et al.* (1995) indicated that consumption of milk represents an important part of nutritional intake and is gradually replaced by the consumption of feed, suggesting that this can help to explain why the mortality rate during this period is largely independent of the feed. Furthermore, Lelkes (1987) indicated that susceptibility to digestive disorders is greater after weaning than before, on account of the many physiological changes which weaning causes.

Table 2.7. Effect of starch on mortality of rabbits from 21 days until weaning (28–35 days of age).

Dietary level of starch (g kg^{-1} DM)	Starch intake (g day^{-1})	Dietary level of ADF (g kg^{-1} DM)	Mortality (%)	No. of litters involved	Reference
130	2.5	221	1.3	Not cited[a]	De Blas *et al.* (1995)
168	3.6	199	0.8		
190	4.3	189	0.7		
246	8.0	167	1.5		
266	6.6	167	1.2		
120		192	0.0	64	Maertens and Luzi (1995)
216		175	0.6		
124	3.0	197	3.5	260	J.J. Pascual and C. Cervera (personal communication)
124	2.8	193	4.0		
198	5.7	199	4.3		
Low[b]		145[c]	0.8	256[d]	Lebas and Maître (1989)
High[b]		136[c]	0.9		
Low[b]		146[c]	0.5	129[e]	
High[b]		136[c]	2.9		

[a]290 rabbit does over a 6-month production cycle.
[b]Around 150 g starch (low) or 250 g starch (high) kg^{-1} air-dry.
[c]CF content, air-dry basis.
[d]In location A; [e]in location B.

References

Abbas, T.A.K., Abdel-Latif, A.M. and Osman, A.M. (1991) Responses of rabbit digestive enzymes amylase and lipase to food composition and intermittent heat exposure. *World Review of Animal Production* 26, 81–86.

AFNOR (1985) *Aliments des Animaux: Méthodes d'Analyse Françaises et Communautaires*. AFNOR, Paris-La Défense.

Alexander, F. and Chowdhury, A.K. (1958) Digestion in the rabbit's stomach. *British Journal of Nutrition* 12, 65–73.

Amber, K.A. (1997) Efecto de la fibra y el almidón del pienso sobre la digestibilidad fecal e ileal en conejos adultos. Doctoral thesis, Universidad Politécnica de Valencia, Spain.

Bedford, M.R. (1995) Mechanism of action and potential environmental benefits from the use of feed enzymes. *Animal Feed Science and Technology* 53, 145–155.

Blas, E. (1986) El almidón en la nutrición del conejo: utilización digestiva e implicaciones prácticas. Doctoral thesis, Universidad de Zaragoza, Spain.

Blas, E., Fandos, J.C., Cervera, C., Gidenne, T. and Perez, J.M. (1990) Effet de la nature et du taux d'amidon sur l'utilisation digestive de la ration chez le lapin au cours de la croissance. In: *Proceedings 5èmes Journées de la Recherche Cunicole*, Vol. 2. Comm. no. 50, INRA-ITAVI, Paris, 9pp.

Blas, E., Cervera, C. and Fernández-Carmona, J. (1994) Effect of two diets with varied starch and fibre levels on the performances of 4–7 weeks old rabbits. *World Rabbit Science* 2, 117–121.

de Blas, C., Taboada, E., Mateos, G.G., Nicodemus, N. and Méndez, J. (1995) Effect of substitution of starch for fiber and fat in isoenergetic diets on nutrient digestibility and reproductive performance of rabbits. *Journal of Animal Science* 73, 1131–1137.

Carré, B. (1992) Factors affecting the digestibility of non starch carbohydrates in monogastric animals. *Poultry Science* 61, 1257–1269.

Champ, M. and Colonna, P. (1993) Importance de l'endommagement de l'amidon dans les aliments pour animaux. *INRA Production Animale* 6, 185–198.

Cheeke, P.R. and Patton, N.M. (1980) Carbohydrate-overload of the hindgut. A probable cause of enteritis. *Journal of Applied Rabbit Research* 3, 20–23.

Corring, T., Lebas, F. and Courtout, D. (1972) Contrôle de l'évolution de l'équipement enzymatique du pancréas exocrine de la naissance à 6 semaines. *Annales de Biologie Animale, Biochimie et Biophysique* 12, 221–231.

de Vizia, B., Ciccimarra, F., De Cicco, N. and Auricchio, S. (1975) Digestibility of starches in infants and children. *Journal of Pediatrics* 86, 50–55.

Duperray, J. (1993) Intérèt d'un aliment périsevrage dans l'optimisation d'un programme alimentaire. *Cuniculture* 20, 79–82.

Duprat, F., Gallant, D., Guilbot, A., Mercier, C. and Robin, J.P. (1980) L'amidon. In: Monties, B. (ed.) *Les Polymères Végétaux*. Gauthiers-Villars, Bordas, Paris, pp. 176–231.

Eggum, B.O., Chwalibog, A., Jensen, N.E. and Boissen S. (1982) Protein and energy metabolism in growing rabbits fed sodium hydroxide treated straw. *Archiv für Tierernärhung* 32, 539–549.

Eliasson, A.C. and Gudmundsson, M. (1996) Starch: physicochemical and functional aspects. In: Eliasson, A.C. (ed.) *Carbohydrates in Food.* Marcel Dekker, Basel, pp. 431–503.

Fernández, C., Merino, J. and Carabaño, R. (1996) Effect of enzyme complex supplementation on diet digestibility and growth performance in growing rabbits. In: Lebas, F. (ed.) *Proceedings of the 6th Congress of World Rabbit Science Association, Toulouse,* Vol. 1. Association Française de Cuniculture, Lempdes, pp. 163–166.

Fraga, M.J., Barreno, C., Carabaño, R., Méndez, J. and de Blas, C. (1984) Efecto de los niveles de fibra y proteína del pienso sobre la velocidad de crecimiento y los parámetros digestivos de los conejos. *Anales del INIA (serie Ganadera)* 21, 91–110.

Gidenne, T. (1992) Effect of fibre level, particle size and adaptation period on digestibility and rate of passage as measured at the ileum and in the faeces in the adult rabbit. *British Journal of Nutrition* 67, 133–146.

Gidenne, T. (1995) Apports de fibres et d'amidon: conséquences digestives chez le lapin en croissance. In: *Proceedings 7ª Jornada Técnica sobre Cunicultura.* Expoaviga, Barcelona, pp. 29–49.

Gidenne, T. (1996) Nutritional and ontogenic factors affecting rabbit caeco-colic digestive physiology. In: Lebas, F. (ed.) *Proceedings of the 6th Congress of World Rabbit Science Association, Toulouse,* Vol. 1. Association Française de Cuniculture, Lempdes, pp. 13–28.

Gidenne, T. (1998) Incidence de la nature de l'amidon sur la digestion du lapin en croissance. *Annales de Zootechnie* (in press).

Gidenne, T. and Jehl, N. (1996) Replacement of starch by digestible fibre in the feed for the growing rabbit. 1. Consequences for digestibility and rate of passage. *Animal Feed Science and Technology* 61, 183–192.

Gidenne, T. and Lebas, F. (1984) Evolution circadienne du contenu digestif chez le lapin en croissance. Relation avec la caecotrophie. In: Finzi, A. (ed.) *Proceedings of the 3rd Congress of World Rabbit Science Association,* Vol. 2. WRSA, Rome, pp. 494–501.

Gidenne, T. and Perez, J.M. (1993a) Effect of dietary starch origin on digestion in the rabbit. 1. Digestibility measurements from weaning to slaughter. *Animal Feed Science and Technology* 42, 237–247.

Gidenne, T. and Perez, J.M. (1993b) Effect of dietary starch origin on digestion in the rabbit. 2. Starch hydrolysis in the small intestine, cell wall degradation and rate of passage measurements. *Animal Feed Science and Technology* 42, 249–257.

Gidenne, T. and Ruckebusch, Y. (1989) Flow and rate of passage studies at the ileal level in the rabbit. *Reproduction, Nutrition and Development* 29, 403–412.

Goodlad, J.S. and Mathers, J.C. (1990) Large bowel fermentation in rats given diets containing raw peas (*Pisum sativum*). *British Journal of Nutrition* 64, 569–587.

Goodlad, J.S. and Mathers, J.C. (1991) Digestion by pigs of non-starch poly-saccharides in wheat bran and raw peas (*Pisum sativum*) fed in mixed diets. *British Journal of Nutrition* 65, 259–270.

Griffiths, M. and Davies, D. (1963) The role of the soft pellets in the production of lactic acid in the rabbit stomach. *Journal of Nutrition* 80, 171–180.

Herrmann, A. (1989) Untersuchungen über die Zusammensetzung des Chymus im Magen-Darm-Kanal von Jungkaninchen in Abhängigkeit vom Rohfaser- und

Stärkegehalt des Futters. Inaugural-Dissertation Doctor Medicinae Veterinariae, Tierärztliche Hochschule, Hannover.

Hizukuri, S. (1996) Starch: analytical aspects. In: Eliasson, A.C. (ed.) *Carbohydrates in Food.* Marcel Dekker, Basel, pp. 347–429.

Hörnicke, H. and Mackiewicz, A. (1976) Production d'amylase, décomposition de l'amidon et formation des acides D- et L- lactiques par les caecotrophes. In: Lebas, F. (ed.) *Proceedings of the 1st Congress of World Rabbit Science Association,* comm. no. 56. WRSA, Dijon, 4 pp.

Inborr, J. (1989) Pig diet: enzymes in combination. *Feed International* 10, 16–27.

INRA (1989) *Alimentation des Animaux Monogastriques: Porc, Lapin, Volailles,* 2nd Edn. INRA, Paris, 282 pp.

INRA (1992) *Nutrition et Alimentation des Volailles.* INRA, Paris, 355 pp.

Jehl, N. and Gidenne, T. (1996) Replacement of starch by digestible fibre in the feed for the growing rabbit. 2. Consequences for microbial activity in the caecum and on incidence of digestive disorders. *Animal Feed Science and Technology* 61, 193–204.

Kozlowski, F. (1994) L'amidon, quel dosage pour quel échantillon? *Cahiers Techniques de l'INRA* 35, 5–22.

Lebas, F. and Fortun-Lamothe, L. (1996) Effects of dietary energy level and origin (starch vs. oil) on performance of rabbit does and their litters: average situation after 4 weanings. In: Lebas, F. (ed.) *Proceedings of the 6th Congress of World Rabbit Science Association, Toulouse,* Vol. 1. Association Française de Cuniculture, Lempdes, pp. 217–222.

Lebas, F. and Maître, I. (1989) Alimentation de présevrage. Etude d'un aliment riche en énergie et pauvre en protéines. Résultats de 2 essais. *Cuniculture* 16, 135–140.

Lee, P.C., Brooks, S.P., Kim, O., Heitlinger, L.A. and Lebenthal, E. (1985) Digestibility of native and modified starches: *in vitro* studies with human and rabbit pancreatic amylases and *in vivo* studies in rabbits. *Journal of Nutrition* 115, 93–103.

Leibholz, J. (1982) Wheat starch in the diet of pigs between 7 and 28 days of age. *Animal Production* 35, 199–207.

Lelkes, L. (1987) A review of rabbit enteric diseases: a new perspective. *Journal of Applied Rabbit Research* 10, 55–61.

Maertens, L. (1992) Rabbit nutrition and feeding: a review of some recent developments. *Journal of Applied Rabbit Research* 15, 889–913.

Maertens, L. and Luzi, F. (1995) The effect of extrusion in diets with different starch levels on the performance and digestibility of young rabbits. In: *Proceedings 9th Symposium on Housing and Diseases of Rabbits, Furbearing Animals and Pet Animals.* DVG, Celle, pp. 131–138.

Mahagna, M., Nir, I., Larbier, M. and Nitsan, Z. (1995) Effect of age and exogenous amylase and protease on development of the digestive tract, pancreatic enzyme activities and digestibility of nutrients in young meat-type chicks. *Reproduction, Nutrition and Development* 35, 201–212.

Maître, I., Lebas, F., Arveux, P., Bourdillon, A., Duperray, J. and Saint Cast, Y. (1990) Taux de lignocellulose (ADF de Van Soest) et performances de croissance du lapin de chair. In: *Proceedings 5èmes Journées de la Recherche Cunicole,* Vol. 2, comm. no. 56. INRA-ITAVI, Paris, 12 pp.

Makkar, H.P.S. and Singh, B. (1987) Comparative enzymatic profiles of rabbit caecum and bovine rumen contents. *Journal of Applied Rabbit Research* 10, 172–174.

Marounek, M., Vook, S.J. and Skrivanová, V. (1995) Distribution of activity of hydrolytic enzymes in the digestive tract of rabbits. *British Journal of Nutrition* 73, 463–469.

Merino, J.M. and Carabaño, R. (1992) Effect of type of fibre on ileal and faecal digestibilities. *Journal of Applied Rabbit Research* 15, 931–937.

Mousset, J.L., Lebas, F. and Mercier, P. (1993) Utilisation d'un aliment de péri-sevrage. *Cuniculture* 20, 83–87.

Padilha, M.T.S., Licois, D., Gidenne, T., Carré, B., Coudert, P. and Lebas, F. (1996) Caecal microflora and fermentation pattern in exclusively milk-fed young animals. In: Lebas, F. (ed.) *Proceedings of the 6th Congress of World Rabbit Science Association, Toulouse,* Vol. 1. Association Française de Cuniculture, Lempdes, pp. 247–251.

Parigi Bini, R., Xiccato, G. and Cinetto, M. (1990) Influenza del contenuto di amido alimentare sulla produttività, sulla digeribilità e sulla composizione corporea di conigli in accrescimento. *Zootecnica e Nutrizione Animale* 16, 271–282.

Peeters, J.E., Orsenigo, R., Maertens, L., Gallazzi, D. and Colin, M. (1993) Influence of two iso-energetic diets (starch vs. oil) on experimental colibacillosis (EPEC) and iota-enterotoxaemia in early weaned rabbits. *World Rabbit Science* 1, 53–66.

Remois, G., Lafargue-Hauret, P. and Rouillere, H. (1996) Effect of amylase supplementation in rabbit feed on growth performance. In: Lebas, F. (ed.) *Proceedings of the 6th Congress of World Rabbit Science Association, Toulouse,* Vol. 1. Association Française de Cuniculture, Lempdes, pp. 289–292.

Rooney, L.W. and Pflugfelder, R.L. (1986) Factors affecting starch digestibility with special emphasis on sorghum and corn. *Journal of Animal Science* 63, 1607–1623.

Thivend, P. (1981) Les constituants glucidiques des aliments concentrés et de leurs dérivés. In: Demarquilly, C. (ed.) *Prévision de la Valeur Nutritive des Aliments des Ruminants.* INRA, Paris, pp. 219–235.

Van der Poel, A.F.B., Den Hartog, L.A., Van Stiphout, W.A.A., Bremmers, R. and Huisman, J. (1990) Effects of extrusion of maize on ileal and faecal digestibility of nutrients and performance of young piglets. *Animal Feed Science and Technology* 29, 309–320.

Vernay, M. (1986) Influence de la caecotrophie sur la production, l'absorption et l'utilisation des acides organiques chez le lapin. *Reproduction, Nutrition and Development* 26, 1137–1149.

Wolter, R., Nouwakpo, F. and Durix, A. (1980) Étude comparative de la digestion d'un aliment complet chez le poney et le lapin. *Reproduction, Nutrition and Development* 20, 1723–1730.

Yoshida, T., Pleasants, J.R., Reddy, B.S. and Wostmann, B.S. (1968) Efficiency of digestion in germ-free and conventional rabbits. *British Journal of Nutrition* 22, 723–737.

Yu, B. and Tsen, H.Y. (1993) An *in vitro* assessment of several enzymes for the supplementation of rabbits' diets. *Animal Feed Science and Technology* 40, 309–320.

3.   Protein Digestion

M.J. Fraga

Departamento de Producción Animal, ETS Ingenieros Agrónomos
Universidad Politécnica, 28040 Madrid, Spain

Some characteristics of main protein sources included in rabbit diets

Proteins are macromolecules made up of long chains of amino acid residues covalently linked by peptide bonds to form polypeptide chains. In each protein, these polypeptide chains are folded in three dimensions to form a characteristic tertiary structure. The properties of each amino acid depend on the structure of its chain (size and electric charge). Eight of them (isoleucine, leucine, lysine, methionine, phenylalanine, threonine, tryptophan and valine) are considered essential from a nutritional perspective because their carbon skeletons cannot be synthesized in higher animals.

The nutritive value of a protein is determined not only by its amino acid composition, but also by its digestibility or proportion of ingested protein that is digested in the gut and absorbed as free amino acids. The main factors involved in protein digestibility in rabbits, as in other non-ruminant species, are chemical structure and properties (the insoluble proteins are more resistant to digestion), and accessibility to enzyme activity.

Plant proteins are divided into two major classes: seed and leaf proteins. The main seed proteins are a part of the reserve material that is necessary for the development of the embryo of the future plant. Thus, the cereal endosperm contains approximately 0.7 of total cereal protein; the remainder is in the germ and in the outer bran. The proportions of the different types of proteins (Table 3.1) differ between cereals: the soluble albumins and globulins derive from the cytoplasm of the cells, and the insoluble prolamins and glutelins are storage proteins. Oat grain contains a considerable proportion of globulins; the grains of sorghum and rice contain low levels of soluble proteins. The bran includes the aleurone layer of endosperm (inner bran) and, because of this, has higher proportions of both crude protein and cell walls than the whole grain. The storage proteins are richer in non-essential amino acids (especially glutamic acid and proline) and lower in lysine, cystine and threonine than cytoplasmic proteins (Table 3.2). In consequence, the amino acid composition of cereals

(eds C. de Blas and J. Wiseman)

Table 3.1. Proportions of the different types of proteins in total protein of cereal and legume grains.[a]

	Cytoplasmic or salt-soluble proteins		Storage or insoluble proteins	
	Albumins	Globulins	Prolamins	Glutelins
Wheat (*Triticum aestivum*)	0.03–0.05	0.10–0.15	0.50–0.65	0.10–0.20
Barley (*Hordeum vulgare*)	0.03–0.04	0.10–0.20	0.45–0.50	0.25–0.35
Oats (*Avena sativa*)	0.01	0.60–0.65	0.10–0.15	0.25–0.30
Maize (*Zea mays*)	Trace	0.05–0.06	0.65-0.75	0.15-0.20
Bean (*Vicia faba*)	0.04	0.67	—	0.29
Peas (*Pisum sativum*)	0.21	0.66	—	0.12
Soya (*Glycine max*)	0.10	0.90	—	0

[a]Larkins (1981), Miflin and Shewry (1981), Boulter and Derbyshire (1978).

Table 3.2. Amino acid composition of wheat proteins.[a]

	Albumins	Globulins	Prolamins	Glutelins
Methionine	1.8	1.7	1.0	1.3
Lysine	3.2	5.9	0.5	1.5
Threonine	3.1	3.3	1.5	2.4
Tryptophan	1.1	1.1	0.7	2.2
Glutamic acid	22.6	15.5	41.1	34.2

[a]Bushuk and Wrigley (1974).

depends on the relative proportions of the different types of proteins.

In general, the grains of legumes and oil seeds contain higher proportions of albumins and globulins than cereal grains (Table 3.1). Thus the proteins of legumes are richer in essential amino acids (especially lysine) and should be more digestible than those of cereals. However, the value of these seeds, when they are used unprocessed, is limited by the presence of various antinutritive factors (e.g. trypsin inhibitors, lectins or tannins) and/or fibrous material.

The proteins of forage plants are concentrated in the leaves. Leaf proteins, unlike storage proteins in the grains, are concerned with the growth and biochemical functions of the cells in the leaf and are found in their cytoplasm. Thus, the major portion of leaf proteins are separated from the cell wall by a membrane, although a comparatively small fraction of insoluble protein remains tightly bound to the cellulose of the cell wall.

Protein digestibility

The main features of protein digestibility are difficult to assess through the chemical analyses commonly used in the evaluation of feedstuffs in which the faecal apparent digestibility coefficient of crude protein (CPD) is obtained. The determination of the proportion of nitrogen bound to acid detergent fibre (N-ADF), which includes heat-damaged protein and nitrogen associated with lignin, permits the estimation of that portion of the nitrogen content of feeds which is indigestible. Although an increase in the non-starch polysaccharide content of feeds has been related to a decrease in its CPD (Longstaff and NcNab, 1991; García *et al.*, 1995b), cell wall content is not always related to the digestibility of proteins.

A few attempts have been made to relate the CPD of different individual feedstuffs to their chemical composition. Martínez and Fernández (1980) obtained a high correlation coefficient ($R^2 = 0.760$) relating CPD (which varied from 0.13 to 0.82) to N-ADF content, using the data of six feed ingredients and two diets with very different chemical compositions. Recently, Villamide and Fraga (1998) have confirmed the high predictive value of N-ADF using only feedstuff data ($R^2 = 0.904$, $n = 11$). However, as this analysis is not frequently undertaken in the assays on feedstuffs for rabbits, it is still not possible to obtain a suitable relationship in which a more homogeneous and representative number of feedstuffs allows the accurate prediction of CPD of the principal raw materials included in rabbit diets.

On the other hand, the validity of grouping dissimilar ingredients for deriving this type of regression is questionable. When some groups of individual classes of similar feeds such as dry forages and protein concentrates are analysed, the best single predictor of CPD was the crude protein (CP) content ($R^2 = 0.405$ and 0.260, respectively). It is commonly accepted that an increase in the CP content of a feedstuff increases its CPD because the proportional contribution of endogenous nitrogen to total faecal nitrogen decreases. In the same way, the structure of proteins of feedstuffs with a high CP content (legume feeds, lucerne leaves) is generally less resistant to digestion.

In relation to complete diets, CPD seems to vary more according to the feed ingredients than to the chemical composition (de Blas *et al.*, 1979, 1984). However, M.J. Villamide and M.J. Fraga (unpublished data), using the information from 72 complete diets in which ADL was analysed, found a reasonable correlation between this variable and CPD ($R^2 = 0.432$). The inclusion of dietary CP in the analysis did not improve the accuracy of prediction, probably because the range of variation in the CP content of complete diets was lower than that observed among the different feedstuffs included in rabbit diets.

The dietary CPD of rabbits is influenced by age. Studies in which digestibility is determined at different ages (from weaning, at 28 days of age, to 11 weeks) show that CPD decreases following weaning to reach a steady value by approximately the 8th–9th week, with the decrease being slower from the 5th week. This effect is common to all dietary components, but the decrease in CPD is higher than that of the digestibility coefficient of organic matter of diets (differences of up to 0.15 and 0.08, respectively; Gidenne *et al.*, 1990).

The effect of age on digestibility may be explained by the increase in feed intake that occurs particularly during the 5th week. This increase may be responsible for the lack of a relationship between the quantities of ingested feed and excreted faeces. Moreover, up to the 9th week, the relative weights of the different portions of the digestive tract of rabbits change, and a proportion of the ingested feed remains in the digestive tract, leading to an overestimate of CPD in young rabbits. Other factors related to CPD, such as caecal ammonia and CP concentrations, together with soft faeces production, are also influenced by the age of rabbits.

However, there is a lack of information about CP digestibilities of individual feedstuffs in young rabbits. According to Blas *et al.* (1990) and Gidenne *et al.* (1990), the CPD of lucerne hay and peas may be lower in recently weaned than in adult rabbits, in comparison with a reference feedstuff of high CPD. This could indicate that the enzymatic processes necessary to digest vegetable protein are not fully developed when weaning is at 28 days of age. This needs to be confirmed in order to select the most appropriate source of protein for weaning diets. In the same way, it would be worth examining the possible hypersensitive reactions caused by the presence of antigens in soybean and in other legume grains. Although information is very limited, Scheele and Bolder (1987) have observed a significant increase in mortality rate of suckling rabbits when 200 g of soybean meal was included in the diet of does as a replacement for other protein concentrates.

Amino acid digestibility

Information regarding the digestibility of amino acids is even more scarce. García *et al.* (1995b) observed that the type of lucerne hay affected both content and digestibility of most of the constituent amino acids, although no significant differences were found for glutamic acid, valine, leucine, tyrosine and arginine digestibilities. Digestibility of those amino acids which are usually limiting in rabbits feeds varied from 0.81 to 0.67, from 0.79 to 0.72, and from 0.77 to 0.67 for lysine, methionine and threonine, respectively. The results lead to the conclusion that there is a positive correlation between the overall protein digestibility and that of the individual amino acids. However, whereas a difference of 0.07 was found between extreme CPD values, a variation of 0.14 was obtained for lysine, although in the remainder of the amino acids the differences were smaller (fluctuating between 0.03 and 0.10 points).

Experiments designed to calculate the minimum requirements of the most limiting amino acid for rabbits provided an opportunity to determine their digestibility. When amino acids were provided by conventional feeds the values obtained for lysine (Taboada *et al.*, 1994), methionine (Taboada *et al.*, 1996) and threonine (de Blas *et al.*, 1996) were 0.74, 0.71 and 0.63, respectively, and the values for the supplemented L-lysine HCl, D,L-methionine and L-threonine were 1.03, 1.04 and 0.94, respectively. These results emphasize the importance of using digestible units instead of total units to express amino acid requirements for rabbits.

Crude protein ileal digestibility

The techniques and methodology for fistulation at the terminal ileum and for caecal cannulation in rabbits developed in the last years (Gidenne and Bouyssou, 1987; Gidenne and Bellier, 1992; Merino, 1994) allow an estimation of the amounts of amino acids and CP that disappear from the small intestine, and *in vivo* studies of caecal fermentation. In other non-ruminant species, ileal digestibility provides a more sensitive predictive system for the detection of variations in the use of dietary amino acids by the animal than faecal digestibility (van Weerden, 1985). In the large intestine, the pattern of non-absorbed amino acid of feed origin is altered mainly because of the net synthesis of microbial protein. The microbial nitrogen contribution to total faecal nitrogen in pigs is about 0.62–0.86 (Sauer *et al.*, 1991); recent results (García *et al.*, 1995a) indicate that this value is lower (nearer to 0.3) in the hard faeces of rabbits. This is due to most of the bacterial nitrogen synthesized in the lower intestinal tract being reingested in the soft faeces of rabbits. Thus, the reingestion of soft faeces and the subsequent digestion of its CP content, in which a significant portion is bacterial nitrogen (from 0.31 to 0.68; García *et al.*, 1995a, 1996), increases the complexity (Merino, 1994) and the interpretation of results of protein digestion studies.

There are few data available on the apparent ileal digestibility coefficient of crude protein (ICPD) in rabbits. Table 3.3 shows the values obtained by Gidenne (1992) and Merino (1994) using diets in which the CP content contribution from lucerne hay varied between 0.40 and 0.70; the other dietary nitrogen sources were more digestible (fish or soybean meal). The principal digestible fraction of CP (more than 0.89) is hydrolysed before the caecum, even in diets where the majority of CP comes from lucerne hay (diet H). On the other hand, in the same diets only 0.6 of organic matter (diets H and L) or 0.75 of DM (diet M) are digested before leaving the ileum.

The high values of ICPD in relation to total CPD could be attributed to the inclusion of the nitrogen that comes from soft faeces in the calculation of ileal digestibility coefficients. However, Merino (1994) also obtained a high ICPD/CPD ratio (0.93) using rabbits fed diet M (Table 3.3) in which coprophagy was prevented. It is possible that the proportion of dietary CP non-digested at the ileal level is poorly used by caecal microorganisms. On the other hand, the results for ICPD suggested that the contribution of

Table 3.3. Effect of type of diet on proportion of digestible crude protein digested before the caecum.

	H[a]	L[a]	M[b]
Diet ingredients (g kg⁻¹) (protein sources)			
Lucerne hay	765	486	490
Fishmeal concentrate	48	100	—
Soybean meal	—	—	120
CP proportion g kg⁻¹ diet	176	182	178
CP proportion from lucerne hay	0.70	0.40	0.45
ICPD	0.585	0.650	0.739
CPD	0.654	0.721	0.722
ICPD/CPD ratio	0.894	0.901	1.023

[a]Gidenne (1992), [b]Merino (1994).

endogenous nitrogen is very important in the last segments of the digestive tract (especially in the caecum).

Using other diets also derived from conventional protein sources (diets AH50 and AH75 in Table 3.4) the ICPD/CPD ratios obtained (0.91 and 1.05) are near to the previously cited range. However, the characteristics of some feedstuffs (especially the presence of antinutritional factors and the content of a high proportion of nitrogen bound to the ADF fraction) can influence the site of dietary protein digestion. Thus, when lucerne hay is partially substituted by grape pomace (Table 3.4), the ICPD decreases from 0.61 to 0.28, with this reduction being higher (more than 100%) than the decrease in total CP digestibility (46%). The low CP digestive utilization of diet GP30 is related to its tannin content (13.5 g catechin equivalents of condensed tannins kg⁻¹ diet; Motta-Ferreira *et al.*, 1996) which may bind to dietary protein, preventing protein digestion. The importance of tannins in CP digestibility has been studied in pigs fistulated at the ileum (Jansman *et al.*, 1993), showing that the effect of tannins on digestibility was higher in the ileal than in the faecal fraction. Moreover, the high proportion of CP linked to ADF in diet GP30 may explain its low digestive utilization.

When lucerne hay was substituted by olive leaves in diet AH75 (diet OL75, Table 3.4) the ICPD decreased by 18% (from 0.79 to 0.67, although 0.66 of dietary protein in that diet came from highly digestible sources such as wheat flour and soya protein isolate). Olive leaves, like other tree leaves, also contained phenolic compounds (hydrolysable tannins) that can be condensed during the drying process. Thus, in ruminants, the CPD of olive leaves varies from 0.11 to 0.30, according to the duration of the conservation process (Gómez-Cabrera *et al.*, 1992). In the same way, 0.45 of the CP content of the olive leaves used in the study of García *et al.* (1996) was found to be bound to the NDF fraction.

In general, from 0.59 to 1.13 of faecal digested protein in rabbits is

Table 3.4. Effect of type of diet on proportion of digestible crude protein digested before the caecum and on related caecal traits.

	AH50[a]	GP30[a]	AH75[b]	OL75[b]
Diet ingredients (g kg⁻¹) (protein sources)				
Lucerne hay	500	200	750	—
Grape pomace	—	300	—	—
Soybean meal	100	70	—	—
Soya isolate	—	—	50	126
Olive leaves	—	—	—	750
CP g kg⁻¹ diet	191(50)[c]	155(23)[c]	203(63)[c]	189(34)[c]
ICPD	0.608	0.284	0.788	0.672
ICPD/CPD ratio	0.911	0.594	1.052	1.127
Caecal traits				
Ammonia (mmol l⁻¹)	4.7	2.7	9.6	2.4
VFA concentration (mmol l⁻¹)	52.1	31.8	64.3	45.2
Microbial N in soft faeces (g kg⁻¹ DM)	—	—	29.9	23.5

[a]Motta-Ferreira *et al.* (1996); Merino and Carabaño (1992).
[b]Garcia *et al.* (1996); R.M. Carabaño (unpublished data).
[c]The amounts of dietary CP coming from lucerne hay, grape pomace, lucerne hay and olive leaves, respectively, are in parentheses.

digested before the caecum. This wide variation enphasizes the importance of evaluating the ileal digestibility of feedstuffs in rabbits, despite the methodological complexity that is implicit in this technique.

Nitrogen metabolism in the caecum

Caecal microorganisms have a low amount of digestible nitrogenous material available to them from digesta compared with that in the rumen, because the major digestible fraction of dietary CP is digested before the caecum. Thus, a significant proportion of the bacterial protein will be synthesized from urea and other endogenous nitrogenous sources (digestive enzymes and cellular desquamation).

Ammonia is the main end product of nitrogen catabolism in the caecum and the main source for microbial protein synthesis. The caecal ammonia concentration in the rabbit fed balanced diets is in the range of 4.5 to 6 mmol l⁻¹ of NH_3, which seems adequate for appropriate protein microbial synthesis when compared with ammonia concentration in the rumen. Some authors have suggested that an increment in the caecal concentration of ammonia (which contributes to increase the caecal pH) could alter the microbial population balance, increasing the risk of digestive disturbances (diarrhoea).

An increment in caecal ammonia concentration was related by different authors (Carabaño *et al.*, 1988, 1989, 1997; Fraga *et al.*, 1991; Motta-Ferreira *et al.*, 1996; García *et al.*, 1995a, 1996) to an increase in the dietary digestible crude protein content. However, as occurs in the rumen (in which the ammonia concentration is related to solubility and rate of degradation of dietary proteins but also is inversely proportional to dietary digestible carbohydrate concentration), a closer relationship can be observed in rabbits between caecal ammonia concentration and dietary digestible energy/digestible protein (DE/DP) ratio. Thus, Fig. 3.1 presents the relationship between dietary E/P and caecal ammonia concentration obtained using the data from the studies previously mentioned. When the protein intake exceeds the nutritional requirements, the urea recycling from blood to the caecum might be increased, leading to an elevation in the caecal ammonia concentration. This supports the view that energy is the most limiting factor for optimum microbial growth in the caecum of rabbits as in pigs and ruminants.

Other dietary factors can affect the caecal ammonia concentration. Thus the acid caecal conditions may lead to the dietary condensed tannins decreasing the proteolytic capacity of caecal microorganisms, as occurs in the rumen (Waghorn *et al.*, 1987). This may in part explain the low caecal ammonia values obtained in diets that contained tannins (e.g. diets GP20, GP25, GP30 and OL75; see Fig. 3.1 and Table 3.4). The decrease in microbial

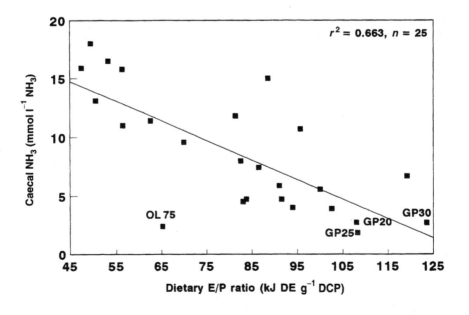

Fig. 3.1. Effect of dietary energy:protein ratio on caecal ammonia concentration (OL75: diet with 750 g of olive leaves kg^{-1}, García *et al.*, 1996; GP30, GP20, GP25: diets with 300, 200 and 250 g of grape pomace kg^{-1}, respectively, Motta-Ferreira *et al.*, 1996; Fraga *et al.*, 1991; other diets see text).

activity may also be estimated by the decrease in caecal VFA concentrations with respect to those obtained with lucerne hay diets. In the same way, the proportion of microbial nitrogen in total nitrogen contained in the soft faeces decreases by 21% in diet OL75 in relation to diet AH75.

It is possible to utilize exogenous sources of non-protein nitrogen (urea) in the caecum for bacterial protein synthesis. However, even in situations where ammonia concentration in the caecum could be the limiting factor for microbial growth (e.g. in very low protein diets) dietary urea supply has not proved to be satisfactory. However, according to the results of Makkar *et al.* (1990), at least a part of the urea administered with a low protein diet enters the caecum directly (together with the blood urea), where it is quickly degraded to ammonia. In this situation, the liveweight gains of rabbits fed the urea-based diet were similar to those obtained with low protein diets.

The CP level of caecal contents is partially influenced by the amount of microbial protein synthesized. According to the view that energy is the limiting factor to bacterial growth, an increase in dietary fibre content has been related to a decrease in the CP of caecal contents. Thus, Carabaño *et al.* (1988) have observed a negative influence of dietary crude fibre (CF) content on CP of caecal contents ($r = 0.81$, $n = 8$). In addition, García *et al.* (1995a) observed that the nitrogen concentration of caecal contents decreases linearly with the increase in dietary NDF proportion. Finally, García *et al.* (1995a, 1996) observed that the efficiency of microbial protein synthesis in the caecum (estimated as microbial nitrogen in soft faeces) decreases when the dietary fibre content increases.

On the other hand, the bacterial activity in the caecum resulted in substantial changes in the amino acid composition of its crude protein when comparing normal with germ-free rabbits (Yoshida *et al.*, 1971). An increased content of five essential amino acids (lysine, methionine, valine, leucine and isoleucine) reflects microbial protein synthesis. The methionine:cystine ratio was close to 1:1 for caecal contents and both hard and soft faeces, a normal ratio for bacterial protein which is predominant in those conditions. The ratio in the germ-free contents was of the order of 1:3, reflecting the high cystine content of pancreatic origin.

Soft faeces and protein digestibility

The main effect of soft faeces reingestion is related to protein reutilization. There are many data on the chemical composition of soft faeces suggesting that it is similar to that of the caecal contents. When comparing the protein concentration of soft faeces with that of the caecal contents of rabbits fed 31 different diets obtained employing the same methodology, the following regression equation was obtained (see Fig. 3.2):

$$y = 100.88 + 6.89\ (\pm 0.8)\ x,\ R^2 = 0.712,\ P < 0.001,\ n = 31,$$

where $y = $ CP (g kg^{-1} soft faeces DM) of soft faeces and $x = $ CP (g kg^{-1} caecal contents DM) of caecal contents.

Having taken into account that the CP concentrations of caecal contents in these studies ranged from 190 to 340 g kg^{-1}, the corresponding CP concentrations in soft faeces ranged from 230 to 335 g kg^{-1}. This suggests that the nitrogen content of the mucosal envelope which covers the caecal contents in the last sections of the large intestine to produce final soft faeces may be near to 55 g kg^{-1}. The amount of endogenous nitrogen secreted daily as mucosal envelope is near to 0.05 g day^{-1} (assuming a mean value of soft faeces production of 20 g DM, with a content of 280 g CP kg^{-1} DM). The relationship between the CF proportions of soft faeces and caecal contents are close ($R^2 = 0.817$); the CF level of soft faeces is an average of 8% lower than that of the caecal contents.

The amount of soft faeces excretion varies with age, physiological status, diet and faeces collection method. Using 25 sets of data from rabbits of approximately 2.0 kg liveweight, in which a wooden collar was added at 08.00 h and removed 24 h later, and with diets in which the NDF content varied between 230 and 550 g kg^{-1} (Carabaño *et al.*, 1988, 1989, 1997; Motta-Ferreira *et al.*, 1996; García *et al.*, 1995a, 1996), the excretion of soft

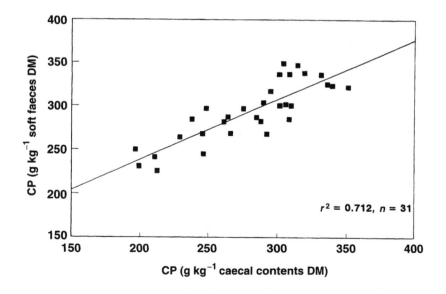

Fig. 3.2. Effect of crude protein (CP) concentration of caecal contents on CP of soft faeces (Carabaño *et al.*, 1988, 1989, 1997; Fraga *et al.*, 1991; Motta-Ferreira *et al.*, 1996; García *et al.*, 1995a).

faeces ranged from 15 to 30 g DM day^{-1}, the mean value was 20.8 g DM day^{-1} (i.e. near to 10 g DM kg^{-1} liveweight). Some authors have reported a constant relationship between the amount of hard and soft faeces excreted. Thus Gidenne and Lebas (1987) have obtained a relationship ($r = 0.639$) between both faeces types from 28 to 133 days of age.

The contribution of soft faeces to total CP intake varies, according to the chemical composition of the diet and principally to the composition of the feed ingredients within diets, from 0.105 to 0.28 (using the studies previously mentioned), the higher values being associated with low digestibility diets with a high proportion of nitrogen coming from forage or from low digestibility by-products. Although higher values have sometimes been obtained (0.55; Falçao e Cunha and Lebas, 1986), in practical diets the protein supply from soft faeces is around 0.18 of total CP intake.

The values for soft faeces production obtained in lactating does fed conventional diets (Lorente *et al.*, 1988) are higher (around 35 g DM day^{-1}) than in growing rabbits, but the contribution of soft faeces to total intake of CP (around 0.16) is maintained in the same range because of higher feed intake of the does.

As a result, the reingestion of soft faeces influenced diet digestibility, especially protein digestibility. When coprophagy is prevented, DM digestibility decreases slightly by about 6%, although in some diets variations are not observed. As an average value, the mean decrease in CPD is near to 20%; the values obtained from the literature range from 4 to 27%. When the dietary protein comes from forage, the decrease is higher than in mixed or non-forage diets.

On the other hand, there is a definite lack of information on the amino acid composition of CP of soft faeces. This composition is influenced by the microbial content of soft faeces and also the differences in the digestibility coefficients of dietary amino acids, and by the contribution of nitrogen of endogenous sources (especially that of its mucosal envelope). Data obtained by Proto (1976) indicate that the soft faeces are a good source of the most frequently limiting amino acids (methionine, lysine and threonine). Results of J.C. de Blas (unpublished data) obtained in lactating does fed diets that meet all their essential nutrient requirements indicated that the contributions of some of the essential amino acids (methionine, lysine, threonine, isoleucine and valine) are higher than the CP contribution of soft faeces to nutrient intake (0.14; see Fig. 3.3). However, the difference was only significant in the case of threonine, which has been identified as the third limiting amino acid in rabbits. From the data mentioned, the main results of digestive processes that determine the amino acid composition of soft faeces with respect to diet composition (in addition to the increase in threonine content) are the increment of methionine/cystine ratio as a consequence of the relatively high value of this ratio in bacterial protein and the decrease in arginine, histidine and phenylalanine.

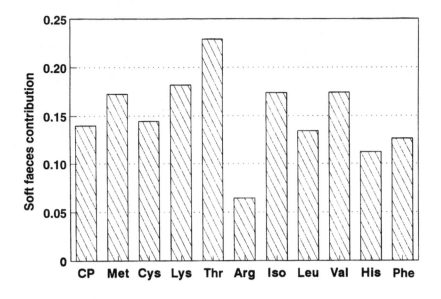

Fig. 3.3. Contribution of soft faeces to total intake of crude protein and some amino acids (J.C. de Blas, unpublished data).

In conclusion, the amino acid supply from soft faeces in conventional diets did not seem to be enough to alter the dietary amino acid pattern in order to meet the essential amino acid requirements of rabbits.

References

Blas, E., Fandos, J.C., Cervera, C., Gidenne, T. and Pérez, J.M. (1990) Effet de la nature et du taux d'amidon sur l'utilization digestive de la ration chez le lapin au cours de la croissance. In: *Proceedings 5èmes Journées de la Recherche Cunicole*, Vol. 2. Comm. no. 1 50, INRA-ITAVI, Paris.

de Blas, J.C., Merino, Y., Fraga, M.J. and Gálvez, J.F. (1979) A note on the use of sodium hydroxide treated straw pellets in diets for growing rabbits. *Animal Production* 29, 427–430.

de Blas, J.C., Fraga, M.J., Rodríguez, J.M. and Méndez, J. (1984) The nutritive value of feeds for growing fattening rabbits. 2. Protein evaluation. *Journal of Applied Rabbit Research* 7, 97–100.

de Blas, J.C., Santomá, G., Carabaño, R. and Fraga, M.J. (1986) Fiber and starch levels in fattening rabbits diets. *Journal of Animal Science* 63, 1897–1904.

de Blas, J.C., Taboada, E., Nicodemus, N., Campos, R., Piquer, J. and Méndez, J. (1996) The response of highly productive rabbits to dietary threonine content for reproduction and growth. In: Lebas, F. (ed.) *Proceedings of the 6th World Rabbit Congress, Toulouse*, Vol. 1. Association Française de Cuniculture, Lempdes, pp. 139–143.

Boulter, D. and Derbyshire, E. (1978) The general properties, classification and distribution of plant proteins. In: Norton. G. (ed.) *Plant Protein.* Butterworths, London, pp. 3–24.

Bushuk, W. and Wrigley, C.W. (1974) In: Inglett, G.E. (ed.) *Wheat Production and Utilization.* AVI Publishing Company, Westport, Colorado, pp. 119–145.

Carabaño, R.M., Fraga, M.J., Santomá, G. and de Blas, J.C. (1988) Effect of diet on composition of caecal contents and on excretion and composition of soft and hard faeces of rabbits. *Journal of Animal Science* 66, 901–910.

Carabaño, R.M., Fraga, M.J. and de Blas, J.C. (1989) Effect of protein source in fibrous diets on performance and digestive parameters of fattening rabbits. *Journal of Applied Rabbit Research* 12, 201–204.

Carabaño, R., Motta Ferreira, W., de Blas, J.C. and Fraga, M.J. (1997) Substitution of sugarbeet pulp for alfalfa hay in diets for growing rabbits. *Animal Feed Science and Technology* 65, 249–256.

Falçao e Cunha, L. and Lebas, F. (1986) Influence chez le lapin adulte de l'origine et du taux de lignine alimentaire sur la digestibilité de la ration et l'importance de la cecotrophie. In: *Proceedings 4èmes Journées de la Recherche Cunicole.* Comm. no. 1 8, INRA-ITAVI, Paris.

Fraga, M.J., Pérez de Ayala, P., Carabaño, R. and de Blas, J.C. (1991) Effect of type of fiber on the rate of passage and on the contribution of soft faeces to nutrient intake of fattening rabbits. *Journal of Animal Science* 69, 1566–1574.

García, J., de Blas, J.C., Carabaño, R. and García, P. (1995a) Effect of type of lucerne hay on caecal fermentation and nitrogen contribution through caecotrophy in rabbits. *Reproduction, Nutrition and Development* 35, 267–275.

García, J., Pérez-Alba, L., Alvarez, C., Rocha, R., Ramos, M. and de Blas, J.C. (1995b) Prediction of the nutritive value of lucerne hay in diets for growing rabbits. *Animal Feed Science and Technology* 54, 33–44.

García, J., Carabaño, R., Pérez-Alba, L. and de Blas, C. (1996) Effect of fibre source on neutral detergent fibre digestion and caecal traits in rabbits. In: Lebas, F. (ed.) *Proceedings of the 6th World Rabbit Congress, Toulouse,* Vol. 1. Association Française de Cuniculture, Lempdes, pp. 175–179.

Gidenne, T. (1992) Effect of fibre level, particle size and adaptation period on digestibility and rate of passage as measured at the ileum and in the faeces in the adult rabbit. *British Journal of Nutrition* 67, 133–146.

Gidenne, T. and Bellier, R. (1992) Etude *in vivo* de l'activité fermentaire caecale chez le lapin. Mise au point et validation d'une nouvelle technique de canulation caecale. *Reproduction, Nutrition and Development* 32, 365–376.

Gidenne, T. and Buoyssou, T. (1987) Mise au point d'une fistule iléale chez le lapin adulte. Caracterisation des digesta prelaves. *Reproduction, Nutrition and Development* 27, 289–290.

Gidenne, T. and Lebas, F. (1987) Estimation quantitative de la caecotrophie chez le lapin en croissance: variations en fonction de l'âge. *Annales de Zootechnie* 36, 225–236.

Gidenne, T., Pérez, J.M., Viudes, P. and Blas, E. (1990) Utilisation digestive de la ration chez le lapin au cours de la croissance: effet de la nature de l'amidon. In: *Proceedings 41ème Réunion Annuelle de la Fédération Européene de Zootechnie.* Toulouse, 6 pp.

Gómez-Cabrera, A., Garrido, A., Guerrero, J.E. and Ortiz, V. (1992) Nutritive value of the olive leaf: effects of cultivar, season of harvesting and system of drying. *Journal of Agricultural Science* 119, 205–210.

Jansman, A.J.M., Huisman, J. and Van der Poel, A.F.B. (1993) Ileal and faecal digestibility in piglet of field beans (*Vicia faba* L.) varying in tannin content. *Animal Feed Science and Technology* 42, 83–96.

Larkins, B.A. (1981) Seed storage proteins: characterization and biosynthesis. In: Stumpf, P.K. and Conn, E.E. (eds) *The Biochemistry of Plants*, Vol. 6. Academic Press, London, pp. 449–489.

Longstaff, M. and McNab, J.M. (1991) The inhibitory effects of hull polysaccharides and tannins of field beans (*Vicia faba* L.) on the digestion of amino acids, starch and lipid and on digestive enzyme activities in young chicks. *British Journal of Nutrition* 65, 199–216.

Lorente, M., Fraga, M.J., Carabaño, R.M. and de Blas, J.C. (1988) Coprophagy in lactating does fed different diets. *Journal of Applied Rabbit Research* 11, 11–15.

Makkar, H.P.S., Singh, B. and Krishna, L. (1990) Effect of feeding urea on some hydrolytic and ammonia assimilation enzymes in rabbit caecum. *Journal of Applied Rabbit Research* 13, 35–38.

Martínez, J. and Fernández, J. (1980) Composición, digestibilidad, valor nutritivo y relaciones entre ambos de diversos piensos para conejos. In: Camps, J. (ed.) *Proceedings of the 2nd World Rabbit Congress*. Asociacíon Española de Cunicultura, Barcelona, pp. 214–223.

Merino, J.M. (1994) Puesta a punto de una técnica de canulación ileal en el conejo para el estudio de el aprovechamiento de los nutrientes de la dieta. MSc thesis, The University Complutense of Madrid, Spain.

Merino, J.M. and Carabaño, R. (1992) Effect of type of fibre on ileal and faecal digestibilities. *Journal of Applied Rabbit Research* 15, 931–937.

Miflin, B.J. and Shewry, P.R. (1981) Seed storage proteins: genetics, synthesis, accumulation and protein quality. In: Bewlwy, J.D. (ed.) *Nitrogen and Carbon Metabolism*. Martinus Nijhoff, The Hague, pp. 195–248.

Motta-Ferreira, W., Fraga, M.J. and Carabaño, R. (1996) Inclusion of grape pomace, in substitution for lucerne hay, in diets for growing rabbits. *Animal Science* 63, 167–174.

Proto, V. (1976) Fisiologia della nutrizione del coniglio con particolare riguardo alla ciecotrofia. *Coniglicoltura* 7, 15–33.

Sauer, W.C., Mosenthin, R., Ahrons, F. and den Hartog, L.D. (1991) The effect of source of fiber on ileal and faecal amino acid digestibility and bacterial nitrogen excretion in growing pigs. *Journal of Animal Science* 69, 4070–4077.

Scheele, C.W. and Bolder, N.M. (1987) Health problems and mortality of young rabbits in relation to dietary composition. In: *Rabbit Production Systems Including Welfare*. CEC, pp. 115–125.

Taboada, E., Méndez, J., Mateos, G.G. and de Blas, J.C. (1994) The response of highly productive rabbits to dietary lysine content. *Livestock Production Science* 40, 329–337.

Taboada, E., Méndez, J. and de Blas, J.C. (1996) The response of highly productive rabbits to dietary sulphur amino acid content for reproduction and growth. *Reproduction, Nutrition and Development* 36, 191–203.

Villamide, M.J. and Fraga, M.J. (1998). Prediction of the digestible crude protein and protein digestibility of feed ingredients for rabbits from chemical analysis. *Animal Feed Science and Technology* 70, 211–224.

Waghorn, G.C., Ulyatt, M.J., John, A. and Fisher, M.T. (1987) The effect of condensed tannins on the site of digestion of amino acids and other nutrients in sheep on *Lotus corniculatus* L. *British Journal of Nutrition* 57, 115–126.

Weerden, E.J., Van (1985) The sensitivity of the ileal digestibility method as compared to the faecal digestibility method. In: Just, A., Jorgensen, H. and Fernández, J.A. (eds) *III International Seminar on Digestive Physiology in the Pig*, Vol. 580. Beretning fra Statens Husdybrugsforsog, p. 392.

Yoshida, F., Pleasants, J.R., Reddy, B. and Wostman, B.S. (1971) Amino acid composition of cecal contents and faeces in germ free and conventional rabbits. *Journal of Nutrition* 101, 1423–1429.

4.   Fat Digestion

G. Xiccato

Dipartimento di Scienze Zootecniche, University of Padua, Agripolis, 35020 Legnaro (Padova), Italy

Chemical structure and physical properties of fats

The word 'fat' is commonly misused to indicate all lipids, a complex group of organic substances composed of C, H and O and characterized by solubility in organic apolar solvents (e.g. chloroform, acetone, ethers, benzene). Lipids can be divided into simple lipids that do not contain fatty acids, and complex lipids which are esterified with fatty acids (Fig. 4.1).

Triglycerides can be considered 'true fats' because they represent the most typical form of energy accumulation in animal and vegetable organisms and therefore only these lipids have real nutritional importance. Triglycerides are the highest energy-yielding components in feeds, with an average of 2.25 times more than other constituents (i.e. protein and starch).

As evident from the name, triglycerides are formed by one glycerol molecule, a trihydric alcohol, to which three fatty acids (FA) are esterified (Fig. 4.2). The physical, chemical and nutritive properties of triglycerides depend on FA characteristics (Fig. 4.3), i.e. the number of carbon atoms and the number and position of unsaturated bonds (double bonds). The number of carbon atoms in triglyceride FAs is usually even, due to the addition or subtraction of a pair of carbon atoms during FA synthesis or oxidation in higher animals and plants, whereas microorganisms are also capable of producing FAs with odd numbers of carbon atoms. There are short-chain FAs formed of 2 (C2) to 8 (C8) carbon atoms, medium-chain FAs, with 10–16 carbon atoms, and long-chain FAs, with 18 carbon atoms and more (up to 22–24).

The number of double bonds is the second distinctive property of FAs: saturated FAs (SFA) contain only single (saturated) bonds between carbon atoms, while unsaturated FAs (UFA) present one or more double (unsaturated) bonds. UFA can be divided into monounsaturated FAs with only one double bond (e.g. oleic acid, C18:1), and polyunsaturated FAs (PUFA) with two (e.g. linoleic acid, C18:2) or more (up to six) double bonds. The position of the double bonds in the carbon chain is another fundamental characteristic of FAs, because it determines the ability of PUFA to act as

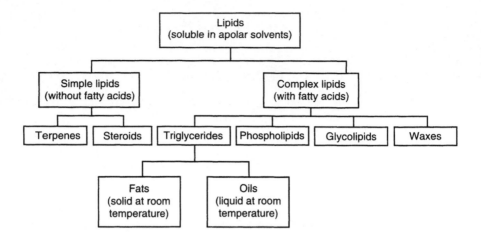

Fig. 4.1. Classification of lipids.

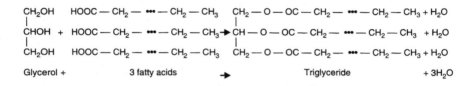

Glycerol +   3 fatty acids   ➜   Triglyceride   + 3H₂O

Fig. 4.2. Synthesis of triglycerides by esterification of glycerol and fatty acids.

Fatty acid	Designation	Chemical structure
Miristic	C14:0	$CH_3{-}(CH_2)_{12}{-}COOH$
Palmitic	C16:0	$CH_3{-}(CH_2)_{14}{-}COOH$
Palmitoleic	C16:1, n–7	$CH_3{-}(CH_2)_5{-}CH{=}CH{-}(CH_2)_7{-}COOH$
Stearic	C18:0	$CH_3{-}(CH_2)_{16}{-}COOH$
Oleic	C18:1, n–9	$CH_3{-}(CH_2)_7{-}CH{=}CH{-}(CH_2)_7{-}COOH$
Linoleic	C18:2, n–6	$CH_3{-}(CH_2)_4{-}CH{=}CH{-}CH_2{-}CH{=}CH{-}(CH_2)_7{-}COOH$
Linolenic	C18:3, n–3	$CH_3{-}CH_2{-}CH{=}CH{-}CH_2{-}CH{=}CH{-}CH_2{-}CH{=}CH{-}(CH_2)_7{-}COOH$

Fig. 4.3. Chemical structure of major fatty acids in fats and oils.

precursors of other essential compounds, such as hormones. Mammals and other higher animals, in fact, are able to elongate the carbon chain (e.g. from C18 to C22), but unable to insert double bonds between the carbon atoms in position 1 (n–1 or ω–1) of the chain (starting from the terminal methyl group, CH₃-) and the carbon in position 9 (n–9 or ω–9). For this reason, animals need an adequate quantity of 'essential FAs' (EFA) in their diet, namely n–3 FAs (with their first double bond in position 3) and n–6 FAs (first double bond in position 6). Dietary EFA are primarily represented by linoleic acid (C18:2, *n*–6) and linolenic acid (C18:3,*n*–3). The former is essential for the synthesis of arachidonic acid (C20:4,*n*–6), the precursor of prostaglandins and prostacyclins (reproductive function) or thromboxanes (homeostasis function); the latter is essential for the synthesis of eicosapentaenoic acid (C20:5,*n*–3), the precursor of several compounds essential for heart, retina and brain functions, and the immune system (Enser, 1984; Sanders, 1988; Sinclair and O'Dea, 1990).

The melting point of fats and oils is influenced by FA chemical structure and falls with a decrease in the number of carbon atoms and with an increase in the number of unsaturated bonds (Table 4.1). For this reason, triglycerides of vegetable origin are liquid at room temperature (oils) being richer in unsaturated bonds, while triglycerides of animal origin are solid (fats).

In addition to influencing physical properties, the degree of unsaturation also affects fat stability, because the double bonds are easily oxidized, thereby forming hydroperoxides that are easily broken down into short-chain compounds, which give fat and feed their typically rancid odour. The rate of oxidation rises as the number of unsaturated bonds increases. As an example, linolenic acid (C18:3) is oxidized 10 times more rapidly than linoleic acid (C18:2), which is oxidized 10 times more rapidly than oleic acid (C18:1) (Enser, 1984).

A chemical index of the degree of unsaturation is the iodine number, i.e. the weight (in g) of iodine capable of reaction with 100 g of triglyceride: in fact, two iodine atoms can react with each double bond. In animal and vegetable lipids, the iodine number represents the average degree of unsaturation of the entire pool of FAs composing the triglycerides (Table 4.1).

Table 4.1. Physical and chemical properties of various fats and oils (Cheeke, 1987).

Type of fat	Iodine number	Melting point (°C)	Fatty acids (g kg^{-1})				
			16:0	18:0	18:1	18:2	18:3
Vegetable oils							
Coconut oil	8–10	20–35	80	28	56	16	—
Corn oil	115–127	<20	120	27	301	547	14
Olive oil	79–80	<20	140	26	740	81	—
Safflower oil	145	<20	123	18	112	743	—
Soybean oil	130–138	<20	115	43	273	497	69
Animal fats							
Butter	26–28	28–36	270	125	350	30	8
Beef tallow	35–45	36–45	262	224	453	16	—
Lard	50–65	35–45	257	56	492	96	11
Poultry fat	80	<30	214	59	395	235	10

Fats in rabbit feeds

The triglycerides usually present in rabbit feed, and pure vegetable and animal fats, contain primarily medium or long-chain FAs (C14–C20), with C16 and C18 FAs being most common (Table 4.1). Rabbits have no specific fat requirements apart from a small amount of EFA (Lang, 1981; INRA, 1989; Lebas, 1989), common to all mammals and other species such as birds and fish (Whitehead, 1984; Sanders, 1988; Sinclair and O'Dea, 1990). This need is easily met by the lipids contained in the conventional raw materials used in the formulation of compound feeds. In addition, rabbit feeding is normally based on moderate energy diets, and therefore pure fats or oils are not added, and the dietary crude fat content does not exceed 30–35 g kg^{-1} on average. Only a part of this chemical constituent is composed of true fat, or triglycerides, given that the larger part is composed of other compounds such as glycolipids, phospholipids, waxes, carotenoids, saponins (Van Soest, 1982; Cheeke, 1987) (Fig. 4.1). All these substances are soluble in ethyl ether or petroleum ether, the solvents utilized to determine the crude fat or ether extract (EE) contents using the Weende feed analysis method. These lipids possess rather low digestibility and metabolic utilization and are therefore considered scarcely relevant from the nutritional point of view.

Triglyceride digestion and utilization

Triglycerides ingested with the diet by rabbits are submitted to rather complex processes of digestion and absorption, but, on the whole, these processes are similar to those observed in other non-ruminants (Brindley, 1984; Freeman, 1984; Cheeke, 1987; Fekete, 1988). In particular, triglycerides are previously emulsified, and then hydrolysed by lipolytic enzymes, before being finally absorbed in the small intestine.

As observed in different species (human, pig, rat, cattle; Freeman, 1984), the digestive process in suckling animals begins in the stomach, where pre-duodenal lipases (oral, and sometimes gastric) hydrolyse the naturally emulsified fat in milk. In rabbits, however, the presence of such lipases has only been hypothesized (Fernández et al., 1994) but never experimentally proven.

When solid feed is given, triglycerides require previous emulsification, and therefore fat digestion occurs only in the small intestine as schematically shown in Fig. 4.4. Fat emulsification is promoted by bile salts secreted by the liver. Bile salts mix with fat droplets and break them down into minute globules that can be easily hydrolysed by pancreatic lipase and other lipolytic enzymes (colipase, sterol ester hydrolase and phospholipase) (Freeman, 1984; Fekete, 1988). The enzymatic hydrolysis of triglycerides leads to the separation of primarily glycerol, free FAs and monoglycerides (glycerol esterified with a single FA) which remain emulsified with bile, forming microscopic micelles. These micelles move to the microvilli of the duodenum and jejunum, which absorb the glycerol, free FAs and monoglycerides leaving the bile salts in the intestinal lumen, which are then absorbed in a lower tract (distal ileum). Fat absorption is passive, i.e. a non-energy consuming process. When absorbed into intestinal cells, glycerol and short-chain FAs (C<12) go directly into the blood, where they circulate as non-esterified fatty acids (NEFA). On the other hand, monoglycerides and medium and long-chain FAs (C>12) are resynthesized triglycerides. Droplets of synthesized triglycerides are then covered by a lipoprotein membrane, forming chylomicrons which pass to the lymph circulation system (Brindley, 1984).

Long-chain FAs esterified in the triglycerides of chylomicrons can be metabolized as energy sources, or either incorporated directly into fat tissue (Wood, 1990) or transferred unchanged to the milk (Seerley, 1984). For this reason, the composition of the dietary fat can significantly influence fat characteristics in the rabbit carcass (Raimondi *et al.*, 1975; Ouhayoun *et al.*,

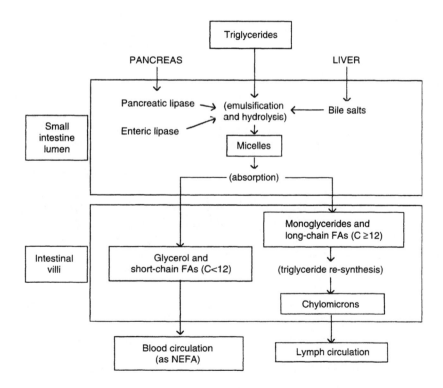

Fig. 4.4. Digestion and absorption of triglycerides in rabbits and other non-ruminants.

1986; Cavani *et al.*, 1996) or the FA composition of the milk fat (Fraga *et al.*, 1989; Christ *et al.*, 1996; Lebas *et al.*, 1996).

FAs which are not digested can pass through the lowest part of the gut and be excreted in the faeces as soaps, or enter the caecum, where UFAs are hydrogenated by the caecal microflora (Fernández *et al.*, 1994; Gidenne, 1996). A *de novo* FA synthesis also occurs with increasing proportions of short- and medium-chain FAs and decreasing levels of C18:2 and C18:3. In addition, an increase in FAs with odd numbers of carbon atoms (C15 and C17) has been observed (Fernández *et al.*, 1994).

Effect of the analytical method on digestibility determination

The precise determination of EE digestibility (EEd) is essential for a correct energy evaluation of complete diets or raw materials for rabbits. According to Jentsch *et al.* (1963) and Maertens *et al.* (1988), the digestible energy content of rabbit feeds can be calculated with good precision (1% residual SD) whenever the digestible EE and other digestible nutrients (crude protein, crude fibre and N-free extract) are known. However, despite the considerable amount of experimental data available on fat digestion efficiency in rabbits, the results often conflict, with variable EEd ranging from 0.40 to 0.95.

Parigi Bini *et al.* (1974) demonstrated the presence of soaps (salts of FAs and Ca^{2+}) in rabbit faeces, which was favoured by the high level of dietary calcium (mostly present in lucerne meal). Such soaps are only partially detected by ether extraction analysis, therefore providing an underestimation of the undigested lipids (faecal lipids) and consequently an overestimation of EEd (Table 4.2). These authors suggested submitting the faecal samples to acid hydrolysis treatment (in 5 M HCl) before extraction with ether in order to remove the FAs from the soap bonds. Using this method, Maertens *et al.* (1986) were able to recalculate the EEd values previously found for different types of fat (0.87–0.88) and obtain the correct values of 0.64 (beef tallow), 0.74 (lard), 0.75 (mixed fat) and 0.83 (soybean oil) after acid hydrolysis pre-treatment.

The problem of analysis methodology for EE determination has recently been examined by a ring-test on the chemical analyses of rabbit diets and faeces conducted by six European laboratories, all members of EGRAN (European Group on Rabbit Nutrition) (Xiccato *et al.*, 1996). This study demonstrated that the reproducibility between laboratories of EE values of feed and faeces was poor, and consequently the EEd calculated in the different laboratories ranged from 0.57 to 0.73 for the same diet.

Table 4.2. Digestibility coefficients of ether extract (EEd) before and after acid hydrolysis of faeces.

Authors	Type of fat	Added fat in the diet (g kg⁻¹)	EE in the diet (g kg⁻¹)	EEd before HCl treatment	EEd after HCl treatment
Parigi Bini *et al.* (1974)	Basal	0	37	0.72	0.52
	Beef tallow	50	92	0.85	0.67
	Beef tallow	100	137	0.89	0.69
Maertens *et al.* (1986)	Basal	0	22	0.69	0.57
	Tallow	60	81	0.88	0.64
	Lard	60	84	0.87	0.74
	Mixed fat	60	84	0.88	0.75
	Soybean oil	60	81	0.87	0.83

Effect of the level and source of fat

Lipid digestibility depends primarily on the level and source of added fats. The EEd of a non-added fat diet, which contains 25–30 g kg⁻¹ of structural lipids linked to vegetable cell walls, is rather low (0.45–0.65). In three diets with 0 (control), 30 and 60 g added fat kg⁻¹, Santomá *et al.* (1987) observed a significant increase in EEd as fat level increased (0.64, 0.75 and 0.79, respectively) without any significant difference between fat sources. These data were confirmed by Fernández *et al.* (1994) using diets with different types of fat (beef tallow, oleins or soybean oil) and inclusion levels (30 and 60 g kg⁻¹). These authors found EEd of 0.48 for a conventional non-added fat diet, 0.71–0.76 for 30 g added fat kg⁻¹ diet, and 0.80–0.83 for 60 g added fat kg⁻¹ diet. The digestibility coefficients (calculated by difference) of the pure fats were 0.86 (beef tallow), 0.90 (oleins) and 0.98 (soybean oil). These values were similar to those listed by Maertens *et al.* (1990) for animal fat (0.90) and soybean oil (0.95).

The increase in EEd with higher levels of dietary fat could also be raised by the reduction of dry matter (DM) intake, which usually happens when feed of a higher dietary energy value is given, as a consequence of the chemostatic regulation of appetite (INRA, 1989; Lebas, 1989; Forbes, 1995; Xiccato, 1996). The decrease of DM intake is associated with a lower transit time of digesta and consequently leads to increased digestion efficiency.

On the other hand, when the inclusion of fat is very high (e.g. >80 g kg⁻¹), EEd often decreases (Table 4.3), probably because both the digestive efficiency and the microflora activity in the caecum are negatively affected by the excessive fat (Maertens *et al.*, 1986; Falcao e Cunha *et al.*, 1996; Gidenne, 1996), as also occurs in ruminants (Van Soest, 1982) and poultry (Wiseman, 1984).

The differences observed in EEd among various sources of fats are mostly attributed to their molecular structure and chemical bonds. As mentioned above, the fat contained in conventional raw materials is linked to plant structures and is therefore poorly digested. Pure added fats are much more easily digestible, and this is also true for the fat contained in heated (or extruded) full-fat oil seeds, such as full-fat soybean (Table 4.4). The EEd of full-fat soybean was found to be very high (0.97), similar to the value of pure soybean oil (Maertens *et al.*, 1990), and only slightly higher than full-fat rapeseed (0.93) (Maertens *et al.*, 1996).

It is not clear whether EEd variation depends on the proportion of the different FAs in the various sources of fat, or not. As in other species, Maertens *et al.* (1986) and Santomá *et al.* (1987) noticed a negative relationship between the degree of saturation and fat digestibility in rabbits: more saturated fats (e.g. beef tallow, lard) are less digestible than unsaturated fats (e.g. sunflower or soybean oils), most probably because these latter are more easily emulsified and therefore digested in the gut. On the basis of such observations, Santomá *et al.* (1987) proposed the following regression equations in order to estimate EEd from the EE content and the linoleic acid (LA) or total UFA contents:

$$EEd = 0.59 + 1.87 \times 10^{-3} \, EE \, (g \, kg^{-1} \, DM) + 0.19 \times 10^{-3} \, LA \, (g \, kg^{-1} \, DM), \, r = 0.59$$
$$EEd = 0.48 + 1.51 \times 10^{-3} \, EE \, (g \, kg^{-1} \, DM) + 0.30 \times 10^{-3} \, UFA \, (g \, kg^{-1} \, DM),$$
$$r = 0.54$$

Table 4.3. Effect of the inclusion level of added fat on the digestibility of ether extract (EEd).

Authors	Type of fat	Inclusion level (g kg⁻¹)	EE in the diet (g kg⁻¹)	EEd of diets
Maertens *et al.* (1986)	Basal diet	0	22	0.57
	Beef tallow	60	81	0.64
	Beef tallow	120	133	0.52
	Lard	60	84	0.74
	Lard	120	136	0.71
Falcao e Cunha *et al.* (1996)	Basal diet	0	46	0.82
	Beef tallow	40	83	0.86
	Beef tallow	80	113	0.84

Table 4.4. Effect of the inclusion of full-fat oilseed on the digestibility of ether extract (EEd).

Authors	Type of full-fat seed	Inclusion level (g kg⁻¹)	EE in the diet (g kg⁻¹)	EEd of diets
Cavani *et al.* (1996)	Basal diet	0	28	0.70
	Soybean	30	32	0.82
	Soybean	60	38	0.85
Maertens *et al.* (1996)	Basal diet	0	29	0.71
	Oilseed rape	300	156	0.89

More recently, Fernández *et al.* (1994) considered that the digestibility of different FAs depends more on the source of fat than the degree of saturation. These authors measured FA digestibility in two diets: diet T, containing 30 g beef tallow kg^{-1}, and diet S containing 30 g soybean oil kg^{-1}. The digestibility coefficients of the two principal SFAs, i.e. C16:0 and C18:0, were higher (0.67 and 0.71) in the T diet than in the S diet (0.57 and 0.31); on the other hand, the PUFA, i.e. C18:2 and C18:3, were more digestible in the more unsaturated diet (0.69 and 0.80 in the T diet vs. 0.84 and 0.84 in the S diet). These authors concluded that the UFA:SFA ratio in dietary fats may not be the most appropriate predictor of fat digestibility. However, due to the higher level of PUFA and their higher digestibility, the more unsaturated diet (S) effectively showed a higher EEd than the T diet (0.81 vs. 0.77).

As mentioned above, the digestibility evaluation of specific FAs in rabbits is probably affected by a systematic bias linked to the influence of caecal microflora. As in the rumen, the microbial population of the caecum may hydrogenate the non-digested UFAs, transforming them into SFAs. Moreover, a *de novo* synthesis of short- and medium-chain, and odd-chain FAs occurs, and this definitely changes the composition of faecal fat and modifies the ratio between digestible SFAs and UFAs (Fernández *et al.*, 1994; Gidenne, 1996).

Effect of age, physiological state and nutritive level

The digestion efficiency for fat, as well as for other nutrients, seems to vary during the life of rabbit.

Rabbit milk contains a high quantity of lipids (100–150 g kg^{-1} depending on lactation period) that are easily digested and absorbed by suckling rabbits. Parigi Bini *et al.* (1991b) estimated the digestibility of milk and solid feed during weaning (from 21 to 26 days of age) by multiple regression. The EEd of milk was found to be practically complete (0.97), while the EEd of pelleted food was much lower (0.74).

Several studies have been conducted to evaluate the variation of digestibility efficiency with age in growing and adult rabbits, but the experimental results are often inconsistent. Digestibility coefficients of different nutrients tend to either decrease or remain constant with age, but EEd seems to follow a different trend. Evans and Jebelian (1982) observed increasing EEd from 0.78 at 5 weeks to 0.82 at 8 weeks. Xiccato and Cinetto (1988) confirmed these results with EEd rising from 0.72 (7 weeks) to 0.79 (12 weeks). On the other hand, Fernández *et al.* (1994) observed higher EEd values in recently weaned rabbits (5 weeks) than in finishing rabbits (10 weeks).

Comparing digestibility efficiency in growing rabbits (male and female) and adult does (pregnant and not pregnant) fed a non-added fat diet, Xiccato *et al.* (1992) observed significantly lower EEd in young rabbits than in adult rabbits (0.58 vs. 0.64), but no differences between sexes in growing rabbits or between physiological states (pregnant or not pregnant) in adult does. These latter results confirmed an absence of any effect ascribable to reproductive status, as was previously observed by Parigi Bini *et al.* (1991a) on does during late pregnancy, early lactation and late lactation.

The influence of age and physiological state on EEd could be attributed to the variation in feed intake, as suggested by Fernández *et al.* (1994). However, this hypothesis does not agree with the results achieved by other studies involving growing rabbits (Xiccato and Cinetto, 1988; Xiccato *et al.*, 1992) and lactating does (Parigi Bini *et al.*, 1992) on either *ad libitum* or restricted feeding.

On the other hand, Parigi Bini (1971) found a very significant negative correlation between dietary crude fibre (CF) level and both DM and EE digestibility:

$$DMd\ (\%) = 0.812 - 1.17 \times 10^{-3}\,g\,kg^{-1}\,DM \qquad r = -0.929$$
$$EEd\ (\%) = 0.993 - 2.08\ \times 10^{-3}\,g\,Cr\,kg^{-1}\,DM \qquad r = -0.936.$$

The decrease of both DMd and EEd is linked to poorer diet quality, and the higher feed intake and faster transit rates that occur when higher fibre levels are given. However, even if this research was unable to distinguish between the simultaneous effects of CF level on DMd and EEd, it appears that the decrease of EEd with increasing CF levels is probably accentuated by a negative interaction between fibre and fat; the latter is always strictly associated with cell walls in non-added fat diets.

References

Brindley, D.N. (1984) Digestion, absorption and transport of fats: general principles. In: Wiseman, J. (ed.) *Fats in Animal Nutrition*. Butterworths, London, pp. 85–103.

Cavani, C., Zucchi, P., Minelli, G., Tolomelli, B., Cabrini, L. and Bergami, R. (1996) Effects of whole soybeans on growth performance and body fat composition in rabbits. In: Lebas, F. (ed.) *Proceedings of the 6th World Rabbit Congress, Toulouse*, Vol. 1. Association Française de Cuniculture, Lempdes, pp. 127–133.

Cheeke, P.R. (1987) *Rabbit Feeding and Nutrition*. Academic Press, Orlando, Florida.

Christ, B., Lange, K. and Jeroch, H. (1996) Effect of dietary fat on fat content and fatty acid composition of does milk. In: Lebas, F. (ed.) *Proceedings of the 6th World Rabbit Congress, Toulouse*, Vol. 1. Association Française de Cuniculture, Lempdes, pp. 135–138.

Enser, M. (1984) The chemistry, biochemistry and nutritional importance of animal fats. In: Wiseman, J. (ed.) *Fats in Animal Nutrition*. Butterworths, London, pp. 23–51.

Evans, E., and Jebelian, V. (1982) Effects of age upon nutrient digestibility by fryer rabbits. *Journal of Applied Rabbit Research* 5, 8–9.

Falcao e Cunha, L., Bengala Freire, J.P. and Gonzalves, A. (1996) Effect of fat level and fiber nature on performances, digestibility, nitrogen balance and digestive organs in growing rabbits. In: Lebas, F. (ed.) *Proceedings 6th World Rabbit Congress, Toulouse,* Vol. 1. Association Française de Cuniculture, Lempdes, pp. 157–162.

Fekete, S. (1988) Recent findings and future perspectives of rabbit's digestive physiology. In: *Proceedings of the 4th World Rabbit Congress, Budapest,* Vol. 2. Sandor Holdas, Hercegalom, Hungary, pp. 327–344.

Fernández, C., Cobos, A. and Fraga, M.J. (1994) The effect of fat inclusion on diet digestibility in growing rabbits. *Journal of Animal Science* 72, 1508–1515.

Forbes, J.M. (1995) *Voluntary Feed Intake and Diet Selection in Farm Animals.* CAB International, Wallingford, UK.

Fraga, M.J., Lorente, M., Carabaño, R.M. and de Blas, J.C. (1989) Effect of diet and remating interval on milk production and milk composition of the doe rabbit. *Animal Production* 48, 459–466.

Freeman, C.P. (1984) The digestion, absorption and transport of fats – non-ruminants. In: Wiseman, J. (ed.) *Fats in Animal Nutrition.* Butterworths, London, pp. 105–122.

Gidenne, T. (1996) Nutritional and ontogenic factors affecting rabbit caeco-colic digestive physiology. In: Lebas, F. (ed.) *Proceedings of the 6th World Rabbit Congress, Toulouse,* Vol. 1. Association Française de Cuniculture, Lempdes, pp. 13–28.

INRA (1989) *L'Alimentation des Animaux Monogastriques: Porc, Lapin, Volailles.* Institut Nationale de la Recherche Agronomique, Versailles.

Jentsch, W., Schiemann, L., Hofmann, L. and Nehering, K. (1963) Die energetische Verwertung der Futterstoffe. 2. Die energetische Verwertung der Kraft-futterstoffe durch Kaninchen. *Archiv für Tierernährung* 13, 133–145.

Lang, J. (1981) The nutrition of the commercial rabbit. Part 1. Physiology, digestibility and nutrient requirements. *Nutrition Abstracts and Reviews, Series B* 51 (4), 197–225.

Lebas, F. (1989) Besoins nutritionnels des lapins. Revue bibliographique et perspectives. *Cuni-Sciences* 5 (2), 1–28.

Lebas, F., Lamboley, B. and Fortun-Lamothe, L. (1996) Effects of dietary energy level and origin (starch vs. oil) on gross and fatty acid composition of rabbit milk. In: Lebas, F. (ed.) *Proceedings of the 6th World Rabbit Congress, Toulouse,* Vol. 1. Association Française de Cuniculture, Lempdes, pp. 223–226.

Maertens, L., Huyghebaert, G. and De Groote, G. (1986) Digestibility and digestible energy content of various fats for growing rabbits. *Cuni-Sciences* 3, 7–14.

Maertens, L., Moermans, R. and De Groote, G. (1988) Prediction of the apparent digestible energy content of commercial pelleted feeds for rabbits. *Journal of Applied Rabbit Research* 11, 60–67.

Maertens, L., Janssen, W.M.M., Steenland, E., Wolfers, D.F., Branje, H.E.B. and Jager, F. (1990) Tables de composition, de digestibilite, et de valeur energetique des matieres premieres pour lapins. In: *Proceedings 5ièmes Journées de la Recherche Cunicole en France, Paris,* Vol. II. Communication No. 57, ITAVI, Paris, pp. 1–9.

Maertens, L., Luzi, F. and Huybrechts, I. (1996) Digestibility of non-transgenic and transgenic oilseed rape in rabbits. In: Lebas, F. (ed.) *Proceedings of the 6th World Rabbit Congress, Toulouse,* Vol. 1. Association Française de Cuniculture, Lempdes, pp. 231–236.

Ouhayoun, J., Kopp, J., Bonnet, M., Demarne, Y. and Delmas, D. (1986) Influence de la composition des graisses alimentaires sur les caractéristiques physico-chimiques des lipides corporels du lapin. In: *Proceedings 4ièmes Journées de la Recherche Cunicole en France, Paris*. Communication no. 6, ITAVI, Paris, pp. 1–13.

Parigi Bini, R. (1971) Ricerche sulla digeribilita ed il valore energetico dei concentrati nel coniglio. *Alimentazione Animale* 15 (3), 17–27.

Parigi Bini, R., Chiericato, G.M. and Lanari, D. (1974) I mangimi grassati nel coniglio in accrescimento. Digeribilita e utilizzazione energetica. *Rivista di Zootecnia e Veterinaria* 2 (3), 193–202.

Parigi Bini, R., Xiccato, G. and Cinetto, M. (1991a) Utilizzazione e ripartizione dell'energia e della proteina digeribile in coniglie non gravide durante la prima lattazione. *Zootecnica e Nutrizione Animale* 17, 107–120.

Parigi Bini, R., Xiccato, G. and Cinetto, M. (1991b) Efficienza digestiva e ritenzione energetica e proteica dei coniglietti durante l'allattamento e lo svezzamento. *Zootecnica e Nutrizione Animale* 17, 167–180.

Parigi Bini, R., Xiccato, G., Cinetto, M. and Dalle Zotte, A. (1992) Energy and protein utilization and partition in rabbit does concurrently pregnant and lactating. *Animal Production* 55, 153–162.

Raimondi, R., De Maria, C., Auxilia, M.T. and Masoero, G. (1975) Effetto della grassatura dei mangimi sulla produzione della carne di coniglio. III – Contenuto in acidi grassi delle carni e del grasso perirenale. *Annali Istituto Sperimentale Zootecnia* 8, 167–181.

Sanders, T.A.B. (1988) Essential and trans-fatty acids in nutrition. *Nutrition Research Reviews* 1, 57–78.

Santomá, G., de Blas, J.C., Carabaño, R. and Fraga, M.J. (1987) The effects of different fats and their inclusion level in diets for growing rabbits. *Animal Production* 45, 291–300.

Seerley, R.W. (1984) The use of fat in sow diets. In: Wiseman, J. (ed.) *Fats in Animal Nutrition*. Butterworths, London, pp. 333–352.

Sinclair, A.J. and O'Dea, K. (1990) Fats in human diets through history: is the western diet out of step? In: Wood, J.D. and Fisher, A.V. (eds) *Reducing Fat in Meat Animals*. Elsiever Applied Science, London, pp. 1–47.

Van Soest, P.J. (1982) *Nutritional Ecology of the Ruminant*. O&B Books, Corvallis, Oregon.

Whitehead, C.C. (1984) Essential fatty acids in poultry nutrition. In: Wiseman, J. (ed.) *Fats in Animal Nutrition*. Butterworths, London, pp. 153–166.

Wiseman, J. (1984) Assessment of the digestible and metabolisable energy of fats for non-ruminants. In: Wiseman, J. (ed.) *Fats in Animal Nutrition*. Butterworths, London, pp. 277–297.

Wood, J.D. (1990) Consequences for meat quality of reducing carcass fatness. In: Wood, J.D. and Fisher, A.V. (eds) *Reducing Fat in Meat Animals*. Elsevier Applied Science, London, pp. 344–397.

Xiccato, G. (1996) Nutrition of lactating does. In: Lebas, F. (ed.) *Proceedings of the 6th World Rabbit Congress, Toulouse,* Vol. 1. Association Française de Cuniculture, Lempdes, pp. 29–50.

Xiccato, G. and Cinetto, M. (1988) Effect of nutritive level and of age on feed digestibility and nitrogen balance in rabbit. In: *Proceedings of the 4th World*

Rabbit Congress, Vol. 3. Sandor Holdas Hercegalom, Budapest, pp. 96–103.

Xiccato, G., Cinetto, M. and Dalle Zotte, A. (1992) Effetto del livello nutritivo e della categoria di conigli sull'efficienza digestiva e sul bilancio azotato. *Zootecnica e Nutrizione Animale* 18, 35–43.

Xiccato, G., Carazzolo, A., Cervera, C., Falcao E Cunha, L., Gidenne, T., Maertens, L., Perez, J.M. and Villamide, M.J. (1996) European ring-test on the chemical analyses of feed and faeces: influence on the calculation of nutrient digestibility in rabbits. In: Lebas, F. (ed.) *Proceedings of the 6th World Rabbit Congress, Toulouse,* Vol. 1. Association Française de Cuniculture, Lempdes, pp. 293–297.

5. Fibre Digestion

T. Gidenne,[1] R. Carabaño,[2] J. García[2] and C. de Blas[2]

*[1]Station de Recherches Cunicoles, INRA Centre de Toulouse, BP 27, 31326
Castanet-Tolosan Cedex, France; [2]Departamento de Producción Animal,
Universidad Politécnica de Madrid, ETS Ingenieros Agronomos, Ciudad
Universitaria, 28040 Madrid, Spain*

Introduction

Dietary fibre (DF) is the main constituent of a commercial rabbit feed and,
depending on the analytical technique adopted, may range from 150 to 500 g
kg^{-1} (Table 5.1). Thus, fibre needs to be evaluated as precisely as possible and
its effect on digestion understood. Initially the definition and structure of the
different classes of fibre and of cell wall constituents will be considered
briefly, followed by a description of some analytical methods employed for
animal or human feeds. Secondly the effects of fibre on rabbit digestion will
be described.

Dietary fibre in animal feeds: definition, physicochemical properties and analysis

Plant cell wall and dietary fibre: definition

The two terms 'cell wall' and 'dietary fibre' are often imprecisely used,
because they refer to a common plant structure. However, they do not exactly
describe the same chemical components and therefore do not have the same
meaning. Accordingly, it is useful to define separately these two terms.

The term 'plant cell walls' (PCWs) must be employed when describing
the structure of the plant cell, which is extremely complex. PCWs are not
uniform; the type, size and shape of the wall is closely linked to the function
of the cell within the plant (skeletal tissue, seeds, etc.). In general, PCWs
consist of a series of polysaccharides often associated and/or substituted with
glycoproteins (extensin), phenolic compounds and acetic acid, together with,
in some cells, the phenolic polymer lignin. Other organic substances, such as
cutin or silica are also found in the walls and/or in the middle lamella. A
growing plant cell is gradually enveloped by a primary wall that contains few
cellulosic microfibrils and some non-cellulosic components such as pectic

(eds C. de Blas and J. Wiseman)

Table 5.1. Current levels of fibre in a complete feed used for the growing rabbit.

Residue analysed	g kg⁻¹ DM
Crude fibre	140–180
Acid detergent fibre (ADF)	160–210
Neutral detergent fibre (NDF)	270–420
Water insoluble cell wall (WICW)	280–470
Total dietary fibre (TDF)	320–510
Other feed constituents	
Starch	100–200
Crude protein	130–180

substances. During plant ageing, some cells develop a thick secondary cell wall consisting of cellulose embedded in a polysaccharide + lignin matrix (Selvendran *et al.*, 1987; McDougall *et al.*, 1996).

Thus, in brief, the wall is formed of cellulose microfibrils (the backbone) embedded in a matrix of lignins, hemicelluloses, pectins and proteins (Fig. 5.1) .

The term 'dietary fibre' refers to the nutrition of animals. Most experts in human nutrition have defined DF as the part of a foodstuff that cannot be digested by the enzymes of the gastrointestinal tract. This empirical 'catch-all' definition describes mainly PCW constituents but also other substances including resistant starch or protein linked to cell walls, etc. Thus for animal nutrition purposes, it is preferable to define DF as the sum of non-starch polysaccharides (NSP) and lignins (Fig. 5.2), because these refer to compounds hydrolysed only by bacterial enzymes and because they have common properties in digestive physiology.

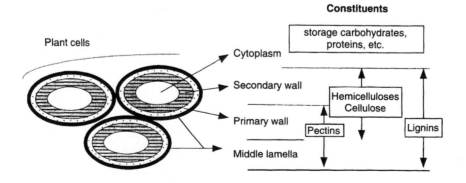

Fig. 5.1. Schematic representation of plant cell walls, and their main constituents.

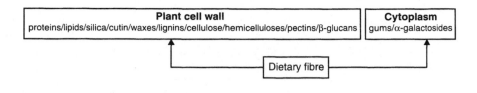

Fig. 5.2. Plant cell wall constituents and dietary fibre.

Biochemical characteristics of dietary fibre

The chemical features of DF are highly variable, depending on many factors such as molecular weight, nature of monomers and types of linkages. Accordingly, chemical features of fibre are one of the main factors responsible for variations in digestibility, and thus it is of importance to describe them.

With the exception of lignin, the cell wall constituents are predominantly polysaccharides composed of neutral and/or acidic sugars. They can be determined using sophisticated extraction techniques, and examples of their concentration in some feedstuffs are given in the Table 5.2.

Among the numerous types of cell wall polymers (Chesson, 1995), it is convenient to select five major classes of fibre, according to their chemical structure and properties (Fig. 5.3):

- one class of various water-soluble non-starch polysaccharides and oligo-saccharides;
- four classes of water-insoluble polymers: lignins, cellulose, hemi-celluloses, pectic substances.

Water-soluble polysaccharides and oligosaccharides include several classes of molecules with a degree of polymerization ranging from about 15 to more than 2000 (β-glucans). Most of them are insoluble in ethanol (80% v/v). Examples include soluble hemicelluloses such as arabinoxylans (in wheat, oat and barley ≈ 20–40 g kg^{-1} DM) and β-glucans (in barley or oat ≈ 10–30 g kg^{-1} DM), oligosaccharides such as α-galactosides (in lupin, pea or soya seeds, 50–80 g kg^{-1} DM), and soluble pectic substances (pulps of fruits or beets, up to 100 g kg^{-1} DM). Because of their highly variable structure, no satisfactory method is at present available to determine precisely these compounds in animal feeds.

Pectic substances are a group of polysaccharides present in the middle lamellae and closely associated with the primary cell wall, especially in the primary cells (young tissues) of dicotyledonous plants, such as in legume seeds (40–140 g kg^{-1} DM in soybean, pea, faba bean, white lupin), and also in fruits and pulps. Pectic substances correspond to several classes of polymers, including pectins (rhamnogalacturonan backbone and side chains of

Table 5.2. Proximate composition of cell-wall constituents (%), according to several methods of analysis, in some raw materials used in rabbit feeds.

Ingredients	Wheat straw	Wheat bran	Dehydrated lucerne	Sugar-beet pulp	Sunflower meal	Soybean hulls	Grape pomace
NDF[a]	80	45	45	46	42	62	64
ADF[a]	54	11	34	22	31	44	54
ADL[a]	16	3	8	2	10	2	34
Crude fibre[b]	40	10	27	19	26	36	26
WICW[c]	84	45	47	50	38	72	69
INSP[d]	55	36	33	64	26	55	36
Rhamnose	<1	<1	<1	1	<1	1	<1
Arabinose	2	8	2	18	3	4	<1
Xylose	18	16	6	2	5	7	8
Mannose	<1	<1	<1	1	1	6	2
Galactose	<1	<1	<1	4	1	2	2
glucose	33	9	19	19	11	29	19
Uronic acids	2	2	7	18	5	6	5
SNSP[e]	1	3	3	10	1	2	1
Crude protein	3	15	16	9	34	11	13

[a]NDF, neutral detergent fibre; ADF, acid detergent fibre; ADL, acid detergent lignin (Van Soest *et al.*, 1991).
[b]According to the Weende method (Henneberg and Stohman, 1864).
[c]Water-insoluble cell wall, including lignin (Carré and Brillouet, 1989).
[d]Insoluble non-starch polysaccharides, not including the lignin, determined by direct monomeric analysis of cell wall polysaccharides (Englyst, 1989; Barry *et al.*, 1990).
[e]Water-soluble non-starch polysaccharides (Brillouet *et al.*, 1988; Englyst, 1989).

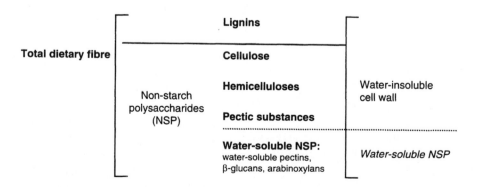

Fig. 5.3. Global classification of dietary fibre.

arabinose and galactose) and neutral polysaccharides (arabinans, galactans, arabinogalactans) frequently associated with pectins. Their extraction requires the use of a chelating agent such as ammonium oxalate or ethylene diamine tetraacetic acid (EDTA) (present in the solution for determining neutral detergent fibre (NDF) so they are not recovered in NDF analysis as described below). Pectins of the middle lamellae serve as an adhesive in plant tissue, cementing plant cells together.

Cellulose is the major structural polysaccharide of the PCW and the most widespread polymer on earth. It is a homopolymer (in contrast to hemicelluloses and pectins), formed from linear chains of $\beta[1\rightarrow4]$ linked D-glucopyranosyl units (whereas starch is formed of $\alpha[1\rightarrow4]$ linked D-glucopyranosyl chains). The degree of polymerization is usually around 8000–10,000. Individual glucan chains aggregate (hydrogen bonding) to form microfibrils, and could serve as the backbone of the plant. Thus, cellulose is only soluble in strong acid solutions (i.e. 72% sulphuric acid) where it is partially hydrolysed. Quantitatively, cellulose represents 400–500 g kg^{-1} DM in hulls of legume and oilseeds, 100–300 g kg^{-1} DM in forages and beet pulps, 30–150 g kg^{-1} DM in oilseeds or legume seeds. Most cereal grains contain small quantities of cellulose (10–50 g kg^{-1} DM) except in oats (100 g kg^{-1} DM).

The hemicelluloses are a group of several polysaccharides, with a lower degree of polymerization than cellulose. They have a $\beta[1\rightarrow4]$ linked backbone of xylose, mannose or glucose residues that can form extensive hydrogen bonds with cellulose. Xyloglucans are the major hemicellulose of primary cell wall in dicotyledonous plants (in vegetables, in seeds), whereas mixed linked glucans ($\beta[1\rightarrow3,4]$) and arabinoxylans are the predominant hemicelluloses in cereals seeds (the latter two include partly water-soluble and water-insoluble polymers, described above). Hemicelluloses include other branched heteropolymers (units linked $\beta[1\rightarrow3]$, $\beta[1\rightarrow6]$, $\alpha[1\rightarrow4]$, $\alpha[1\rightarrow3]$) such as highly branched arabinogalactans (in soybean), galactomannans (seeds of legumes), or glucomannans. Polymers formed of linear chains of pentose (linked $\beta[1\rightarrow4]$) such as xylans (in secondary walls), or hexose such as mannans (in palm kernel meal) are also classed as hemicelluloses. Pentosans such as xylans and arabinoxylans are soluble in weak basic solutions (5–10%), or in hot dilute acids (5% sulphuric acid). Hexosans such as mannans, glucomannans or galactans can only be dissolved in strong basic solutions (17–24%). Quantitatively, hemicelluloses constitute 100–250 g kg^{-1} of the DM in forages and agro-industrial by-products (brans, oilseeds and legume seeds, hulls and pulps) and about 20–120 g kg^{-1} DM of grains and roots.

Lignins are the only non-saccharidic polymer of the cell wall. They can be described as very branched and complex three-dimensional networks (high molecular weight), built up from three phenylpropane units (coniferilic, coumarilic and sinapyilic acid). Lignin networks tend to fix the other polymers in place, exclude water and make the cell wall more rigid and resistant to various agents, such as bacterial enzymes. Most concentrate feeds and young forages contain less than 50 g lignin kg^{-1}. With ageing, the degree of lignification of the PCW may reach 120 g kg^{-1} in forages.

Other constituents are also present in cell walls, but in smaller quantities. Minerals, such as silica, are essentially in graminaceous leaves. Phenolic acids are chemically linked to hemicelluloses and lignin in gramineous plants. Some proteins are linked to cell walls through intermolecular bonds from amino acids such as tyrosine, and thus resist standard extractions. In addition, plant epidermal cells may be covered by a complex lipid (cutin for aerial parts, suberin for underground structures) which could encrust and embed the cell walls, making them impermeable to water.

Other phenolic compounds can also be mentioned, i.e. condensed tannins, which may exist in higher plants. They form cross-linkages with protein and other molecules. They could be included in the sum of indigestible polysaccharides + lignin. However, condensed tannins, lignins and indigestible proteins are closely related because indigestible complexes of these substances are common in plants (Van Soest *et al.*, 1987).

Methods for estimating the dietary fibre (DF) content of animal feeds

Estimating the fibre content of feeds for ruminant and non-ruminant animals is important, because the fraction is highly related to their organic matter digestibility. For instance, a new method was established to predict the metabolizable energy content in poultry feeds, based on only one chemical fibre criterion: the water-insoluble cell wall content, 'WICW' (Carré, 1991, and Fig. 5.4), because insoluble fibres are indigestible for poultry.

Estimating the DF is also necessary to describe a feed or raw material. This is particularly important in rabbit feeding because a deficiency in the fibre supply leads to serious digestive problems. Consequently, it is of importance to evaluate the concentration of the different classes of fibres and their effects on the digestive processes.

Unfortunately, because of the diversity of fibre classes (described above), currently no method is adequately able to fractionate DF. However, methods have been developed to estimate the DF content in animal feeds, but none of them correspond to a precise DF fraction. Detailed reviews have been recently published on this subject (Asp and Johansson, 1984; Carré, 1991; Giger-Reverdin, 1995). Accordingly only those techniques currently used in rabbit feeding will be described (Fig. 5.4).

Initially, the Weende method (Henneberg and Stohmann, 1864) must be mentioned because it is quick, simple, cheap and frequently used all over the world. This technique extracts a fibre residue, 'crude fibre' (after an acidic followed by a basic hydrolysis), which corresponds partly to a mixture of cellulose, lignins, cutin and suberin. The main limitations of this method are that crude fibre is too global a criterion and that the fibre residue contains in

RESIDUE OF   CLASS OF POLYMER

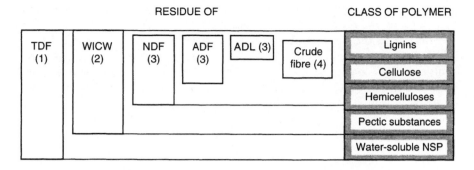

Fig. 5.4. Gravimetric methods for the determination of dietary fibre and identification of the residue of analysis.1: Lee *et al.* (1992); Li (1995); 2: Carré and Brillouet (1989); 3: Van Soest *et al.* (1991); 4: according to the Weende technique (Henneberg and Stohmann, 1864).

fact various proportions of different fibre classes (depending on the raw material analysed): 0.30–1.0 cellulose, 0.14–0.20 pentosans, 0.16–0.90 lignin (e.g. 0.30 for a wheat straw).

The technique of Van Soest and Wine aims to fractionate the cell wall and to obtain a fibre residue without contaminants such as proteins, through the combined activity of detergents (in a neutral and then an acidic environment). The method was developed initially for forages and has been subsequently modified by several authors in order to be used for 'concentrates' (raw materials and compound feeds used in non-ruminant feeding such as cereals and various by-products) (Van Soest *et al.*, 1991), through the use of pre-extraction steps with proteolytic and amylolytic enzymes. The main advantage of this method is that three fibre residues are obtained (see Fig. 5.4) from which it is possible to evaluate the lignins (ADL), the cellulose (ADF-ADL) and the hemicelluloses (NDF-ADF). The main limitations of this method are that the CP content of the NDF fibre is highly variable (10–200 g kg^{-1} for concentrates; NDF could also contain starch or residual pectins), the pectic substances are removed from the NDF residue, and the pre-treatment with enzymes is not totally standardized. However, the Van Soest method is very widely used for animal feeds, because it gives a lignocellulose residue without the main contaminants found for crude fibre, and because this technique is relatively quick and simple and improves noticeably the fractionation of the CW.

Considerable developments have occurred in fibre analysis methods during the last 10 years because of a redefinition of the term 'dietary fibre' for humans and non-ruminant animals to include lignin and NSP resistant to mammalian enzymes. New methods to replace NDF have been developed for measuring total dietary fibre (TDF) (Fig. 5.4), which corresponds to soluble and insoluble NSP (including pectins and β-glucans) + lignins (Lee *et al.*,

1992; Li, 1995). However, the methods have at present only been applied to human foods and remain to be validated for animal feeds.

In conclusion, the determination of the fibre content of a compound feed (Table 5.1) or a raw material (Table 5.2) is highly variable, depending on the analytical method of estimation. The choice of which definition is to be used by the nutritionist thus depends on the type of information required (to relate to digestive processes, to predict the nutritive value).

Physicochemical properties of fibre related to digestion

Particle size

As described below, part of the fibre requirements of rabbits are related to the effect of the large size fibre particles on the passage rate of digesta through the gut.

Particle size can be measured by dry or wet sieving and varies largely depending on the fibre source. Table 5.3 shows the distribution of particles by size of several commercial sources of fibre. Part of this variation can be explained by differences in chemical composition, as cellulose tends to produce larger and thiner particles than lignin. For the six fibrous feeds presented in Table 5.3, the correlations between proportion of particles larger than 0.315 mm and the cellulose and lignin contents of NDF were +0.37 and –0.35, respectively. The low values obtained indicate that other factors are also involved.

Fibre particle size is also modified during the feed manufacturing process (see Chapter 13). Particle size is reduced in successive millings, even with using sieves of the same diameter. In this way, Morisse (1982) by milling a feed one or three times with the same sieve size (4 mm) observed an increase in the proportion of fine particles (<0.25 mm) from 0.308 to 0.738.

On the other hand, fibre composition is not homogeneous among particles of different size within the same feed: the proportion of lignin tends to increase with particle size, because the force required to shear the fibre particles increases with lignification (Van Soest, 1994).

Table 5.3. Proportional distribution of particles by size of some commercial sources of fibre (García *et al.*, 1996).

Particle size (mm)	Paprika meal	Olive leaves	Lucerne hay	Soybean hulls	Treated barley straw	Sunflower husk
< 0.160	0.8414	0.5383	0.5373	0.2446	0.2299	0.0236
0.160–0.315	0.0906	0.0834	0.1774	0.2240	0.2340	0.2384
0.315–0.630	0.0519	0.0898	0.1557	0.2596	0.2108	0.3875
0.630–1.250	0.0160	0.1942	0.1086	0.2351	0.2161	0.3055
> 1.250	0.0000	0.0942	0.0209	0.0367	0.1090	0.0449

Water-holding capacity

Some cell wall constituents, such as β-glucans, pentosans and pectins, are hydrophilic, tending to form gels in solution. On the other hand, hydration is negatively related to the size of the fibre particles (Van Soest, 1994).

A high dietary content of fibre with a high water-holding capacity might slow transit time in the small intestine of rabbits, as occurs in poultry, and reduce the digestibility of other nutrients. It might also increase the weight of the stomach and caecal contents, which would reduce carcass yield and negatively affect feed intake.

Buffering capacity

The cation exchange capacity of fibre is dependent on its concentration of carboxyl, amino and hydroxyl groups (Van Soest *et al.*, 1991). Accordingly, buffering capacity is high in feeds containing pectins (e.g. 70 mEq 100 g^{-1} for beet pulp) and is also significantly higher in legumes than in grasses (50 vs. 13 mEq 100 g^{-1}).

The type of fibre would then interact with the acidity of caecal contents, although little is currently known about this effect in rabbits.

Degradation of dietary fibre in the rabbit gut

Precaecal digestion of fibre

Traditionally, fermentation of dietary fibre has been considered to be a post-ileal activity of the endogenous microflora. However, there is increasing evidence that some components of structural carbohydrates disappear or are degraded prior to entering the caecum of rabbits. This has also been observed in other non-ruminant species such as pigs and poultry.

The extent of precaecal fibre digestion in rabbits varies from 0.07 to 0.19 for crude fibre (Yu *et al.*, 1987), from 0.05 to 0.43 for NDF (Gidenne and Ruckebush, 1989; Merino and Carabaño, 1992) and from 0 to 0.17 for NSP (Gidenne, 1992). In the latter study, arabinose and uronic acids, typical monomers of pectic substances, were largely digested before the ileum (from 0.2 to 0.4). These results imply that from 0.2 to 0.8 of total digestible fibre (including water-soluble NSP) is degraded before the caecum.

The presence of microbial hydrocarbonases in the stomach and in the small intestine of rabbits, especially pectinases and xylanases (Marounek *et al.*, 1995), could partially explain these results. However, it is important to note that partial fibre digestion in the ileum could be important. Using a technique of *in vitro* simulation of the precaecal digestion on 27 complete rabbit diets (the samples were treated with a solution of pepsin–hydrochloric acid and then with a solution containing pancreatic enzymes), Ramos (1995) observed an average NDF disappearance of 0.11, with values ranging from

0 to 0.255. This suggests that some components of fibre could be solubilized, filtered and then be considered as digested, analytical errors which could lead to overestimation of the actual fibre digestion in the ileum.

Caecal digestion of fibre

Fibre degradation is ultimately determined by microbial activity, digesta retention time in the caecum and by fibre chemical composition.

Microbial activity

The caecum is the major organ where microbial activity takes place in rabbits. As was mentioned in Chapter 1, the caecal microbial population secretes enzymes capable of hydrolysing the main components of dietary fibre. Greater enzymatic activity for degrading pectins and hemicellulose than for degrading cellulose has been detected in several studies (Jehl *et al.*, 1995; Marounek *et al.*, 1995). These results parallel those for faecal digestibility of the corresponding dietary fibre constituents in rabbits (Gidenne, 1996), and are also consistent with the smaller counts of cellulolytic bacteria in the rabbit caecum compared with xylanolytic or pectinolytic bacteria (Boulahrouf *et al.*, 1991).

Fermentation time

The retention time of digesta in the caecum can be estimated from the difference in ileo-rectal mean retention time (i-r MRT, h) and minimal transit time (TTm, h) obtained using ileally cannulated animals. The latter value is relatively constant with a range from 3.5 to 4.5 h, averaging 3.7 h (Gidenne, 1994; García, 1997). Several studies (Gidenne *et al.*, 1991b; Gidenne and Perez, 1993; Gidenne, 1994; García, 1997) have measured the i-r MRT for diets based on lucerne hay, wheat bran and fibrous by-products. The results show this trait was linear and negatively correlated with dietary NDF content, which varied from 220 to 470 g kg^{-1} (DM basis). The regression equation obtained was:

i-r MRT = 26.5($\pm$4.9) – 0.0368($\pm$0.015) NDF (DM);
$R^2 = 0.35$; $n = 13$; $P = 0.03$.

According to this equation, the i-r MRT of an average rabbit diet containing 360 g NDF kg^{-1} diet DM would be 13.2. The time of fermentation (i.e. retention in the caecum and proximal colon) could be estimated as 9.5 h (13.2–3.7).

Ileo-rectal MRT is also related to the weight of the caecal contents (CCW as a proportion of body weight; $P = 0.04$), according to studies where both traits were determined (Gidenne, 1992; García, 1997). CCW is easier to determine than retention time, and the amount of information available in the literature is much larger. Some results are shown in Fig. 5.5, where CCW has been related to dietary NDF content. Sampling time significantly affects this trait (see Chapter 1), so that only results obtained using the same methodology have been chosen. A stepwise regression analysis was undertaken by relating CCW to chemical composition and level of dietary fibre. The equation obtained was:

$$CCW = 18.0 - 0.064 \, NDF + 0.000081 \, NDF^2 - 0.0028 \, ADL;$$
$$(\pm 2.1) \quad (\pm 0.12) \quad\quad (\pm 0.0015) \quad\quad (\pm 0.013)$$

$$n = 40; \, R^2 = 0.52; \, P < 0.001$$

This response shows that, when a wide range of level and sources of fibre is considered, dietary NDF content affects CCW quadratically. From this equation, it can be calculated that CCW would reach a minimal value for a dietary NDF content of 395 g kg^{-1} DM. The significant effect of the degree of lignification of NDF on CCW indicates an additional influence of source of fibre. Diets containing low lignified beet pulp or high lignified sunflower husk tended to give, respectively, higher and lower CCW values at the same NDF level.

Another factor related to fermentation time is particle size. As was observed by Björnhag (1972), the particle size of fibre influences the entry of digesta in the caecum. Also, Gidenne (1993) observed that particles smaller than 0.3 mm were retained for longer (10 h more, on average) than particles larger than 0.3 mm.

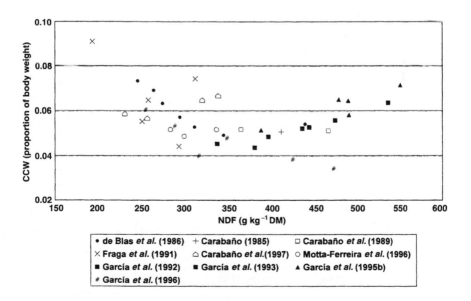

Fig. 5.5. Effect of dietary NDF content on the weight of caecal contents.

Digestion rate

Rate of fermentation of cell wall constituents is a primary factor influencing their digestion efficiency in rabbits, because of the relatively short mean retention time (about 10 h) within the fermentative region.

Caecal microbial NDF degradation rate may be derived from *in situ* rumen measurements, as is shown in Fig. 5.6 for six fibrous feedstuffs used in rabbit diets. From these data, it can be concluded that the relative value of fibre is highly dependent on time of fermentation. For instance, paprika meal has a relatively high degradation rate at 10 h and low at 72 h, whereas the opposite occurs for NaOH-treated wheat straw.

Cell wall composition is the main factor affecting degradation rate. Lignin and cutin are considered almost totally undegradable. Cellulose and hemicellulose are potentially degradable but cellulolytic bacteria require some time for attachment to the cell wall before the degradation starts. Furthermore, digestion is negatively related to lignin concentration. Moreover, because of the crystalline structure of cellulose more time is required to degrade this linear polymer. Pectins, β-glucans, pentosans and galactosans are the components more easily fermented (see Table 5.4).

Faecal NDF digestibility of several fibrous feedstuffs is presented in Table 5.5. The highest value (0.845) was obtained for beet pulp. Fibre in this feed is poorly lignified and would have a lengthy fermentation time in the caecum as it is not made up of long molecules. The lowest NDF digestibility

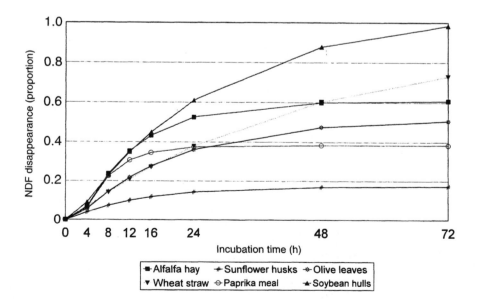

Fig. 5.6. Effect of type of fibre on NDF disappearance in the rumen (Escalona *et al.*, 1995).

Table 5.4. Mean apparent faecal digestibility coefficient of dietary fibre.

Class of dietary fibre	Mean	Range
Lignin (ADL)	0.1–0.15	−0.13 to +0.50
Cellulose (ADF-ADL)	0.15–0.18	0.04–0.37
Hemicellulose (NDF-ADF)	0.25–0.35	0.11–0.60
Pectins (total uronic acids)	0.70–0.76	nd

nd, not determined.

Table 5.5. Neutral detergent fibre digestibility (NDFd) of several feedstuffs in rabbits.

Feedstuff	NDFd	Reference
Dehydrated lucerne	0.15–0.18	Gidenne *et al.* (1991)
Dehydrated lucerne (*n* = 12)	0.255–0.407	Perez (1994)
Lucerne hay (*n* = 6)	0.175–0.276	García *et al.* (1995a, 1996)
Beet pulp	0.845	Gidenne (1987)
Paprika meal	0.351	García *et al.* (1996)
Soybean hulls	0.282	García *et al.* (1996)
Sunflower husk	0.100	García *et al.* (1996)
Barley straw NaOH-treated	0.167	García *et al.* (1996)

(0.10) has been found for sunflower husk, a highly lignified (210 g kg^{-1}) source of fibre with a low proportion (0.262) of fine particles (<0.315 mm). Similar NDF digestibilities were observed for paprika meal and soybean hulls. The lower lignin content in the former (150 vs. 210 g kg^{-1}, respectively) was compensated for by its lower proportion of fine particles (0.932 vs. 0.469, respectively) and by a shorter fermentation time.

Fibre digestibility is not significantly affected by the dietary level of fibre. In fact, it may be concluded that the quantity of fibre entering the caecum is not a limiting factor for the fermentation processes, as the digesta retention time in the caecum is relatively short, allowing, predominantly, degradation of the more easily digestible fibre fractions such as pectins or hemicelluloses. Moreover, as indicated above, the retention time in the caecum increased proportionally to the reduction of the fibre intake, and could then compensate for an eventual limitation of the quantity of fibre entering the caecum. Overall, transit in the proximal part of the tract is regulated according to the fibre intake. An increase in fibre intake stimulates transit and thus increases the rate of passage in the overall tract, while the retention time is slightly prolonged in the stomach and slightly shortened in the small intestine.

Fermentation pattern
Volatile fatty acids (VFA). These are the main products of carbohydrate microbial fermentation. Consequently, their concentration in the fermentative

areas can be used as an indirect estimation of microbial activity. VFA are rapidly absorbed in the hindgut and provide a regular source of energy for the rabbit. Butyrate seems to be a preferential source of energy for the hindgut, whereas acetate is mainly metabolized in the liver for lipogenesis and cholesterogenesis (Vernay, 1987). Furthermore, VFA have been suggested as being a stimulative factor of colon mucosal growth (Chiou *et al.*, 1994). Although VFA have been proposed as a protective factor against pathogen microflora (*Escherichia coli*) infections (Prohaszka, 1980; Wallace *et al.*, 1989), recent work (Padilha *et al.*, 1996) has demonstrated a lack of relationship between these two parameters.

Carbohydrate uptake by intestinal flora includes most of the cell wall constituents, besides starch not digested in the small intestine, and endogenous mucopolysaccharides. Additionally, protein residues of ileal digesta (undigested dietary protein, mucosal-cell protein, enzymes) can be utilized (after deamination) as an energy source for the microbial population. The relative contribution of these sources to total caecal VFA production is unknown.

Some studies (Fraga *et al.*, 1984; García *et al.*, 1995b, 1996; Motta-Ferreira *et al.*, 1996; Carabaño *et al.*, 1997) have determined the composition of caecal contents using different diets and the same methodology. Caecal VFA concentration averaged 57.2 mmol l^{-1}, and ranged from 31.8 to 88.5 mmol l^{-1}. It tended ($P = 0.09$) to increase linearly with dietary NDF digestible content by 1.42 mmol l^{-1} per each 10 g unit increment. Consequently, source of fibre affected VFA concentration, as poorly digested highly lignified sources of fibre (grape pomace, wheat straw, sunflower hulls) produced the lowest VFA concentration (Carabaño *et al.*, 1988; García *et al.*, 1996; Motta-Ferreia *et al.*, 1996). Chiou *et al.* (1994), using isolated components of dietary fibre (cellulose, pectins and lignin) and lucerne hay, also observed a negative effect of lignin on this trait. Furthermore, the intercept of this relationship (43.9 ± 8.2 mmol l^{-1}) indicates that NDF fermentation only accounts for around 0.25 of the average total VFA concentration. The remainder would be produced from pectins (not included in NDF) and water-soluble DF_l, and also from endogenous materials and/or other dietary components. In this way, Vernay and Raynaud (1975) observed a caecal VFA concentration of 17.8 mmol l^{-1} in fasted rabbits, suggesting that a significant amount of endogenous materials can be fermented in the caecum.

The caecal VFA profile is specific to the rabbit, with a predominance of acetate (C2 = 60–80 mmol 100 ml^{-1}) followed by butyrate (C4 = 8–20 mmol 100 ml^{-1}) and then by propionate (C3 = 3–10 mmol 100 ml^{-1}). These molar proportions are affected by fibre level. For instance, the proportion of butyrate generally increases significantly when fibre level decreases.

Caecal pH. This is frequently determined in digestion studies as it gives an estimation of the extent of the fermentation, and because of its possible negative relationship with diarrhoea incidence and health status of the rabbits.

However, its relationship with the extent of the fibre degradation is not clear. Previous work (Pote *et al.*, 1980; Champe and Maurice, 1983; Gidenne *et al.*, 1991a; García *et al.*, 1995b) has shown that, when lucerne hay is used as the main source of fibre, neither type of lucerne hay nor dietary fibre concentration has an effect on caecal pH. In other studies caecal pH was modified significantly by varying both type and level of fibre (Fig. 5.7). In this way, the inclusion of increasing amounts of wheat straw and lucerne hay to replace both barley grain and wheat bran increased caecal pH (de Blas *et al.*, 1986). The same effect was obtained when wheat straw, lucerne hay and wheat bran replaced wheat grain (Bellier and Gidenne, 1996). On the other hand, the substitution of sugar-beet pulp for barley grain or lucerne hay increased acidity of caecal contents (García *et al.*, 1992, 1993; Carabaño *et al.*, 1997). In conclusion, level of fibre can increase or decrease caecal pH depending on the source of fibre used.

The effect of type of fibre on caecal pH has been investigated in several studies, indicating that inclusion of large-sized, low-pectin fibre from wheat straw would tend to increase caecal pH, whereas the opposite would occur when including beet pulp in the diet. Finally, the use of balanced sources of fibre, such as lucerne hay, would not modify caecal acidity. These results agree with a recent study (García *et al.*, 1996) using six semipurified diets based on fibrous feedstuffs differing both in chemical and physical characteristics. In this study, caecal pH was mainly explained by the uronic acid concentration of the

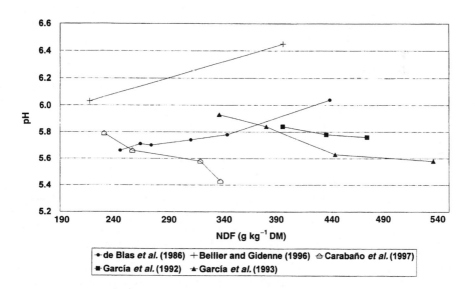

Fig. 5.7. Effect of dietary NDF content on caecal pH.

feed and by the proportion of fine particles (<0.315 mm) ($R^2 = 0.53$; $P < 0.001$; $n = 58$). Both variables were negatively correlated with pH.

From a chemical point of view, caecal pH would be expected to be related to the main sources of H^+ and OH^-, AGV and $N\text{-}NH_3$, respectively. However, a poor or no relationship has been found between these variables (Bellier, 1994; García *et al.*, 1996). This could be due to the presence of buffer substances in the caecum from endogenous or feed origin, which would also explain the stability of caecal pH among animals fed different diets.

Effect of fibre level on rabbit digestion

When the fibre level increases, the feed digestibility normally decreases. This effect depends on the fibre fraction added in the feed. As an example, one unit of lignocellulose addition led to a −1.2 to −1.5 point decrease in organic matter digestibility (OMD). Thus, lignocellulose is considered to have a negative effect on feed digestion in the rabbit, while NDF level (which includes hemicelluloses, a more digestible fraction) only had a dilution effect on OMD (one point decrease in OMD per 10 g unit increment in NDF). In this way, the inclusion of purified lignin in the diet by the substitution of lucerne hay produced a reduction in villi height either in the jejunum or in the ileum (Chiou *et al.*, 1994), that could impair both enzymatic activity and nutrient absorption. Furthermore, the inclusion of lignified sources of fibre (sunflower hulls, olive leaves and paprika meal) elicited a reduction of saccharase activity in the ileum compared with diets based on lucerne, soybean hulls or NaOH-treated straw (1070 vs. 1560 mmol glucose g^{-1} protein, respectively) (García *et al.*, unpublished). Other lignified sources of fibre, such as grape by-products, contained in the lignin fraction (ADL) and phenolic compounds, such as tannins, decrease protein utilization in the ileum (Merino and Carabaño, 1992).

References

Asp, N.G. and Johansson, C.G. (1984) Dietary fiber analysis. *Nutrition Abstracts and Reviews* 54, 735–752.
Barry, J.L., Hoebler, C., David, A., Kozlowski, F. and Gueneau, S. (1990) Cell wall polysaccharides determination: comparison of detergent method and direct monomeric analysis. *Sciences des Aliments* 10, 275–282.
Bellier R. (1994) Contrôle nutritionel de l'activité fermentaire caecale chez le lapin. PhD Thesis. Ecole Nationale Supérieure Agronomique, Institut National Polytechnique de Toulouse.
Bellier, R. and Gidenne, T. (1996) Consequences of reduced fibre intake on digestion, rate of passage and caecal microbial activity in the young rabbit. *British Journal of Nutrition* 75, 353–363.

Björnhag, G. (1972) Separation and delay of contents in the rabbit colon. *Swedish Journal of Agricultural Research* 7, 105–114.

de Blas, J.C., Santomá, G., Carabaño, R. and Fraga, M.J. (1986) Fiber and starch levels in fattening rabbit diets. *Journal of Animal Science* 63, 1897–1904.

Boulahrouf, A., Fonty, G. and Gouet, P. (1991) Establishment, counts and identification of the fibrolytic bacteria in the digestive tract of rabbit. Influence of feed cellulose content. *Current Microbiology* 22, 1–25.

Brillouet, J.M., Rouau, X., Hoebler, C., Barry, J.L., Carré, B. and Lorta, E. (1988) A new method for determination of insoluble cell-walls and soluble non-starchy polysaccharides from plant materials. *Journal of Agricultural and Food Chemistry* 36, 969–979.

Carabaño, R. (1985) Influencia de la composición química de la dieta sobre distintos parámetros cecales de conejos en crecimiento y su relación con el aporte nutritivo de las heces coprófagas. PhD Thesis, Universidad Politécnia de Madrid.

Carabaño, R., Fraga, M.J., Santoma, G. and de Blas, J.C. (1988) Effect of diet on composition of caecal contents and on excretion and composition of soft and hard feces of rabbits. *Journal of Animal Science* 66, 901–910.

Carabaño, R., Fraga, M.J. and De Blas, J.C. (1989) Effect of protein source in fibrous diets on performance and digestive parameters of fattening rabbits. *Journal of Applied Rabbit Research* 12, 201–204.

Carabaño, R., Motta-Ferreira, W., de Blas, J.C. and Fraga, M.J. (1997) Substitution of sugarbeet pulp for alfalfa hay in diets for growing rabbits. *Animal Feed Science and Technology* 65, 249–256.

Carré, B. (1991) The chemical and biological bases of a calculation system developed for predicting dietary energy values: a poultry model. In: Fuller, M.F. (ed.) In Vitro *Digestion for Pigs and Poultry*. CAB International, Wallingford, UK, pp. 67–85.

Carré B. and Brillouet J.M. (1989) Determination of water-insoluble cell-walls in feeds: interlaboratory study. *Journal of the Association of Official Analytical Chemists* 72, 463–467

Champe, V.A. and Maurice, D.V. (1983) Response of early weaned rabbits to source and level of dietary fibre. *Journal of Animal Science* 56, 1105–1114.

Chesson, A. (1995) Dietary fiber. In: Stephen, A.M. (ed.) *Food Polysaccharides and their Applications*. Marcel Dekker, New York, pp. 547–576.

Chiou, P.W.S., Yu, B. and Lin, C. (1994) Effect of different components of dietary fiber on the intestinal morphology of domestic rabbits. *Comparative Biochemistry and Physiology* 108A, 629–638.

Englyst, H. (1989) Classification and measurement of plant polysaccharides. *Animal Feed Science and Technology* 23, 27–42.

Escalona, B., Rocha, R., García, J., Carabaño, R. and de Blas, C. (1995) Efecto del tipo de fibra sobre la extensión y velocidad de degradación de la fibra neutro detergente. *ITEA* 16, 45–47.

Fraga, M.J., Barreno, C., Carabaño, R., Mendez, J. and de Blas, J.C. (1984) Efecto de los niveles de fibra y proteína del pienso sobre la velocidad de crecimiento y los parámetros digestivos de los conejos. *Anales INIA Serie Ganadera* 21(6), 91–110.

Fraga, M.J., Pérez, P., Carabaño, R. and de Blas, J.C. (1991) Effect of type of fiber on the rate of passage and on the contribution of soft feces to nutrient intake of finishing rabbits. *Journal of Animal Science* 69, 1566–1574.

García, J. (1997) Estudio de distintas fuentes de fibra en la alimentación del conejo. PhD Thesis, Universidad Politécnica de Madrid.

García, G., Gálvez, J.F. and de Blas, J.C. (1992) Substitution of barley grain by sugar-beet pulp in diets for finishing rabbits. 2. Effect on growth performance. *Journal of Applied Rabbit Research* 15, 1017–1024.

García, G., Gálvez, J.F. and de Blas, J.C. (1993) Effect of substitution of sugarbeet pulp for barley in diets for finishing rabbits on growth performance and on energy and nitrogen efficiency. *Journal of Animal Science* 71, 1823–1830.

García, J., Pérez-Alba, L., Alvarez, C., Rocha, R., Ramos, M., and de Blas, C. (1995a) Prediction of the nutritive value of lucerne hay in diets for growing rabbits. *Animal Feed Science and Technology* 54, 33–44.

García, J., de Blas, J.C., Carabaño, R. and García, P. (1995b) Effect of type of lucerne hay on caecal fermentation and nitrogen contribution through caecotrophy in rabbits. *Reproduction, Nutrition and Development* 35, 267–275.

García, J., Carabaño, R., Pérez-Alba, L. and de Blas, C. (1996) Effect of fibre source on neutral detergent fibre digestion and caecal traits in rabbits. In: Lebas, F. (ed.) *Proceedings of the 6th World Rabbit Congress*. Association Française de Cuniculture, Lempdes, pp. 175–180.

Gidenne, T. (1987) Effet de l'addition d'un concentré riche en fibres dans une ration à base de foin, distribuée à deux niveaux alimentaires chez la lapine adulte. 2. Mesures de digestibilité. *Reproduction, Nutrition and Development* 27, 801–810.

Gidenne, T. (1992) Effect of fibre level, particle size and adaptation period on digestibility and rate of passage as measured at the ileum and in the faeces in the adult rabbit. *British Journal of Nutrition* 67, 133–146.

Gidenne, T. (1993) Measurement of the rate of passage in restricted-fed rabbits: effect of dietary cell wall level on the transit of fibre particles of different sizes. *Animal Feed Science and Technology* 42, 151–163.

Gidenne, T. (1994) Effets d'une réduction de la teneur en fibres alimentaires sur le transit digestif du lapin. Comparaison et validation de modéles d'ajustement des cinétiques d'excrétion fécale des marqueurs. *Reproduction, Nutrition and Development* 34, 295–307.

Gidenne, T. (1996) Nutritional and ontogenic factors affecting rabbit caeco-colic digestive physiology. In: Lebas, F. (ed.) *Proceedings of the 6th World Rabbit Congress*. Association Française de Cuniculture, Lempdes, pp. 13–28.

Gidenne, T. and Perez, J.M. (1993) Effect of dietary starch origin on digestion in the rabbit. 2. Starch hydrolysis in the small intestine, cell wall degradation and rate of passage measurements. *Animal Feed Science and Technology* 42, 249–257.

Gidenne, T. and Ruckebusch, Y. (1989) Flow and passage rate studies at the ileal level in the rabbit. *Reproduction, Nutrition and Development* 29, 403–412.

Gidenne, T., Scalabrini, F. and Marchais, C. (1991a) Adaptation digestive du lapin à la teneur en constituants pariétaux du régime. *Annales de Zootechnie* 40, 73–84.

Gidenne, T., Carre, B., Segura, M., Lapanouse, A. and Gomez, J. (1991b) Fibre digestion and rate of passage in the rabbit: effect of particle size and level of lucerne meal. *Animal Feed Science and Technology* 32, 215–221.

Giger-Reverdin, S. (1995) Review of the main methods of cell wall estimation:

interest and limits for ruminants. *Animal Feed Science and Technology* 55, 295–334.

Henneberg, W. and Stohmann, F. (1864) *Beiträge zur Begründung einer rationnellen Fütterung der Wiederkäuer*, Vol. 2. Schwetschke, Braunschweig, Germany. (Cited by Giger-Reverdin, 1995.)

Jehl, N., Martin, C., Nozière, P., Gidenne, T. and Michalet-Doreau, B. (1995) Comparative fibrolytic activity of differnt microbial populations from rabbit caecum and bovine rumen. *Annales de Zootechnie* 44, 186 (abstract).

Lee, S.C., Prosky, L. and De Vries, J.W. (1992) Determination of total, soluble and insoluble dietary fiber in foods – enzymatic-gravimetric method, MES-TRIS buffer: collaborative study. *Journal of the Association of Official Analytical Chemists* 75, 395–416.

Li, B.W. (1995) Determination of total dietary fiber in foods and food products by using a single-enzyme, enzymatic-gravimetric method: interlaboratory study. *Journal of the AOAC International* 78, 1440–1444.

Marounek, M., Vovk, S.J. and Skrinova, V. (1995) Distribution of activity of hydrolytic enzymes in the digestive tract of rabbits. *British Journal of Nutrition* 73, 463–469.

McDougall, G.J., Morrison, I.M., Stewart, D. and Hillman, J.R. (1996) Plant cell walls as dietary fibre: range, structure, processing and function. *Journal of the Science of Food and Agriculture* 70, 133–150.

Merino, J. and Carabaño, R. (1992) Effect of type of fibre on ileal and faecal digestibilities. *Journal of Applied Rabbit Research* 15, 931–937.

Morisse, J.P. (1982) Taille des particules de l'aliment utilisé chez le lapin. *Revue Médecine Vétérinaire* 133, 635–642.

Motta-Ferreira, W., Fraga, M.J. and Carabaño, R. (1996) Inclusion of grape pomace, in substitution for lucerne hay, in diets for growing rabbits. *Animal Science* 63, 167–174.

Padilha, M.T.S., Licois, D., Gidenne, T., Carré, B., Coudert, P. and Lebas, F. (1996) Caecal microflora and fermentation pattern in exclusively milk-fed young rabbit. In: Lebas, F. (ed.) *Proceedings of the 6th World Rabbit Congress*. Association Française de Cuniculture, Lempdes, pp. 247–251.

Perez, J.M. (1994) Digestibilité et valeur energetique des luzernes deshydratées pour le lapin: influence de leur composition chimique et de leur technologie de preparation. In: *Vièmes Journées de la Recherche Cunicole, La Rochelle*, Vol. 2, pp. 355–364.

Pote, L.M., Cheeke, P.R. and Patton, N.M. (1980) Utilization of diets high in lucerne meal by weanling rabbits. *Journal of Applied Rabbit Research* 3, 5–10.

Prohaszka, L. (1980) Antibacterial effect of volatile fatt acids in enteric *E. coli* infections of rabbits. *Zentrabl Veterinaermed Reihe B* 27, 631–639.

Ramos, M.A. (1995) Aplicación de técnicas enzimáticas de digestión *in vitro* a la valoración nutritiva de piensos para conejos. PhD Thesis, Universidad Complutense de Madrid.

Selvendran, R.R., Stevens, B.J.H. and Du Pont, M.S. (1987) Dietary fiber: chemistry, analysis and properties. In: Chixhester, C.O. (ed.) *Advances in Food Research*. Academic Press, New York, pp. 117–212.

Van Soest, P.J. (1994) *Nutritional Ecology of the Ruminant*, 2nd Edn. Cornell University Press, Ithaca, New York.

Van Soest P.J., Conklin, N.L. and Horvath, P.J. (1987) Tannins in foods and feeds. In: *Proceedings of the Nutrition Conference for Feed Manufacturers, 26–28 October, Syracuse*, pp. 115–122. (Cited by Giger-Reverdin, 1995.)

Van Soest, P.J., Robertson, J.B. and Lewis, B.A. (1991) Methods for dietary fiber, neutral detergent fiber and non-starch polysaccharides in relation to animal nutrition. *Journal of Dairy Science* 74, 3583–3597.

Vernay, M. (1987) Origin and utilization of volatile fatty acids and lactate in the rabbit: influence of the faecal excretion pattern. *British Journal of Nutrition* 57, 371–381.

Vernay, M. and Raynaud, P. (1975) Répartition des acides gras volatils dans le tube digestif du lapin domestique. II. Lapins soumis au jeûne. *Annales Recherches Vétérinaires* 6, 369–377.

Wallace, R.J., Falconer, M.L. and Bhargava, P.K. (1989) Toxicity of volatile fatty acids at rumen pH prevents enrichment of *Escherichia coli* by sorbitol in rumen contents. *Current Microbiology* 19, 277–281.

Yu, B., Chiou, P.W.S., Young, Ch.L. and Huang, H.H. (1987) A study of rabbit T-type cannule and its ileal digestibilities. *Journal of the Chinese Society of Animal Science* 16, 73–81.

6. Feed Evaluation

M.J. Villamide[1], L. Maertens[2], C. de Blas[1] and J.M. Perez[3]

[1]Departamento de Producción Animal, ETS Ingenieros Agrónomos, Universidad Politécnica de Madrid, 28040 Madrid, Spain; [2]Agriculture Research Centre–Ghent, Rijksstation voor Kleinveeteelt, Burg. Van Gansberhelaan 92, 98290 Merelbeke, Belgium; [3]Station de Recherches Cunicoles, INRA Centre de Toulouse, BP 27, 31326 Castanet-Tolosan Cedex, France

Units for feed evaluation

Energy

The most commonly used unit for expressing energy value in rabbit diets is digestible energy (DE). However, the use of DE leads to some systematic errors in feed evaluation, especially for certain groups of ingredients.

For instance, DE overestimates the energy content of protein concentrates, as it does not take into account either the higher energy losses in urine, or the energy cost of urea synthesis in the liver associated with the use of an excess of this type of ingredient in the diet. Similarly, the DE content of feedstuffs containing significant amounts of digestible fibre, e.g. beet or citrus pulps or soybean hulls, also overvalues their relative energy concentration, as the use of DE does not consider the energy losses (methane and heat of fermentation) linked to the microbial digestion of fibre in the caecum. On the other hand, the relative energy content of fats and feeds with a high fat content is underestimated by DE, because dietary fatty acids are retained in the body more efficiently than other nutrients.

These disadvantages explain the current interest in replacing DE by metabolizable (ME) or net energy (NE). However, most of the information available at present has been obtained as DE. As a consequence, DE should be retained as the unit of expression of the energy value of feedstuffs in rabbits for the foreseeable future. However, some important points should be taken into account in order to minimize errors.

Thus, to prevent the use of protein concentrates as energy sources, a maximum protein content of compound feeds should be established. This restriction is also required to control the incidence of diarrhoea (de Blas et al., 1981) and to reduce environmental pollution by animal excreta (Maertens and Luzi, 1996). An alternative is to use ME corrected to N equilibrium (MEn), as in poultry nutrition, instead of DE. Values of MEn can be derived for each ingredient by subtracting 4.8 kJ g^{-1} digestible protein from its DE content (Maertens, 1992).

Digestible fibre and fat content of compound feeds should also be limited, to allow for a maximal energy intake (García *et al.*, 1993) and for technological reasons (rabbit feed must be pelleted), respectively. These restrictions limit the errors associated with the use of DE to less than 5% in extreme practical diets (de Blas *et al.*, 1985; Ortiz *et al.*, 1989; de Blas and Carabaño, 1996; Fernández and Fraga, 1996). In any case, the use of correction factors (-2.51 and $+5.85$ kJ DE kg^{-1}) for ingredients or diets containing a high proportion of digestible neutral detergent fibre (NDF) and ether extract (EE), respectively, might be envisaged.

Protein and amino acids

The use of crude protein (CP) and amino acids to formulate diets for rabbits is still very common. However, protein digestibility varies from 0.15 (grape pomace) to 0.85 (soybean meal), according to the review of Santomá *et al.* (1989). Furthermore, recent work has shown large variations in essential amino acid digestibilities for lucerne hays harvested at different stages of maturity (García *et al.*, 1995), or between the average lysine digestibility in a basal diet and that of L-lysine (Taboada *et al.*, 1994). A similar situation has been described for D,L-methionine (Taboada *et al.*, 1996) and L-threonine (de Blas *et al.*, 1996).

These results indicate that the use of digestible, instead of crude, units for expressing protein value would considerably improve the accuracy of feed evaluation. However, more information is needed on amino acid digestibility of the most commonly used ingredients.

Fibre

Fibre is one of the main components of rabbit diets, which usually contain 300–380 g of NDF kg^{-1}. Diets must supply a minimum of fibre to allow adequate digestive transit and to prevent incidence of diarrhoea. Accordingly, present feeding standards (e.g. de Blas *et al.*, 1986; INRA, 1989; Maertens, 1992) include a minimum content of crude or acid detergent fibre (ADF). They also include a minimum content of indigestible crude fibre (CF), as low-lignified highly digestible sources of fibre might lead to a longer retention time and to a decrease in performance. However, recent studies have shown that substitution of lucerne hay by beet pulp in isofibrous diets does not affect either rate of passage in the whole digestive tract (Gidenne *et al.*, 1987; Fraga *et al.*, 1991) or performance in the fattening period (Motta, 1990). Other work (Björnhag, 1972; Gidenne, 1993) has shown that particle size is a main factor affecting rate of passage. According to these authors, the most effective fibre fraction in terms of promoting transit time is that longer than 0.3 mm. In this respect, Fraga *et al.* (1991) observed a relatively long retention time for rice hulls. This feedstuff is characterized by a very low fibre digestibility, but also by a small particle size. Consequently, it could be envisaged that future recommendations might include two minimal values, one for total and the other for long fibre, as occurs in dairy cows.

Methodology of feed evaluation

Complete diets

The nutritive value of complete diets is usually determined by digestibility assays. The standardization of the procedures used in these assays is the first step in reducing the variability of the results. In this way, a European reference method for *in vivo* determination of diet digestibility in rabbits has been proposed by the European Group on Rabbit Nutrition (Pérez *et al.*, 1995a). The most relevant variables to control in a digestibility assay are the length of experimental period and the number of animals used. The recommended values are at least 7 days of adaptation period and 4 days of collection period (which imply 5 days of control) using ten rabbits per treatment. No advantage in accuracy was found by Pérez *et al.* (1996a) when increasing the adaptation period from 7 to 14 days. The number of replicates can be decreased when the length of the collection period increases. Villamide and Ramos (1994) found the same variability for DM digestibility using ten rabbits and 4 days, eight rabbits and 7 days or seven rabbits and 10 days of collection period. Similarly, Lebas *et al.* (1994) obtained the same accuracy in digestibility determinations with ten cages of one rabbit or with four cages of four rabbits each, although in the latter case there is a greater risk of missing data if any of the animals of the group have health problems.

Other sources of variation in digestibility assays are rabbit breed, sex, litter, age and physiological state. However, no differences have been found for meat breeds of rabbit of the same size (Maertens and De Groote, 1982; Dessimoni, 1984). Furthermore, because of the low sexual dimorphism of this species, the effect of sex is not relevant (Xiccato *et al.*, 1992; Pérez *et al.*, 1995b) and, for examining the possible effect of litter, rabbits of the same litter should be distributed evenly across the treatments.

Digestibility measurements are usually performed with growing rabbits, but the effect of age during this period, and the validity of the extrapolation of the result obtained with young animals to reproductive females is not clear. It seems that the effect of age on energy and protein digestibility from 7 weeks to slaughter is limited (Maertens and De Groote, 1982; Xiccato and Cinetto, 1988). However, as the caecal content weight increases during the first weeks after weaning (by 70% from 30 to 40 days of age; Peeters *et al.*, 1992), a significant part of the ingested feed remains in the digestive tract, leading to an overestimation of the digestibility in young rabbits (Blas *et al.*, 1991; Fernández *et al.*, 1994). On the other hand, comparisons between digestibility of growing rabbits and breeding does are contradictory. Maertens and De Groote (1982) and Perez *et al.* (1996b) found higher digestibility values in growing rabbits than in breeding does, whereas Xiccato *et al.* (1992) obtained higher digestibility of fibrous fractions but lower values for ether extract for growing rabbits and no differences in other components, and de Blas *et al.* (1996) found a higher digestibility in breeding does, especially for NDF.

Digestibility determinations in rabbits have a higher variability than in other non-ruminant species. The average coefficient of variation (CV) for DE estimations are 2.8, 1.9 and 1.7% in rabbit, pigs and for AMEn in poultry, respectively (Villamide, 1996). The accuracy also depends on the component studied. In a recent inter-laboratory study (Perez *et al.*, 1995b) the repeatability (standard deviation within laboratories) of energy, CP and NDF digestibility was 0.017, 0.031 and 0.027, respectively. However a very good reproducibility (standard deviation among laboratories) was obtained in this study, when all the laboratories used the reference method (0.009, 0.019 and 0.057 for energy, CP and NDF digestibility, respectively). Thus, the lowest CV in digestibility assays corresponds to energy (2.8%) and the highest to the fibrous fractions (from 21.3 to 33.8% for NDF and ADF, respectively).

Feedstuffs

The nutritive value of feedstuffs can be determined directly or can be estimated using the substitution method. Some relatively balanced and palatable feedstuffs, like lucerne hay and wheat bran, can be evaluated directly, thus using these feedstuffs as sole feeds in a digestibility assay. Maertens and De Groote (1981) did not find differences between lucerne hay evaluated by substitution (0.2, 0.4 and 0.6 substitution) or directly. Recently, García *et al.* (1995) determined energy, NDF, CP and amino acid digestibility of five samples of lucerne hay using the direct method and obtained a good accuracy (mean CV 2.1, 1.3, 4.9, 2.7 and 2.7% for energy, CP, NDF, lysine and methionine digestibility, respectively).

Most of the feedstuffs are not nutritionally balanced in relation to the requirements of rabbits and, when they are fed as a sole diet, the digestive process can be different from that in normal diets. This problem is partially corrected using the substitution method, where a basal diet of known nutritive value is substituted by the test feedstuff. The evaluation of the feedstuff is undertaken by difference. The correct use of this method implies that there is no interaction between the basal diet and the test ingredient. Therefore, both the substitution rate of the test ingredient and the basal diet should be designed to prevent interactions and to obtain accurate estimates of its nutritive value, taking into account that the higher the substitution rate the lower the error (Fig. 6.1) but also the greater the probability of interaction between the test ingredient and the basal diet.

An increase in substitution rate from 0.15 to 1 leads to a linear increase (P = 0.008) in the estimation of the DE of sugar beet pulp from 10.5 to 14.5 MJ kg^{-1} DM (Fig. 6.2). This effect might be related to a longer retention time of digesta in the caecum of diets with high levels of soluble fibre (de Blas and Carabaño, 1996). An effect of basal diet was also detected for citrus and sugar beet pulps (Table 6.1). Nutritive values of both feedstuffs were significantly lower when they were estimated from the lowest energy basal diet, probably because the high levels of indigestible fibre in this diet produced a lower entry rate of the potentially digestible fibre of pulps into the caecum.

The opposite effect occurs when high subtitution rates (>0.2) of feedstuffs with high levels of indigestible fibre are evaluated. The use of basal diets with a high proportion of wheat straw leads to an underestimation of the test feedstuff (Villamide *et al.*, 1991). Another dietary interaction has been observed with fats. Maertens *et al.* (1986) obtained higher DE values for tallow at 0.06 than at 0.12 substitution. Similarly, Santomá *et al.* (1987) and Fraga *et al.* (1989) found an increase in the digestibility coefficient of 0.058 of all nutrients when 30–60 g fat was added to the diets.

When interactions between feedstuffs are expected, or low rates of inclusion (<0.2) have to be used (because of technological or nutritional problems), substitution of the basal diet at several rates is recommended (Villamide, 1996). The linearity between the dietary nutritive value and the substitution rate is analysed and, if it is established, estimation of the nutritive value of the test feedstuff is undertaken by regression and extrapolation to total substitution. When the relation is non-linear, a nutritive value of the test feedstuff can be

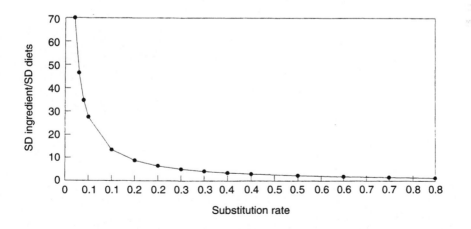

Fig. 6.1. Effect of substitution rate on the standard deviation (SD) of the estimated nutritive value of ingredients in relation to the standard deviation of complete diets (Villamide, 1996).

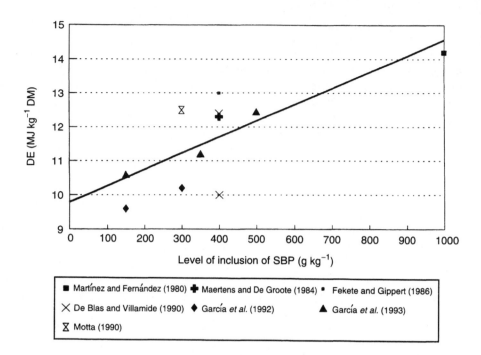

Fig. 6.2. Effect of level of inclusion of sugar beet pulp in the experimental diets on DE estimations (de Blas and Carabaño, 1996).

Table 6.1. Effect of type of basal diet on the nutritive value of citrus and beet pulps (de Blas and Villamide, 1990).

	Citrus pulp				Beet pulp			
Basal diet DE (MJ kg^{-1} DM)	10.01	12.32	SE	*P*	10.01	12.32	SE	*P*
Digestible energy (MJ kg^{-1} DM)	11.27	13.09	0.73	*	9.97	12.37	0.54	*
Protein digestibility	0.173	0.885	0.15	*	0.178	0.764	0.07	*
ADF digestibility	0.677	0.827	0.04	*	0.385	0.717	0.05	*

*Significant at $P < 0.01$.

assigned for a recommended range of inclusion. Outside this range the nutritive value should be calculated from second or third degree equations.

The above methodology has been extensively applied to energy value determinations, and can be used for protein evaluation of feedstuffs (at least, with medium content of CP), but the results obtained for fibre digestibility are erratic and highly variable. Maertens and De Groote (1984) evaluated 14 feedstuffs and obtained three negative values of CF digestibility, one of them for soybean meal, while the highest CF digestibility (0.71) corresponded to full-fat soybean. Likewise, soybean meal ADF digestibility determined at three substitution rates and with two different basal diets (Villamide *et al.*, 1991) varied from −0.14 to +0.51, although the values calculated by extrapolation of the regression equations were 0.38 and 0.32 for the two basal diets. In this study, the CV varied from 27 to 114%, with the average value being 50%. More repeatable and accurate values were determined for feedstuffs with high or medium fibre content. De Blas *et al.* (1989) found an ADF digestibility of wheat straw of 0.33 and 0.145 for substitution rates of 0.2 and 0.4, respectively, and a mean CV for the estimation of 15%. However, for this feedstuff the CP digestibility showed erratic values (from 0.42 to 2.23). Similarly, Villamide *et al.* (1989) obtained a relatively good accuracy for ADF digestibility (0.096 ± 0.039; 0.277 ± 0.060; 0.583 ± 0.056) of wheat bran, corn gluten feed and distilled dried grains and soubles (DDGS), respectively. The reason for this discrepancy between fibre digestibility values and for their low accuracy is the high variability in fibre digestibility of complete diets. This problem also affects the evaluation of other components when they are present at very low concentrations (i.e. CP of wheat straw), but the influence of an incorrect evaluation of CP digestibility of these feedstuffs on overall dietary values is small.

Another method of possible use in analysis of fibre digestibility is the utilization of semipurified diets where all the fibre comes from the test feedstuff. The results obtained with this method increase the precision of the estimates, although sometimes the results differ from those obtained using the substitution method (García *et al.*, 1996). The question is whether these fibrous sources maintain their value when they are in combination with another kind of fibre or when they are included at lower levels in commercial diets.

The effect of errors in DE determinations of experimental diets on the DE estimate of ingredients is shown in Table 6.2. Small differences in the nutritive value of experimental diets result in large differences

Table 6.2. Error in the estimation of mean DE of an ingredient (12.55 MJ kg^{-1} DM, DE of basal diet 11.3 MJ kg^{-1}) when the variables were measured with ±1% error using substitution rates of 0.2 and 0.4. Figures express % of error and the difference between the actual and measured DE of ingredient (kJ kg^{-1} DM) (Villamide, 1996).

	0.2 substitution rate		0.4 substitution rate	
	%	kJ kg^{-1}	%	kJ kg^{-1}
Error in basal diet				
DM intake or excreted, GE faeces	2.25	284	0.85	109
GE basal diet	5.85	736	2.20	276
Error in substituted diets				
DM intake or excreted, GE faeces	2.73	343	1.35	171
GE test diets	7.33	920	3.67	460

Table 6.3. Composition of some feeds commonly used for rabbits (data in g kg⁻¹ on as-fed basis).

	DM	Ash	CP	EE	CF	NDF	ADF	ADL	ST	Sugar	Lys	Met	SAA	Thr	Trp	Ca	P	Na	Cl	Mg	K	CPd	DE (MJ kg⁻¹)
Cereals																							
Barley	880	22	108	25	46	175	55	9	520	25	3.7	1.4	3.7	3.3	1.3	0.6	3.6	0.2	1.4	1.3	5.1	0.67	12.9
Corn	880	12	92	38	19	100	25	5	640	15	2.6	1.8	4.0	3.4	0.6	0.2	2.5	0.1	0.5	1.1	3.2	0.65	13.1
Oats	880	26	106	51	111	280	135	22	370	13	4.4	1.9	5.3	3.7	1.3	0.6	3.0	0.3	0.7	1.3	4.0	0.73	10.9
Triticale	880	18	116	17	23	115	31	9	580	45	4.3	1.9	4.8	3.7	1.4	0.5	3.4	0.1	0.5	1.2	4.2	0.75	12.9
Wheat	880	16	110	22	22	105	31	9	610	30	3.2	1.8	4.5	3.4	1.3	0.4	3.5	0.2	0.6	1.2	4.1	0.77	13.1
Cereal by-products																							
Corn gluten feed	900	67	215	43	78	312	94	12	190	20	7.1	4.1	9.0	8.0	1.6	1.7	8.6	1.2	2.2	3.8	9.7	0.70	11.4
DDGS	900	60	253	90	81	316	89	12	90	10	6.6	5.1	8.9	8.9	1.9	1.4	7.3	0.5	2.0	2.9	9.7	0.70	12.7
Malt sprouts	900	61	232	19	126	378	139	18	100	100	10.8	3.1	6.0	8.1	2.3	2.1	6.6	0.6	4.0	1.5	4.7	0.75	10.8
Rice bran	900	90	135	153	81	211	101	36	200	30	5.9	2.1	4.4	5.3	1.4	1.2	16.0	0.6	0.8	10.0	16.0	0.65	12.4
Wheat bran	880	53	158	44	102	428	128	35	190	50	6.4	2.4	5.7	5.2	2.4	1.5	10.9	0.3	0.8	4.4	11.0	0.73	10.1
Wheat feed	880	40	140	40	59	271	77	24	270	90	5.0	2.5	7.0	5.0	2.0	1.0	9.0	0.2	0.9	4.0	10.2	0.76	12.1
Wheat shorts	880	51	167	41	85	366	110	30	240	50	7.3	2.6	6.0	5.5	2.5	1.4	10.5	0.3	0.8	4.2	13.0	0.75	10.8
Other energy concentrates																							
Beet molasses	750	86	105	0	0	0	0	0	0	450	0.4	0.5	1.0	0.6	1.0	2.2	0.2	10.0	10.8	2.0	39.1	0.70	10.7
Cane molasses	750	98	45	0	0	0	0	0	0	470	0.2	0.2	0.4	0.5	—	7.4	0.9	1.5	20.0	4.2	29.2	0.60	10.1
Cassava 60	880	57	26	7	48	124	77	21	600	25	1.0	0.3	0.7	0.8	0.3	3.0	2.0	0.4	1.1	1.4	12.0	0.50	12.0
Cassava 65	880	57	26	7	44	95	68	20	650	30	1.0	0.3	0.7	0.8	0.3	2.5	1.7	0.3	0.7	1.1	7.5	0.50	12.5
Cassava 70	880	35	26	7	31	80	50	14	700	35	1.0	0.3	0.7	0.8	0.3	2.0	1.5	0.3	0.7	0.9	4.4	0.50	13.1
Legume and oil seeds																							
Faba bean	880	33	257	13	77	123	89	8	390	35	16.8	1.8	5.0	9.2	2.4	1.2	5.3	0.2	0.7	1.5	12.4	0.80	13.0
Lupin	880	35	326	70	128	210	155	15	0	60	15.9	2.5	7.3	11.6	2.6	2.3	3.2	0.5	0.4	1.7	8.5	0.80	12.7
Peas	880	34	220	12	57	130	70	4	450	50	16.3	2.2	5.4	8.4	1.8	1.0	4.0	0.2	0.4	1.2	10.5	0.85	13.2
Rapeseed	900	41	189	396	81	211	130	49	0	50	11.5	4.2	9.2	8.7	2.4	4.0	6.0	0.3	0.6	2.4	7.9	0.78	20.9
Soybean	900	47	369	193	56	117	73	8	0	75	23.3	5.2	11.4	14.4	4.8	2.5	5.6	0.1	0.3	3.0	17.0	0.85	18.0

Feed																							
Oil meals																							
Rapeseed meal	900	68	361	25	121	277	189	86	0	90	19.4	7.6	16.2	15.7	4.3	7.0	10.0	0.7	0.3	4.5	12.5	0.76	11.3
Soybean meal 44	900	68	432	18	77	161	100	8.0	0	80	27.2	6.0	12.5	16.8	5.9	2.9	6.0	0.2	0.4	2.5	18.0	0.82	13.3
Soybean meal 46	900	63	450	18	63	132	82	6.0	0	80	28.4	6.3	13.1	17.6	6.0	2.9	6.1	0.2	0.4	2.7	19.5	0.83	13.9
Soybean meal 48	900	61	468	18	50	104	65	5.0	0	80	29.5	6.6	13.6	18.3	6.3	2.9	6.4	0.2	0.4	2.8	20.5	0.84	14.7
Sunflower meal 28	900	68	279	27	252	428	302	101	0	50	10.0	6.7	12.0	10.3	3.6	3.5	10.0	0.3	1.5	5.0	11.0	0.73	9.6
Sunflower meal 32	900	68	306	23	225	383	270	90	0	50	11.2	7.4	13.1	11.3	4.0	3.0	9.5	0.3	1.5	5.0	10.0	0.76	10.3
Sunflower meal 36	900	68	342	19	180	306	216	72	0	50	12.5	8.2	14.7	12.7	4.4	2.5	9.0	0.3	1.6	5.0	11.0	0.80	11.1
Oils and fats																							
Animal fat	995	0	0	990	0	0	0	0	0	0	0	0	0	0	0	0	0	0	0	0	0	0	33.5
Olein	995	0	0	990	0	0	0	0	0	0	0	0	0	0	0	0	0	0	0	0	0	0	31.4
Rapeseed oil	995	0	0	990	0	0	0	0	0	0	0	0	0	0	0	0	0	0	0	0	0	0	35.1
Soybean oil	995	0	0	990	0	0	0	0	0	0	0	0	0	0	0	0	0	0	0	0	0	0	35.6
Sunflower oil	995	0	0	990	0	0	0	0	0	0	0	0	0	0	0	0	0	0	0	0	0	0	35.6
Fibrous feedstuffs																							
Lucerne meal 12	900	90	126	23	297	475	371	83	0	30	5.4	1.9	3.4	5.2	2.1	14.0	2.6	0.6	3.5	2.0	19.0	0.56	6.7
Lucerne meal 15	900	99	153	32	261	418	326	73	0	30	6.6	2.3	4.1	6.3	2.5	15.0	2.6	0.7	4.8	2.7	21.0	0.60	7.4
Lucerne meal 18	900	99	180	36	216	346	270	60	0	30	7.7	2.7	4.9	7.4	3.0	16.0	2.7	0.8	4.9	3.0	25.0	0.64	8.3
Beet pulp	900	72	90	10	180	428	212	18	0	60	5.3	1.9	3.1	4.4	0.9	7.6	1.0	2.0	1.0	2.3	4.9	0.50	10.4
Citrus pulp	900	67	59	27	133	220	155	16	0	230	2.0	0.7	1.5	2.0	0.6	15.9	1.2	1.0	0.6	1.4	7.1	0.60	11.3
Cocoa hulls	900	80	164	50	183	390	300	140	0	—	7.5	1.5	3.5	6.0	1.0	3.0	3.5	0.8	1.5	4.0	25.0	0.50	8.3
Flax chaff	900	76	84	35	380	550	370	130	0	—	3.0	0.5	1.0	1.5	—	18.0	3.0	0.6	0.9	1.0	9.0	0.40	3.9
Grape seed meal	900	36	99	14	441	730	650	550	0	—	4.0	1.5	3.5	2.0	0.9	6.0	1.2	0.1	0.1	1.0	6.0	0.10	2.8
Grape pomace	900	81	117	54	280	560	480	300	0	20	4.9	1.7	3.5	3.7	0.7	7.0	2.0	0.1	0.1	1.2	16.0	0.15	4.5
Grass meal	900	126	144	36	225	460	250	50	0	80	6.0	2.0	3.5	5.5	1.5	7.0	4.0	1.0	0.8	2.0	25.0	0.55	8.1
Soybean hulls	900	46	122	20	355	588	426	21	0	10	7.0	1.4	3.4	4.6	1.5	5.0	1.6	0.2	0.3	2.0	12.6	0.50	7.2
Sunflower hulls	900	34	54	40	468	693	562	202	0	10	2.3	1.2	2.5	2.3	0.7	4.0	2.0	1.0	1.0	1.7	10.5	0.15	4.3
Wheat straw	900	61	36	12	395	750	474	80	0	—	—	—	—	—	—	3.8	0.8	1.6	4.6	0.9	9.5	0.20	2.7
Treated wheat straw	900	73	32	8	365	694	444	75	0	—	—	—	—	—	—	4.3	0.6	8.6	4.3	0.7	8.9	0.25	3.7

CPd, crude protein digestibility; DE, digestible energy; —, no analytical data available.

(proportionally to substitution rate) in nutritive value of the test feedstuff, so very careful determinations (large numbers of animals and replicates in the chemical analyses of diets) of the nutritive value of diets must be performed.

Composition and nutritive value of feedstuffs for rabbits

Table 6.3 shows the chemical composition and nutritive value of 47 feeds commonly used in rabbit nutrition. The data are expressed on an as-fed basis, with a common DM content for each group of feedstuffs, in view of practical utilization. Chemical composition includes DM, ash, CP, ether extract (EE), CF, NDF, ADF, acid detergent lignin (ADL), starch (ST), lysine (Lys), methionine (Met), methionine + cystine (SAA), threonine (Thr), tryptophan (Trp), calcium (Ca), phosphorus (P), sodium (Na), chlorine (Cl), magnesium (Mg) and potassium (K). The chemical composition is based mostly on the data of the INRA (1989), CVB (1994) and FEDNA (1997).

The nutritive value is based on a literature compilation. The tables proposed by Maertens *et al.* (1990) were revised taking into account recently published experimental data (Villamide *et al.*, 1991; Perez, 1994; García *et al.*, 1996, Fernández-Carmona *et al.*, 1996). However, for some feedstuffs the data concerning the CP digestibility (CPd) and DE were very divergent. The values proposed in Table 6.3 were retained after judging the methodology used and were considered as the most accurate at normal levels of dietary inclusion. They may be used for least cost diet formulation using linear programming. Digestibility data for CF and the cell wall constituents of feedstuffs are not presented because the reliability of the data was judged to be insufficient due to the methodology and the analytical procedures used.

References

Björnhag, G. (1972) Separation and delay of contents in the rabbit colon. *Journal of Agricultural Research* 7, 105–114.
Blas, E., Fando, J.C., Cervera, C., Gidenne, T. and Perez, J.M. (1991) Effet de la nature de l'amidon sur l'utilisation digestive de la ration chez le lapin au cours de la croissance. In: *5émes Journées de la Recherche Cunicole*, communication no. 50. Paris.
de Blas, C. and Carabaño, R. (1996) A review on the energy value of sugar beet pulp for rabbits. *World Rabbit Science* 4, 33–36.
de Blas, C. and Villamide, M.J. (1990) Nutritive value of beet and citrus pulps for rabbits. *Animal Feed Science and Technology* 31, 239–246.
de Blas, C., Pérez, E., Fraga, M.J., Rodríguez, M. and Gálvez, J. (1981) Effect of diet on feed intake and growth of rabbits from weaning to slaughter at different ages and weights. *Journal of Animal Science* 52, 1225–1232.

de Blas, C., Fraga, M.J. and Rodríguez, M. (1985) Units for feed evaluation and requirements for commercially grown rabbits. *Journal of Animal Science* 60, 1021–1028.

de Blas, C., Santomá, G., Carabaño, R. and Fraga, M.J. (1986) Fiber and starch levels in fattening rabbit diets. *Journal of Animal Science* 63, 1897–1904.

de Blas, C., Villamide, M.J. and Carabaño, R. (1989) Nutritive value of cereal by-products for rabbits. 1. Wheat straw. *Journal of Applied Rabbit Research* 12, 148–151.

de Blas, C., Taboada, E., Nicodemus, N., Campos, R. and Méndez, J. (1996) The response of highly productive rabbits to dietary threonine content for reproduction and growth. In: *Proceedings of the 6th World Rabbit Congress, Toulouse*, Vol. 1. Association Française de Cuniculture, Lempdes, pp. 139–143.

CVB (1994) *Veevoedertabel*. Central Veevoederbureau, Lelystad, the Netherlands.

Dessimoni, R. (1984) Efecto de razas y de diferentes niveles de proteina y fibra bruta sobre la digestibilidad de nutrientes en raciones de conejos. In: *Proceedings of the 3rd World Rabbit Congress*, Vol. 1. Rome, pp. 314–322.

FEDNA (1997) *Normas FEDNA para la Formulación de Piensos Compuestos*. Fundación Española para el Desarrollo de la Nutrición Animal, Madrid.

Fekete, S. and Gippert, T. (1986) Digestibility and nutritive value of nineteen feedstuffs. *Journal of Applied Rabbit Research* 9, 103–108.

Fernández, C. and Fraga, M.J. (1996) Effect of fat inclusion in diets for rabbits on the efficiency of digestible energy and protein utilization. *World Rabbit Science* 4, 19–23.

Fernández, C., Cobos, A. and Fraga, M.J. (1994) The effect of fat inclusion on diet digestibility in growing rabbits. *Journal of Animal Science* 72, 1508–1515

Fernández-Carmona, J., Cervera, C. and Blas, E. (1996) Prediction of the energy value of rabbit feeds varying widely in fibre content. *Animal Feed Science and Technology* 64, 61–75.

Fraga, M.J., Lorente, M., Carabaño, R.M. and de Blas, J.C. (1989) Effect of diet and of remating interval on milk production and milk composition of the doe rabbit. *Animal Production* 48, 459–466.

Fraga, M.J., Pérez, P., Carabaño, R. and de Blas, C. (1991) Effect of type of fiber on the rate of passage and on the contribution of soft feces to nutrient intake of finishing rabbits. *Journal of Animal Science* 69, 1566–1574.

García, G., Gálvez, J. and de Blas, C. (1992) Substitution of barley grain by sugar beet pulp in diets for finishing rabbits. 1. Effect of energy and nitrogen balance. *Journal of Applied Rabbit Research* 15, 1008–1016.

García, G., Gálvez, J. and de Blas, C. (1993) Effect of substitution of sugarbeet pulp for barley in diets for finishing rabbits on growth performance and on energy and nitrogen efficiency. *Journal of Animal Science* 71, 1823–1830.

García, J., Pérez, L., Alvarez, C., Rocha, R., Ramos, M. and de Blas, C. (1995) Prediction of the nutritive value of lucerne hay in diets for growing rabbits. *Animal Feed Science and Technology* 54, 33–44.

García, J., Villamide, M.J. and de Blas, C. (1996) Nutritive value of sunflower hulls, olive leaves and NaOH-treated barley straw for rabbits. *World Rabbit Science* 4, 205–209.

Gidenne, T. (1993) Measurement of the rate of passage in restricted-fed rabbits: effect of dietary cell wall level on the transit of fibre particles of different sizes.

Animal Feed Science and Technology 42, 151–163.

Gidenne, T., Poncet, C. and Gómez, L. (1987) Effet de l'addition d'un concentré riche en fibres, distribuées à deux niveaux alimentaires chez le lapin adulte. *Reproduction, Nutrition and Development* 27, 733–743.

INRA (1989) *L'Alimentation des Animaux Monogastriques: Porc, Lapin, Volailles*, 2nd Edn. Institut National de la Recherche Agronomique, Paris, 282 pp.

Lebas, F., Perez, J.M., Juin, H. and Lamboley, B. (1994) Incidence du nombre d'individus par cage sur la precision des coefficients de digestibilité mesurés chez le lapin en croissance. In: *6émes Journées de la Recherche Cunicole*. ITAVI, Paris, pp. 317–323.

Maertens, L. (1992) Rabbit nutrition and feeding: a review of some recent developments. In: Cheeke, P.R. (ed.) *Proceedings of the 5th World Rabbit Congress*. University of Corvallis, Oregon, pp. 889–913.

Maertens, L. and De Groote, G. (1981) L'énergie digestible de la farine de luzerne déterminée par des essais de digestibilité avec des lapins de chair. *Revue de l'Agriculture* 34, 79–92.

Maertens, L. and De Groote, G. (1982) Étude de la variabilité des coefficients de digestibilité des lapins suite aux differences d'âge, de sexe, de race et d'origine. *Revue de l'Agriculture* 35, 2787–2797

Maertens, L. and De Groote, G. (1984) Digestibility and digestible energy content of a number of feedstuffs for rabbits. In: Finzi, A. (ed.) *Proceedings of the 3rd World Rabbit Congress, Rome*, Vol. 1, pp. 244–251.

Maertens, L. and Luzi, F. (1996) Effect of dietary protein dilution on the performance and N-excretion of growing rabbits. In: Lebas, F. (ed.) *Proceedings of the 6th World Rabbit Congress, Toulouse*. Association Française de Cuniculture, Lempdes, pp. 231–235.

Maertens, L., Huyghebaert, G and De Groote, G. (1986) Digestibility and digestible energy content of various fats for growing rabbits. *Cuni-Sciences* 3, 7–14.

Maertens, L., Janssen, W.M.A., Steenland, E.M., Wolters, D.F., Branje, H.E.B. and Jager, F. (1990) Tables de composition, de digestibilité et de valeur énergétique des matières premières pour lapins. In: *5èmes Journées Recherche Cunicole*. Communication 57, ITAVI, Paris.

Martinez, J. and Fernández, J. (1980) Composition, digestibility, nutritive value and relations among them of several feeds for rabbits. In: Camps. J. (ed.) *Proceedings of the 2nd World Rabbit Congress*. Asociación Española de Cunicultura, Barcelona, pp. 214–223.

Motta, W. (1990) Efectos de la sustitución parcial de heno de alfalfa por orujo de uva o pulpa de remolacha sobre la utilización de la dieta y los rendimientos productivos en conejos en crecimiento. PhD Thesis, Universidad Politécnica de Madrid, Spain.

Ortiz, V., de Blas, C. and Sanz, E. (1989) Effect of dietary fiber and fat content on energy balance in fattening rabbits. *Journal of Applied Rabbit Research* 12, 159–162.

Peeters, J.E., Maertens, L. and Geeroms, R. (1992) Influence of galacto-oligosaccharides on zootechnical performance, cecal biochemistry and experimental colibacillosis 0103/8+ in weanling rabbits. In: Cheeke, P.R. (ed.) *Proceedings of the 5th World Rabbit Congress*. Oregon State University, Corvallis, Oregon, pp. 1129–1136.

Perez, J.M. (1994) Digestibilité et valeur energetique des luzernes deshydratées pour le lapin. In: *6èmes Journées de la Recherche Cunicole.* ITAVI, Paris, pp. 355–364.

Perez, J.M., Lebas, F., Gidenne, T., Maertens, L., Xiccato, G., Parigi Bini, R., Dalle Zotte, A., Cossu, M.E., Carazzolo, A., Villamide, M.J., Carabaño, R., Fraga, M.J., Ramos, M.A., Cervera C., Blas, E., Fernández, J., Falcao e Cunha, L. and Bengala Freire, J. (1995a) European reference method for *in vivo* determination of diet digestibility in rabbits. *World Rabbit Science* 3, 41–43.

Perez, J.M., Cervera C., Falcao e Cunha, L. Maertens, L., Villamide, M.J. and Xiccato, G. (1995b) European ring-test on *in vivo* determination of digestibility in rabbits: reproducibility of a reference method in comparison with domestic laboratory procedures. *World Rabbit Science* 3, 171–178.

Perez, J.M., Bourdillon, A., Lamboley B. and Naour J. (1996a) Length of adaptation period: influence on digestive efficacy in rabbits. In: Lebas, F. (ed.) *Proceedings of the 6th World Rabbit Congress, Toulouse.* Association Française de Cuniculture, Lempdes, pp. 263–266.

Perez, J.M., Fortun-Lamothe L. and Lebas, F. (1996b) Comparative digestibility of nutrients in growing rabbits and breeding does. In: Lebas, F. (ed.) *Proceedings of the 6th World Rabbit Congress, Toulouse.* Association Française de Cuniculture, Lempdes, pp. 267–270.

Santomá, G., de Blas, C., Carabaño, R. and Fraga, M.J. (1987) The effects of different fats and their inclusion level in diets for growing rabbits. *Animal Production* 45, 291–300.

Santomá, G., de Blas, C., Carabaño, R. and Fraga, M.J. (1989) Nutrition of rabbits. In: Haresign, W. and Cole, D.J.A. (eds) *Recent Advances in Animal Nutrition.* Butterworths, London, pp. 109–138.

Taboada, E., Méndez, J., Mateos, G.G. and de Blas, C. (1994) The response of highly productive rabbits to dietary lysine content. *Livestock Production Science* 40, 329–337.

Taboada, E., Méndez, J. and de Blas, C. (1996) The response of highly productive rabbits to dietary sulphur amino acid content for reproduction and growth. *Reproduction, Nutrition and Development* 36, 191–203.

Villamide, M.J. (1996) Methods of energy evaluation of feed ingredients for rabbits and their accuracy. *Animal Feed Science and Technology* 57, 211–223.

Villamide, M.J. and Ramos, M.A. (1994) Length of collection period and number of rabbits in digestibility assays. *World Rabbit Science* 2, 29–35.

Villamide, M.J., de Blas, C. and Carabaño, R. (1989) Nutritive value of cereal by-products for rabbits. 2. Wheat bran, corn gluten feed and dried distillers grain and solubles. *Journal of Applied Rabbit Research* 12, 152–155.

Villamide, M.J., Fraga, M.J. and de Blas, C. (1991) Effect of type of basal diet and rate of inclusion on the evaluation of protein concentrates with rabbits. *Animal Production* 52, 215–224.

Xiccato, G. and Cinetto, M. (1988) Effect of nutritive level and of age on feed digestibility and nitrogen balance in rabbit. In: *Proceedings of the 4th World Rabbit Congress, Budapest,* Vol. 3. Sandor Holdas, Hercegalom, pp. 96–104.

Xiccato, G., Cinetto, M. and Dalle Zotte, A. (1992) Effeto del livello nutritivo e della categoria di conigli sull'eficienza digestiva e sul bilancio azotato. *Zootechnia Nutritione Animale* 18, 35–43.

7. Energy Metabolism and Requirements

R. Parigi Bini and G. Xiccato

Dipartimento di Scienze Zootecniche, University of Padua, Agripolis, 35020 Legnaro (Padova), Italy

Energy units and their measurement

Energy is the potential ability to produce work and the joule (J) is the international unit used to measure all forms of energy. Since this unit is very small, the multiples kilojoules (1 kJ = 1000 J) and megajoules (1 MJ = 1000 kJ) are used much more widely in animal nutrition and feeding.

Joules can easily be converted into other energy units (e.g. calories) by using appropriate conversion factors. The standard calorie (cal) is equivalent to the energy cost of increasing the temperature of 1 gram of distilled water by 1°C (or more precisely, from 14.5 to 15.5°C). In practice, kilocalories (1 kcal = 1000 cal) and megacalories (1 Mcal = 1000 kcal) are still very commonly used and converted into kJ and MJ by multiplying by 4.184.

In the nutrition and feeding of rabbits, as in other species, the following energy parameters are used to express energy requirements and nutritive value of feeds: gross energy (GE), digestible energy (DE), metabolizable energy (ME), and net energy (NE) (Fig. 7.1).

GE, or heat of combustion, is the quantity of chemical energy lost as heat when organic matter is completely oxidized, forming water and carbon dioxide. In food, the GE content depends on the organic matter chemical composition: the energy values of the single components are about 22–24 MJ kg^{-1} for crude protein, 38–39 MJ kg^{-1} for fats (ether extract), and 16–17 MJ kg^{-1} for carbohydrates (fibre and starch). Therefore, the GE value depends essentially on the dietary fat concentration, because fat has more than twice the GE content of other constituents. However, the GE concentration in complete diets or raw materials does not provide any useful information on the availability and utilization of dietary energy by the animal and for this reason is not a relevant unit in the energy evaluation of feeds or animal energy requirements.

DE can be measured *in vivo*, by subtracting from GE the quantity of energy recovered in the faeces (FE), in other words, the energy of undigested nutrients:

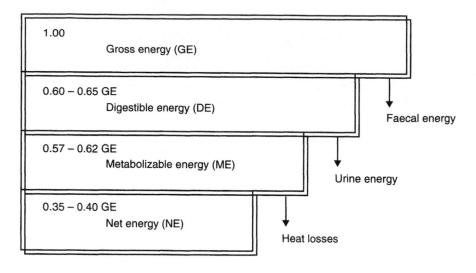

Fig. 7.1. Energy utilization of feeds in rabbits (from Xiccato, 1989).

$DE = GE - FE$.

In compound feeds for rabbits, DE represents a portion that usually varies from 0.50 to 0.80 of the GE. The DE is a good estimation of the energy value of feeds, because the latter depends on nutrient digestibility and can be measured quite easily. The energy system in rabbits is currently based on DE, and a standardized method (Perez *et al.*, 1995) capable of offering repeatable and reproducible measurements for the *in vivo* determination of dry matter (DM) digestibility (and consequently energy digestibility) has been recently proposed and adopted by the scientific community.

ME is calculated from DE by subtracting the energy loss associated with urine (UE) and intestinal fermentation gases (GasE), primarily methane:

$ME = DE - (UE + GasE)$.

In ruminants, GasE accounts for a significant proportion of DE, whereas in rabbits it is practically negligible, as are the heat losses from caecal fermentation (Gidenne, 1996). On the other hand, the energy loss associated with urine is substantial and depends on feed protein concentration (Maertens, 1992). Nitrogen losses increase (and consequently the energy associated with urea and other nitrogen catabolites) as dietary protein increases. The UE can be calculated from the quantity of N excreted daily in the urine (UN) (Parigi Bini and Cesselli, 1977):

$UE \ (kJ \ day^{-1}) = 51.76 \ UN \ (g \ day^{-1}) - 3.01$ r.s.d. = 3.93; $r = 0.90$.

ME is undoubtedly more precise for estimation of energy requirement and feed energy evaluation because it is not affected by the energy losses due to fermentative processes and the metabolic cost of protein utilization (Close, 1990; Maertens *et al.*, 1990). On the other hand, in practical feeding, UE losses are closely linked to the total intake of digestible protein (DP). For this reason, when common compound diets with 120–150 g DP kg^{-1} are fed to rabbits, the DE and ME are closely correlated and the ME is about 0.94–0.96 of DE (Parigi Bini and Cesselli, 1977; Partridge *et al.*, 1986b; Ortiz *et al.*, 1989; Santomá *et al.*, 1989; Xiccato, 1989). A value of 0.5 as an average proportional UE loss in rabbits (ME = 0.95 DE) can be proposed.

The NE is the proportion of GE that is actually utilized by the animal for maintenance and productive purposes. Therefore, the NE is the most precise estimate of feed energy value and energy requirements. The feed value in NE is related to specific energy utilization, i.e. NE for maintenance (NE_m), growth (NE_g), milk production (NE_l), and so on. The NE requirements in growing rabbits have been studied previously by Parigi Bini *et al.* (1974, 1978). A number of equations estimating NE of compound feeds for rabbits have also been proposed: some based on contents of digestible nutrient (Jentsch *et al.*, 1963) and others on organic matter digestibility (Parigi Bini and Dalle Rive, 1978).

Even though the NE system is generally considered the most suitable method for estimating energy requirement and feed evaluation because it is related to the effective utilization of dietary energy, its experimental determination is extremely complicated and expensive. Therefore, the real choice is between DE and ME systems.

ME is the system preferred for poultry, because birds excrete urine and faeces together. For rabbits, however, it is very difficult to collect and measure the urine energy values. As mentioned above, ME and DE values are highly correlated: in normal diets only a small amount of energy is lost in urine, and ME represents almost a constant fraction of DE (about 0.95 DE). For these reasons, DE values are still commonly used both in studies on energy metabolism and in practical rabbit feeding.

Methods for estimating energy requirements

Without going into details which are covered comprehensively in specialized energy metabolism texts (Blaxter, 1989; Webster, 1989; Close, 1990), the principal methods used for the measurement of energy requirements are as follows.

1. Long-term feeding experiments, carried out to establish the feed needed to maintain constant live weight or, conversely, to measure the variations in live weight (or milk production or fetal growth) associated with a certain quantity of feed. This method necessitates the keeping of a large number of animals for long periods under conditions similar to those on farms. However, the method does not provide any useful information on body composition changes which are very often found in rabbits during growth, lactation or pregnancy;

2. Calorimetric methods, which measure the heat lost by the animals. Measurement is direct (direct calorimetry) when calorimeters are used, and indirect (indirect calorimetry) when based on the gaseous exchanges determined in respiration chambers of various types (open or closed circuit). These methods allow the direct measurement of ME intake (MEI) and energy lost as heat (HE). The retained energy (RE) in either the body or the products (milk, fetal body, wool) is calculated by difference (RE = MEI – HE). Calorimetric methods require very complex and expensive equipment and can be utilized only on a few animals and in short-term experiments. Moreover, although the measurements obtained are highly accurate and repeatable on the same animal in time, they are hardly comparable to those obtained under practical rearing conditions. In addition, as in feeding experiments, calorimetric methods do not permit the identification of the origin of heat lost from different physiological functions (e.g. whether from feed digestion or body tissue utilization) or the partition of RE (e.g. energy retained in maternal or fetal body);

3. Comparative slaughter technique, which measures the variation of the energy contained in the body. Unlike calorimetric methods, this technique allows the direct measurement of MEI and RE, while HE is calculated by difference (HE = MEI – RE). This method constitutes the basis of the California net energy system developed for beef cattle (Lofgreen and Garrett, 1968) and applied to rabbit nutrition by others (Parigi Bini *et al.*, 1974, 1978, 1990a, 1992; Parigi Bini and Xiccato, 1986; Partridge *et al.*, 1989; Xiccato *et al.*, 1992b, 1995; Fortun *et al.*, 1993; Fortun, 1994; Nizza *et al.*, 1995). Using this method, body energy change is measured by first analysing the empty body (EB = live body – gut content) of a reference group of animals (initial slaughter group). A second group is subjected to the feeding experiment and then the empty bodies of these animals are analysed (final slaughter group). The rabbits are given a diet with a DE (or ME) content measured experimentally. In growing rabbits, the RE is calculated by subtracting the body energy found in the final group from the body energy in the initial group, that is the RE_g (RE for growth). Similarly, the energy excreted in the milk of lactating does (E_{milk}) and/or retained in the fetuses in pregnant does (RE_{fetus}) can be also measured during the entire lactation or pregnancy. The comparative slaughter technique is based on the assumption that the body composition of the initial slaughter group is very similar to the body composition of the final slaughter group at the beginning of the experiment. Moreover, the difference in body composition at the beginning and the end of experiment is a good estimate of RE only if the animals in the initial and the final groups are homogeneous, their number is relatively high and the length of experiment is long enough (e.g. a complete growing period or an entire pregnancy or lactation) (Close, 1990). As described below, comparative slaughter permits the estimation of the variation in chemical composition in the EB and therefore allows the partition of RE between the energy retained (or lost) as protein (RE_p) and fat (RE_f);

4. Non-destructive methods for measurement of body composition, allows measurement of the variation of body composition and then body RE without slaughter. Several methods have been proposed for rabbits, including dilution methods, nuclear magnetic resonance (NMR), computerized tomography (CT), total body electrical conductivity (Tobec) (see the review of Fekete, 1992). These methods often need very expensive equipment and their efficacy is not completely proven.

Energy metabolism and requirements

Several factors influence energy metabolism and consequently the energy requirements in rabbits. The most important are:

- body size, which depends on breed, age, sex;
- vital and productive functions, such as maintenance, growth, lactation, pregnancy;
- environment, i.e. temperature, humidity, air speed.

Only those aspects of energy metabolism related to vital and productive functions will be described here.

Voluntary feed and energy intake

The energy requirements for rabbits are usually listed as a proportion of the diet (i.e. in terms of MJ kg^{-1} of feed). Additional information that might provide a more precise definition of energy requirements, such as the feed intake or quantitative/qualitative performance, is not frequently given.

Animals in good health normally consume sufficient feed to meet their energy requirements. This is particularly true for growing rabbits, whereas reproducing does often demonstrate an energy deficit due to the high energy requirements for pregnancy, lactation or both activities that are not covered by adequate voluntary intake.

Appetite regulation in rabbits is mostly controlled by chemostatic mechanisms, which explains why the total quantity of energy ingested daily tends to be constant. Voluntary intake is proportional to metabolic live weight ($LW^{0.75}$). In growing rabbits, voluntary intake is about 900–1000 kJ DE day^{-1} kg^{-1} $LW^{0.75}$ and the chemostatic regulation appears only with a DE concentration of the diet higher than 9–9.5 MJ kg^{-1} (Lebas *et al.*, 1984; Partridge, 1986; Cheeke, 1987; Parigi Bini, 1988; Lebas, 1989; Santomá *et al.*, 1989), below which level a physical-type regulation is prevalent and linked to gut fill (Fig. 7.2). Less is known about the voluntary intake of

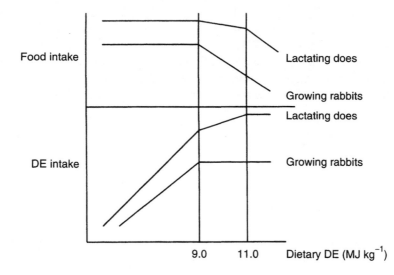

Fig. 7.2. Influence of dietary DE concentration on the voluntary food and energy intake in rabbits (from Xiccato, 1993).

reproducing females, where the energy consumption (in terms of metabolic weight) is lower in growing rabbits than in lactating females, which can consume 1100–1300 kJ DE day^{-1} kg^{-1} LW$^{0.75}$, with the lowest value recorded by primiparous females (Maertens and De Groote, 1988; Lebas, 1989; Parigi Bini *et al.*, 1990b, 1992; Xiccato *et al.*, 1992b, 1995). A second factor, which differentiates between voluntary intake of reproducing does in comparison with growing rabbits, is the energetic limit of chemostatic regulation. Some research has demonstrated that an increase in DE concentration over the normal values of 10–10.5 MJ kg^{-1} permits a further increase in the daily energy intake of the lactating females (Maertens and De Groote, 1988; Fraga *et al.*, 1989; Castellini and Battaglini, 1991; Xiccato *et al.*, 1995). In these animals, the regulating limit probably varies between 10.5 and 11 MJ kg^{-1}. This limit also depends on the dietary energy source and tends to be higher in added-fat diets than in high-starch diets (Fraga *et al.*, 1989; Castellini and Battaglini, 1991; Xiccato *et al.*, 1995).

Energy for maintenance and efficiency of energy utilization

By varying the quantity of DE or ME intake (e.g. feeding low-energy diets or in restricted quantities), it is possible to modify both the quantity of RE in the bodies of growing rabbits (Fig. 7.3) and the quantity of energy excreted in the milk or the quantity of energy retained in the fetal body.

The MEI promoting maintenance of energy equilibrium in the body (RE = 0) is called ME requirement for maintenance (ME$_m$). The slope of regres-

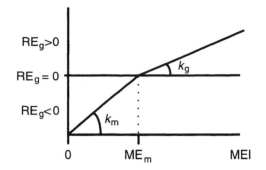

Fig. 7.3. Variation of RE_g as a function of ME intake (MEI).

sion of RE on MEI ($\Delta RE/\Delta MEI$) is defined as the efficiency of utilization of ME and commonly indicated by the letter k. When MEI is below the maintenance level (MEI < ME_m), RE becomes negative or, in other words, the body loses energy, with a specific slope on MEI (k for maintenance, k_m). When MEI is above the maintenance level (MEI > ME_m), RE is positive, or rather the body gains energy with a specific efficiency (k for growth, k_g). The efficiency of energy utilization for growth is lower than that for maintenance ($k_g < k_m$).

When the energy system is based on DE, RE can be related to DE intake (DEI) instead of MEI. In different studies on rabbit energy metabolism, the efficiencies of utilization of DEI for maintenance, growth, lactation or pregnancy ($\Delta RE/\Delta DEI$) have been estimated instead of the k coefficients relating to MEI (de Blas *et al.*, 1985; Parigi Bini and Xiccato, 1986; Partridge *et al.*, 1989; Xiccato *et al.*, 1995). Assuming a constant ratio ME = 0.95 DE, the efficiency of DEI utilization can be easily transformed into k values by dividing by 0.95. For example, using data from de Blas *et al.* (1985):

$$k_g = \Delta RE_g/\Delta DEI\ /\ 0.95 = 0.53/0.95 = 0.56.$$

Energy requirements for maintenance

In rabbits, as in all animals, energy losses for maintenance (basal metabolism and voluntary activity) are related to metabolic weight and physiological state.

As reviewed by Parigi Bini (1988) and Lebas (1989), different estimates of DE requirements for the maintenance (DE_m) of growing rabbits have been found, varying from 381 kJ day^{-1} kg^{-1} LW$^{0.75}$ in New Zealand White rabbits (Partridge *et al.*, 1989) to 552 kJ DE day^{-1} kg^{-1} LW$^{0.75}$ in Giant Spanish growing rabbits (de Blas *et al.*, 1985). Such a wide range can be attributed to both the breeds, characterized by highly contrasting daily gain and body composition, and to the methodology used (calorimetric methods usually give lower DE_m than comparative slaughter). Table 7.1 shows a review of

Table 7.1. Energy requirements for maintenance of energy equilibrium (RE = 0) in New Zealand White or hybrid growing rabbits.

Authors	DE_m (kJ day^{-1} kg^{-1} LW$^{0.75}$)	ME_m[a] (kJ day^{-1} kg^{-1} LW$^{0.75}$)
Isar (1981)	470	446
Scheele *et al.* (1985) (at 17°C)	413	392
Parigi Bini and Xiccato (1986)[b]	425–454	404–431
Partridge *et al.* (1989)	381	362
Nizza *et al.* (1995)	441–454	419–432
Average	431	409

[a]Calculated from DE_m by assuming ME = 0.95 DE.
[b]Recalculated values from original data expressed on metabolic empty body weight (EBW$^{0.75}$).

studies based on growing New Zealand White pure-bred or hybrid rabbits. An average DE_m of 430 kJ DE day^{-1} kg^{-1} LW$^{0.75}$ might be proposed, a value that is slightly higher than the requirement proposed by Lebas (1989) (400 kJ DE day^{-1} kg^{-1} LW$^{0.75}$).

Another important factor to be considered is the difference between the requirement for the maintenance of energy equilibrium (RE = 0) and the maintenance of live body weight (live weight gain, LWG = 0). Parigi Bini and Xiccato (1986) observed a DE_m of 425 kJ DE day^{-1} kg^{-1} LW$^{0.75}$ at RE = 0 and only of 273 kJ DE day^{-1} kg^{-1} LW$^{0.75}$ when LWG = 0. As shown in Table 7.3, when LWG = 0 the rabbit is in energy deficit, as a consequence of the loss of fat primarily compensated by water gain. Similar results were observed by de Blas *et al.* (1985) who estimated the DE_m for LW maintenance to be 10% lower than the DE_m required for body energy equilibrium.

In reproducing does, experimental estimates of DE_m are often inconsistent (Table 7.2). In adult non-pregnant does, the DE_m was found to be 398 kJ day^{-1} kg^{-1} LW$^{0.75}$ by Parigi Bini *et al.* (1990a, 1991a), a value somewhat lower than the DE_m measured in growing rabbits but higher than the estimate of Partridge *et al.* (1986b; 326 kJ day^{-1} kg^{-1} LW$^{0.75}$).

In pregnant does, the estimated DE_m ranged from 352 kJ day^{-1} kg^{-1} LW$^{0.75}$ when measured by Partridge *et al.* (1986b) and 452 kJ day^{-1} kg^{-1} LW$^{0.75}$ by Fraga *et al.* (1989), while Parigi Bini *et al.* (1990a, 1991a) gave intermediate requirements (431 kJ day^{-1} kg^{-1} LW$^{0.75}$).

In lactating does, the DE_m was estimated from 413 to 500 kJ day^{-1} kg^{-1} LW$^{0.75}$ by Partridge *et al.* (1983, 1986b) and 473 kJ day^{-1} kg^{-1} LW$^{0.75}$ by Fraga *et al.* (1989). This considerable difference between the DE_m in pregnant and lactating does can be ascribed to the fact that studies did not take into account any body energy change during lactation. Parigi Bini *et al.* (1991b, 1992) demonstrated that primiparous does are almost always in energy deficit and utilize body tissues (protein and especially fat) as a source of energy to compensate for the insufficient energy intake. These latter studies estimated a

Table 7.2. DE requirements for maintenance in rabbit does (kJ day^{-1} kg^{-1} LW$^{0.75}$).

Author	Non-lactating does		Lactating does	
	Non-pregnant	Pregnant	Non-pregnant	Pregnant
Partridge *et al.* (1983)			413–446	
Partridge *et al.* (1986b)	326	352	500	
Fraga *et al.* (1989)		452	473	
Parigi Bini *et al.* (1990a)	398	431		
Parigi Bini *et al.* (1991a)			432	468
Xiccato *et al.* (1992b)				470
Lebas (1989)	400		460	
Maertens (1992)	420		460	
Xiccato (1996)	400	430	430	470

DE$_m$ equal to 432 kJ day^{-1} kg^{-1} LW$^{0.75}$ in lactating does and 468 kJ day^{-1} kg^{-1} LW$^{0.75}$ in concurrent pregnant and lactating does. Xiccato *et al.* (1992b) confirmed higher DE$_m$ in does submitted to an intensive remating system (470 kJ day^{-1} kg^{-1} LW$^{0.75}$).

In a review, Lebas (1989) proposed a DE$_m$ equal to 400 and 460 kJ day^{-1} kg^{-1} LW$^{0.75}$ for non-reproducing and lactating does, respectively. Xiccato (1996) proposed 400 kJ day^{-1} kg^{-1} LW$^{0.75}$ for non-reproducing does, 430 kJ day^{-1} kg^{-1} LW$^{0.75}$ for pregnant or lactating does, and 460 kJ day^{-1} kg^{-1} LW$^{0.75}$ for concurrent pregnant and lactating does.

Energy requirements for growth

Figure 7.4, taken from Partridge *et al.* (1989), shows the response in terms of daily gain and energy intake to the increase of diet DE concentration when the DP to DE ratio is maintained constant and protein contains the major amino acids in satisfactory equilibrium. This typical growth-response curve shows that the maximum average daily growth is achieved when the dietary DE concentration is between 11 and 11.5 MJ kg^{-1} DM or 10 and 10.5 MJ kg^{-1} as fed.

An increase in the level of dietary energy intake can also affect composition of body gain and the partition of energy retained as protein and fat. The body composition changes are not linearly correlated with DEI, because some constituents (e.g. fat) tend to increase more than proportionally. In five comparative slaughter experiments involving 180 growing New Zealand White rabbits, Parigi Bini and Xiccato (1986) derived quadratic regression equations which estimated the daily gains (g day^{-1} kg^{-1} LW$^{0.75}$) of empty body (EBG), water (WG), protein (WG), fat (PG) and ash (AG) from DEI (MJ day^{-1} kg^{-1} LW$^{0.75}$):

$$EBG = -12.61 + 48.50 \text{ DEI} - 8.15 \text{ DEI}^2 \qquad \text{r.s.d.} = 1.64, r = 0.93$$
$$WG = -6.52 + 33.59 \text{ DEI} - 10.69 \text{ DEI}^2 \qquad \text{r.s.d.} = 0.61, r = 0.96$$

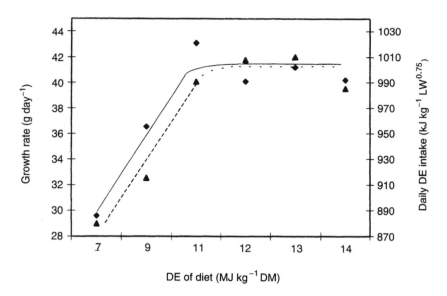

DE of diet (MJ kg^{-1} DM)

Fig. 7.4. Effect of dietary energy concentration on growth rate ◆ and total DE intake ▲ (Partridge *et al.*, 1989).

$$PG = -2.42 + 10.01 \text{ DEI} - 1.70 \text{ DEI}^2 \qquad \text{r.s.d.} = 0.38, r = 0.91$$
$$FG = -2.76 + 1.69 \text{ DEI} + 5.75 \text{ DEI}^2 \qquad \text{r.s.d.} = 0.45, r = 0.94$$
$$AG = -0.91 + 3.21 \text{ DEI} - 1.51 \text{ DEI}^2 \qquad \text{r.s.d.} = 0.12, r = 0.67$$

The same equations were transformed to formulate other equations which estimated retained energy as protein (RE_p) and as fat (RE_f) and total RE (RE_g) as a function of DEI (all data are expressed as MJ day^{-1} kg^{-1} LW$^{0.75}$):

$$RE_p = -0.057 + 0.234 \text{ DEI} - 0.040 \text{ DEI}^2 \qquad \text{r.s.d.} = 0.009, \qquad r = 0.91$$
$$RE_f = -0.098 + 0.060 \text{ DEI} + 0.204 \text{ DEI}^2 \qquad \text{r.s.d.} = 0.016, \qquad r = 0.94$$
$$RE_g = -0.155 + 0.294 \text{ DEI} + 0.164 \text{ DEI}^2 \qquad \text{r.s.d.} = 0.012, \qquad r = 0.96$$

In fact, body protein and fat have specific caloric values, namely 23.347 and 35.564 MJ kg^{-1} respectively, as determined by bomb calorimetry (Parigi Bini and Dalle Rive, 1978). Recently, Nizza *et al.* (1995) found similar caloric values (23.1 and 35.7 kJ g^{-1}), thereby confirming that the caloric value of rabbit fat is lower than that of other animal fats (on average 38–40 MJ kg^{-1}) (Close, 1990). The caloric value of fat was found to be higher in reproducing does (36–37 MJ kg^{-1}) than in growing animals, probably related to the lower content of phospholipids in older and fatter animals (Cambero *et al.*, 1991; Hulot *et al.*, 1992; Fortun, 1994).

As presented in Fig. 7.5 and Table 7.3, the previous equations show an increase in EBG as DEI increases and a strong modification of the chemical composition of EBG (and consequently its energy value). When DEI = 0, substantial losses of body weight and body tissues occur and the body loses

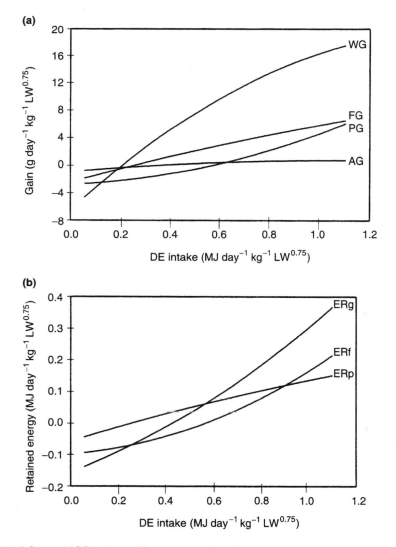

Fig. 7.5. Influence of DE intake on EB gain composition (a) and partition of energy retained as protein and fat (b) (from Parigi Bini and Xiccato, 1986).

155 kJ day^{-1} kg^{-1} LW$^{0.75}$. This loss is called 'fasting metabolism', which means the loss of body energy at fasting.

As mentioned above, when DEI is equal to 273 kJ day^{-1} kg^{-1} LW$^{0.75}$, the EB weight is maintained (EBG = 0) but losses of fat (−1.9 g day^{-1} kg^{-1} LW$^{0.75}$) and energy (−62 kJ day^{-1} kg^{-1} LW$^{0.75}$) are observed.

As discussed above, when DEI = DE$_m$, the energy equilibrium is reached as a consequence of a gain of energy (36 kJ day^{-1} kg^{-1} LW$^{0.75}$) as protein (+1.5 g day^{-1} kg^{-1} LW$^{0.75}$) and an equivalent loss of energy as fat (−1.0 g day^{-1} kg^{-1}

Table 7.3. Empty body gain (EBG) composition and RE partition as influenced by DE intake (DEI) (all data are in kJ or g day^{-1} kg^{-1} LW$^{0.75}$) (recalculated from Parigi Bini and Xiccato, 1986).

		EBG composition						
DEI (kJ)	EBG (g)	WG (g)	PG (g)	FG (g)	AG (g)	RE$_p$ (kJ)	RE$_f$ (kJ)	RE$_g$ (kJ)
0	−12.6	−6.5	−2.4	−2.8	−0.9	−57	−98	−155
273	0.0	1.8	0.2	−1.9	−0.1	4	−66	−62
425	6.5	5.8	1.5	−1.0	0.2	36	−36	0
900	24.4	15.0	5.2	3.4	0.8	122	122	244
1000	27.8	16.4	5.9	4.7	0.8	137	166	303

LW$^{0.75}$). At the same time, EBG is positive (6.5 g day^{-1} kg^{-1} LW$^{0.75}$) primarily due to water retention.

With increasing DEI, protein gain (in weight) always remains higher than fat gain, but, at DEI = 900 kJ day^{-1} kg^{-1} LW$^{0.75}$, RE as protein and fat become equal (RE$_p$ = RE$_f$ = 122 kJ day^{-1} kg^{-1} LW$^{0.75}$). When DEI reaches the highest values (about 1000 kJ day^{-1} kg^{-1} LW$^{0.75}$), which correspond to voluntary intake, RE$_f$ > RE$_p$ and total RE reaches the maximum level (RE$_g$ = 303 kJ day^{-1} kg^{-1} LW$^{0.75}$).

These equations can be used to calculate composition of daily gain throughout the entire growing period (e.g. from 0.65 to 2.5 kg LW). During this time, the average LW$^{0.75}$ is about 1.4 kg. With *ad libitum* feeding (DEI = 1000 kJ day^{-1} kg^{-1} LW$^{0.75}$), EBG is 38.9 g day^{-1} (27.8 g day^{-1} kg^{-1} LW$^{0.75}$ × 1.4), and LW gain is 44.7 g day^{-1} (assuming EB weight = 0.87 LW). Daily EBG is then composed of 22.9 g (0.589) of water, 8.3 g (0.213) of protein, 6.6 g (0.170) of fat and 1.1 g (0.028) of ash. This chemical composition of daily growth is typical of a young, rapidly growing rabbit.

The above listed regression equations relating RE to DEI are not linear. From the same data set, the following linear regression equation can be estimated (data are expressed in kJ day^{-1} kg^{-1} LW$^{0.75}$):

$$RE_g = -235 + 0.527 \text{ DEI} \qquad\qquad \text{r.s.d} = 22, \ r = 0.97$$

which indicates a DE$_m$ = 446 kJ day^{-1} kg^{-1} LW$^{0.75}$ and an efficiency of utilization of DE for growth of 0.53. This efficiency is close to that found by de Blas *et al.* (1985) and Partridge *et al.* (1989) (Table 7.4).

The efficiency of energy utilization for growth is clearly influenced by the composition of growth, because energy is retained as protein less efficiently than energy is retained as fat (Blaxter, 1989; Close, 1990). Estimates of efficiency of DE utilization for protein and fat deposition are given in Table 7.4. Efficiencies of DE utilization for energy deposition as protein and fat appear to be 0.38–0.44 and 0.60–0.70, respectively.

Using the factorial method and the above-mentioned coefficients of

Table 7.4. Efficiency of DE utilization for RE_g and RE as protein (RE_p) and fat (RE_f).

Authors	$\Delta RE_g/\Delta DEI$	$\Delta RE_p/\Delta DEI$	$\Delta RE_f/\Delta DEI$
De Blas *et al.* (1985)	0.52	0.38	0.65
Parigi Bini and Xiccato (1986)	0.53	0.44	0.70
Partridge *et al.* (1989)	0.47	0.39	0.60
Nizza *et al.* (1995)	0.51[a]	0.39–0.41	0.64–0.66
Average	0.51	0.40	0.65

[a]Calculated value.

energy utilization and DE_m values, the DE requirement and energy and chemical balance in growing rabbits can be estimated. An example regarding rabbits from 0.8 to 2.4 kg is provided in Box 7.1.

Energy requirements for reproduction and lactation

The energy metabolism of rabbit does in different reproductive states has been investigated (see review of Partridge, 1986). According to the results of these studies, the dietary DE appears to be utilized very efficiently by lactating does, i.e. more than 0.80. This very high overall efficiency of energy utilization may be due to body fat mobilization for milk synthesis, particularly in mid-lactation (when milk production is high) or when pregnancy is concurrent with lactation.

Information on the changes of body composition in reproducing does, and on the partition and utilization of dietary energy for maternal and fetal tissue synthesis and/or for milk production was given in a successive series of experiments (described below) using the comparative slaughter technique. A further objective of these experiments was a more precise definition of the energy requirements and the utilization and partition of dietary energy in primiparous rabbit does.

Pregnancy

Parigi Bini *et al.* (1990a, 1991a) observed that primiparous does undergo wide variations in body composition, tissue deposition, and energy retention (Table 7.5). During early and mid gestation (0–21 days), the live weight increase is similar to that of non-pregnant does. During late pregnancy (21–30 days), the empty body weight decreases as a result of a protein and fat loss and a transfer of energy to the rapidly growing fetuses. At the same time, non-pregnant does continue to gain weight and retain body energy, primarily in the form of fat.

Milisits *et al.* (1996) have recently confirmed these results by comparing the variations in the body composition of pregnant and non-pregnant does using CT. The total balance of body tissue showed a net loss in fat during the

Box 7.1. Estimate of DE requirements and efficiency of utilization of DE in growing rabbits from weaning (35 days) to slaughter (75 days).

Reference data
  LW at weaning = 0.8 kg
  LW at slaughter = 2.4 kg
  EBG = 0.87 LW gain
  EBG composition = water 0.61; protein 0.21; fat 0.15; ash 0.03
  Caloric value of body protein = 23.2 kJ g^{-1}
  Caloric value of body fat = 35.6 kJ g^{-1}
  Efficiency of DE utilization for RE as protein = 0.40
  Efficiency of DE utilization for RE as fat = 0.65
  Dietary DE concentration = 10 MJ kg^{-1}
  Maximum DE intake = 950 kJ day^{-1} kg^{-1} $LW^{0.75}$

Calculated data
  Daily LW gain = 40 g day^{-1}
  EBG = 34.8 g day^{-1}
  Protein gain = 7.3 g day^{-1}
  Fat gain (FG) = 5.2 g day^{-1}
  Average metabolic LW ($LW^{0.75}$) = $((0.800 + 2.400)/2)^{0.75}$ = 1.42 kg
  RE as protein (RE_p) = 169 kJ day^{-1}
  RE as fat (RE_f) = 185 kJ day^{-1}
  Total RE (RE_g) = 354 kJ day^{-1}

DE requirement and efficiency for growth
  DE_m = 430 kJ day^{-1} kg^{-1} $LW^{0.75}$ = 611 kJ day^{-1}
  DE requirement for RE_p = 169 / 0.40 = 423 kJ day^{-1}
  DE requirement for RE_f = 185 / 0.65 = 285 kJ day^{-1}
  DE requirement for growth (DE_g) = 708 kJ day^{-1}
  Efficiency of DE utilization for growth = 354 / 708 = 0.50
  Total DE requirement = DE_m + DE_g = 1319 kJ day^{-1} (929 kJ day^{-1} kg^{-1} $LW^{0.75}$)
  Required feed intake = 132 g day^{-1}

entire pregnancy that was evenly distributed throughout the different fat deposits (intrascapular, perirenal and pelvic fat).

Data concerning various blood plasma metabolites confirm the significant modification of energy metabolism in late pregnancy (Parigi Bini et al., 1990a). In the last period of pregnancy, the glucagon level increased, but glucose and triglyceride levels decreased. Glucagon is involved in the control of catabolic utilization of body reserves and a similar change in level was reported by Jean-Blain and Durix (1985). The transfer of energy from the body of the doe to the fetuses leads to an energy deficit that is particularly concentrated in the last 10 days of pregnancy, as reported for sows (Noblet and Close, 1980) and ewes (Rattray et al., 1980).

Table 7.5. Variation of body composition in non-pregnant and pregnant (1st pregnancy) does (from Parigi Bini and Xiccato, 1993).

Days on trial	0	21	30
Non-pregnant does			
Empty body weight (kg)	2.70	3.16	3.27
Water	0.62	0.60	0.58
Protein	0.22	0.21	0.20
Fat	0.13	0.16	0.19
EB energy (MJ kg^{-1})	9.8	10.6	11.4
Pregnant does			
Days of pregnancy	0	21	30
Empty body weight[a] (kg)	2.70	3.15	2.98
Water	0.62	0.60	0.60
Protein	0.22	0.22	0.22
Fat	0.13	0.16	0.15
EB energy (MJ kg^{-1})	9.8	10.7	10.6

[a]Excluding pregnant uterus.

The efficiency of utilization of DE for maternal tissue accretion in pregnant or non-pregnant does was estimated to be 0.49 (Parigi Bini *et al.*, 1991a) and this value is similar to the above-mentioned efficiency in growing rabbits. On the other hand, Partridge *et al.* (1986) found a higher efficiency of DE for body growth (0.64) in non-lactating does (pregnant and non-pregnant), but with a concomitant lower DE_m requirement (352 kJ day^{-1} kg^{-1} LW$^{0.75}$).

The efficiencies of utilization of dietary DE for fetal growth ($\Delta E_{fetus}/\Delta DEI$) (Parigi Bini *et al.*, 1991a, 1992; Xiccato *et al.*, 1992b), namely 0.31 in pregnant nulliparous does and 0.27 in lactating and pregnant does, appear to be low (see Table 7.6). Similar or lower efficiencies for fetal growth were observed in experiments on pigs (0.20–0.30), as reported by Walach-Janiak *et al.* (1986). Explanation for the high energy cost of fetal growth may come from the very high protein content of the fetal body and its extremely rapid turnover (Young, 1979).

Lactation and concurrent pregnancy

The energy output in the milk (E_{milk}) during lactation is exceptionally high in rabbits, compared with other species, due to the rate of milk production (200–300 g day^{-1}) and the high concentration of DM (300–350 g kg^{-1}), protein (100–150 g kg^{-1}) and fat (120–150 g kg^{-1}) (Lebas, 1971; Partridge *et al.*, 1983, 1986b; Fraga *et al.*, 1989; Parigi Bini *et al.*, 1992). The chemical composition of rabbit milk changes substantially during lactation, as described by Lebas (1971). In particular, the DM content decreases in the first 1–3 days, as colostrum becomes milk, then remains constant for 2–3 weeks, and finally increases as the milk yield decreases. On the other hand, the

R. Parigi Bini and G. Xiccato

Table 7.6. Efficiency of utilisation of DE and body energy of does (RE_d) for fetal growth (E_{fetus}) and energy milk production (E_{milk}).

Authors	$\Delta E_{fetus}/\Delta DEI$	$\Delta E_{milk}/\Delta DEI$	$\Delta E_{milk}/\Delta RE_d$
Partridge et al. (1983, 1986b)		0.68–0.84	0.94
Partridge et al. (1986a)		0.62	
Fraga et al. (1989)		0.71	
Parigi Bini et al. (1991a,b, 1992)	0.27–0.31	0.63	0.76–0.81
Xiccato et al. (1992b, 1995)	0.30	0.63	0.76

composition of milk DM tends to remain unchanged, except for a constant reduction in lactose, and therefore the caloric value of milk is strictly dependent on the variation of DM content (Parigi Bini et al., 1991b; Xiccato et al., 1995). Different measurements of milk energy concentration and variation during lactation are listed in Table 7.7.

The average caloric value of 8.4 MJ kg^{-1} is very close to the value of 8.53 MJ kg^{-1} reported by Blaxter (1989) and is about 2.8 times higher than that of cow milk (2.97 kJ g^{-1}). At the same time, if the daily excretion of energy as milk is expressed in terms of metabolic weight, the average milk energy output is higher in rabbits than in cows. For example, a 4 kg doe producing 250 g day^{-1} of milk excretes 752 kJ E_{milk} $day^{-1} kg^{-1}$ $LW^{0.75}$, while a 600 kg cow producing 25 kg day^{-1} of milk excretes only 612 kJ E_{milk} $day^{-1} kg^{-1}$ $LW^{0.75}$.

With respect to the utilization of DE for milk production (Table 7.6), the estimate of 0.63 in both lactating non-pregnant and lactating and concurrent pregnant does (Parigi Bini et al., 1991a,b, 1992) agrees with the estimate made by Lebas (1989), but is much lower than the results of Partridge et al. (1983, 1986b), who reported an efficiency of ME utilization for milk production varying from 0.74 to 0.94. Nevertheless, these coefficients were ascribed to the particular diets used (high fat, high protein) and a value of

Table 7.7. Energy concentration (MJ kg^{-1}) and variation of rabbit milk during lactation.

Authors	1st week	2nd and more weeks	Final week	Average
Lebas (1971)[a]	8.10	7.11	10.22	8.02
Partridge et al. (1983)	8.42	9.01	10.25	9.17
Partridge et al. (1986a)[a]		7.13–7.79		7.46
Partridge et al. (1986b)		8.39	10.0–14.6	
Fraga et al. (1989)		7.98		7.98
Parigi Bini et al. (1991b)		7.75	9.84	8.27
Maertens (1992)		8.0	9.0–12.0	
Xiccato et al. (1995)		8.06–8.76		8.42
Average	8.3	8.0	10.8	8.4

[a]Estimated energy values from chemical composition (Partridge et al., 1986a).

metabolizable energy efficiency of 0.65 for normal diets was proposed (Partridge, 1986; Partridge *et al.*, 1986a). This value corresponds to about 0.61–0.62 in terms of DE efficiency.

Finally, the efficiency of utilization of energy retained in the body (RE_d) for milk production was 0.81 in lactating does (Parigi Bini *et al.*, 1991a,b) and 0.76 in lactating and pregnant does (Parigi Bini *et al.*, 1992; Xiccato *et al.*, 1992b); the values found for utilization of both the dietary energy and the maternal energy for lactation agree with the results of experiments conducted on other animal species (cattle and pigs) and the theoretical calculations reported by Blaxter (1989).

Similarly to the estimate for growing rabbits, a calculation of the energy requirement and body balance can be made for reproducing does. Box 7.2 provides an example regarding lactating non-pregnant does.

Energy and chemical balance during reproduction
The considerable energy excretion through milk in lactating does, which is even more pronounced in selected 'hybrid' does, is not completely compensated for by voluntary DE intake, and this causes a consistent deficit in both body tissues and energy. During the first lactation, the body of the doe is subjected to a marked reduction in energy reserves following the mobilization of fat deposits, while the body protein level remains unchanged (Parigi Bini *et al.*, 1990b, 1991b, 1992; Xiccato *et al.*, 1992b, 1995) (Fig. 7.6). Unlike other species, this energy loss remains constant throughout lactation (Parigi Bini *et al.*, 1990b) and no recovery is observed during the final phase due to milk production, which remains high even after 25–30 days of lactation (Fortun, 1994).

In *post partum* (PP) mated does (therefore concurrently pregnant and lactating), a rapid reduction in milk production is observed after 20 days of lactation (Lebas, 1972; Lebas *et al.*, 1984; Maertens and De Groote, 1988; Parigi Bini *et al.*, 1992; Xiccato *et al.*, 1995) (see Fig. 7.9). Nevertheless, energy requirements remain high due to the rapid development of the fetuses and uterine tissues. The simultaneous condition of pregnancy accentuates these chemical modifications and is responsible for a further reduction in fat content and body energy levels. The overlapping of pregnancy and lactation prevents the return to normal body conditions (Fortun *et al.*, 1993) and increases the protein requirements in response to the elevated demand for protein by the fetuses and the rapid turnover in fetal protein (Parigi Bini *et al.*, 1992; Xiccato *et al.*, 1992b, 1995).

Fortun (1994) compared the energy balance in pregnant non-lactating does (P) and in pregnant and lactating does (LP) in their second pregnancy (Table 7.8). In this study, a positive energy balance in P does (+12.51 MJ at 28 days of pregnancy) was estimated, whereas LP does were found to be in negative balance (–11.78 MJ). Moreover, in LP does the energy deficit appears higher during the second half of pregnancy.

Box 7.2. Estimate of DE requirements and body energy and fat balance in lactating does (assuming protein balance in equilibrium).

Reference data
  Average LW = 4.25 kg
  Days of lactation = 30 days
  Milk production = 220 g day^{-1}
  Caloric value of milk = 8.4 MJ kg^{-1}
  Caloric value of fat in the maternal body = 36.5 MJ kg^{-1}
  Efficiency of DE utilization for milk energy (E_{milk}) = 0.63
  Efficiency of body energy (RE_d) utilization for E_{milk} = 0.78
  Dietary DE concentration = 11 MJ kg^{-1}
  Maximum DE intake (DEI) = 1250 kJ day^{-1} kg^{-1} LW$^{0.75}$
  Protein balance = 0

Calculated data
  Average LW$^{0.75}$ = 2.96 kg
  E_{milk} = 220 × 8.4 = 1848 kJ day^{-1}

DE requirement
  DE_m = 430 kJ day^{-1} kg^{-1} LW$^{0.75}$ = 1273 kJ day^{-1}
  DE requirement for milk production (DE_{milk}) = 1848 / 0.63 = 2933 kJ day^{-1}
  Total DE requirement = DE_m + DE_{milk} = 4206 kJ day^{-1} (1420 kJ day^{-1} kg^{-1} LW$^{0.75}$)

Body energy and tissue balance
  Maximum DEI = 1250 × 2.96 = 3700 kJ day^{-1}
  Maximum feed intake = 336 g day^{-1}
  DEI deficit = –506 kJ day^{-1}
  E_{milk} from dietary energy = 1848 – 506 × 0.63 = 1529 kJ day^{-1}
  E_{milk} from body energy = 1848 – 1529 = 319 kJ day^{-1}
  Body energy (RE_d) retained = –319 / 0.78 = –409 kJ day^{-1}
  Fat loss = –409 / 36.5 = –11.2 g day^{-1}
  Total body fat loss during lactation = –336 g
  Total body energy loss = –12.27 MJ

Rabbit does in their second pregnancy and maintained in different nutritional and lactation conditions were compared (Fortun, 1994). The energy balances of these experimental groups are shown in Table 7.9, where it is apparent that a moderate feeding restriction during the second half of pregnancy helps to partially reduce the energy gain in the maternal body and limit overfattening (RNL does). A severe feeding restriction (M1 and R1 does) during pregnancy, on the other hand, can induce an energy deficit similar to that observed in does lactating for ten pups. The reduction of the number of suckling pups to only four in the FL rabbits did not consistently decrease the energy deficit; in fact, milk production remained high because

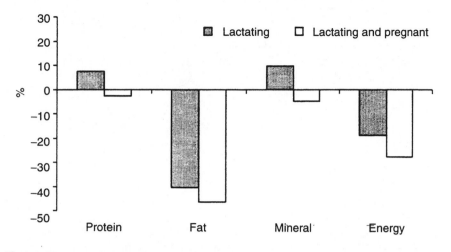

Fig. 7.6. Material and energy balance of primiparous does in different physiological states (Parigi Bini and Xiccato, 1993; Xiccato, 1996).

each pup suckled more milk and the energy requirement for pregnancy concurrent with lactation remained unchanged.

The nutritional deficit provoked by lactation or feeding restriction also appears to be responsible for the decreased reproductive efficiency of lactating and concurrently pregnant does, and consequently a reduction in fetal development and viability (Viudes-De Castro *et al.*, 1991; Parigi Bini *et al.*, 1992; Fortun *et al.*, 1993; Fortun and Lebas, 1994; Fortun-Lamothe and Bolet, 1995).

The stimulation of energy intake is necessary to reduce the energy deficit in the lactating does (Cheeke, 1987). In fact, the genetic selection and the crossbreeding programmes of the most common European hybrid lines are based on the increase in litter size at birth and daily milk production as their main objectives, relegating the goal of increasing voluntary feed intake and maintaining the body conditions of the females to secondary importance (De Rochambeau, 1990; Maertens, 1992; Xiccato, 1996).

Table 7.8. Energy balance in pregnant non-lactating does (P) and concurrently pregnant and lactating does (LP) at 28 days of their second pregnancy (Fortun, 1994). All values are in MJ.

	P does	LP does
No. of does	72	79
DE intake	54.72	87.35
DE requirements		
Maintenance	33.21	35.51
Milk production		57.11
Fetal growth	9.00	6.51
Total balance	+12.51	−11.78

Table 7.9. Effect of the nutritive level and the reduction of the number of suckled pups on the energy balance of does during the second pregnancy (Fortun, 1994). All values are in MJ.

Feeding level	Restricted		Ad libitum	
Groups[a]	RLN does	M1 does	R1 does	FL does
No. of does	27	20	19	27
DE intake	51.58	32.53	23.50	80.32
DE requirements				
Maintenance	34.85	32.96	32.25	35.78
Milk production				46.64
Fetal growth	8.15	5.96	5.70	6.85
Total balance	+8.58	−6.39	−14.45	−8.95

[a]RNL does; non-lactating and restricted (0.75 ad libitum) from 15 to 28 days of pregnancy; M1 does: non-lactating and restricted at maintenance level; R1 does: non-lactating and restricted at 0.75 of maintenance level; FL does: lactating for four pups and fed ad libitum.

It has been demonstrated that the limiting factor on doe productivity is not milk production but voluntary feed intake. For this reason, as DEI increases, milk production also tends to increase, thereby cancelling, at least partially, the effect of increased DEI. Figure 7.7 illustrates the energy balance achieved in 104 breeding does in different experiments (Xiccato, 1996). An increase of 1 kJ in the DEI leads to a proportional increase in milk energy

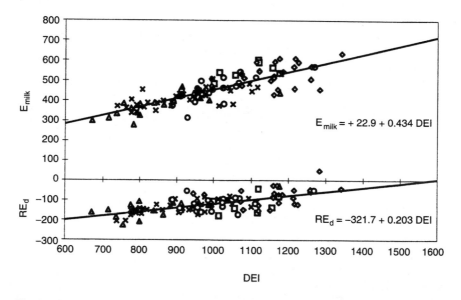

Fig. 7.7. Effect of DEI on energy milk production (E_{milk}) and energy retained in the doe body (RE_d) (data expressed in kJ day^{-1} kg^{-1} LW$^{0.75}$) (Xiccato, 1996).

output (+0.434 kJ) and a more limited reduction in the energy deficit (−0.203 kJ). This trend, linear throughout the time period evaluated, shows a DEI capable of maintaining the body energy equilibrium of a doe ($RE_d = 0$) equal to 1585 kJ day^{-1} kg^{-1} LW$^{0.75}$. At this DEI level, the energy milk output is 711 kJ day^{-1} kg^{-1} LW$^{0.75}$, which corresponds to about 250 g day^{-1} of milk in a 4.25 kg rabbit (assuming a milk energy concentration of 8.4 MJ kg^{-1}). Using a diet with 10.5 MJ kg^{-1} of DE, this female must be able to ingest at least 150 g day^{-1} kg^{-1} LW$^{0.75}$, i.e. about 440–450 g day^{-1}. Such an average voluntary intake during the entire period of lactation is very unusual in primiparous and second parity does.

The above-mentioned equations indicate that any intervention designed to stimulate energy intake will only very rarely provide a substantial reduction in the body energy deficit. In some cases, a simultaneous increase in daily energy intake and milk production does not produce improvements in the nutritional state of does. Figure 7.8 shows the change in protein and energy deficit in lactating and concurrent pregnant does fed three different diets: one moderate-energy diet (M diet, 10.1 MJ kg^{-1}) and two high-energy diets (H diet, with increased starch concentration, 11.0 MJ kg^{-1}; and F diet, with added animal fat, 10.7 MJ kg^{-1}) (Xiccato *et al.*, 1995). The high-energy diets, especially the F diet, stimulated DEI and milk production proportionally, in this way positively affecting litter growth but without improving protein and energy balance.

In addition to the stimulation of milk production, any increase in DEI is generally associated with increased feed intake. This leads to a faster digestive transit and a consequent reduction in the digestive utilization of the dietary energy, which makes the objective of solving the energy deficit even more difficult to achieve (Xiccato *et al.*, 1992a; de Blas *et al.*, 1995; Xiccato, 1996).

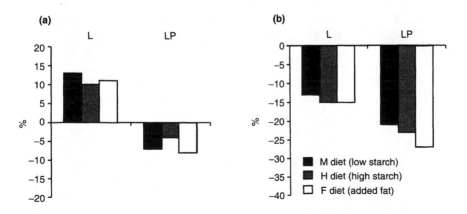

Fig. 7.8. Protein (a) and energy (b) balance in lactating (L) and concurrently lactating and pregnant does (LP) (Xiccato *et al.*, 1995).

The problem of body energy deficit occurs most frequently during the first lactation, but assumes less importance in successive lactations, as reported by many studies (Simplicio *et al.*, 1988; Parigi Bini *et al.*, 1989; Battaglini and Grandi, 1991; Castellini and Battaglini, 1991) which observed an increase of 10–20% in feed intake from the first to the second lactation, 7–15% from the second to the third, and 3–7% from the third to the fourth lactation, before finally reaching a stable level. Milk production also increases with parity order, but less markedly, and this permits better maintenance of the energy balance.

Breeding rhythm

As described from the comparison between pregnant and non-pregnant does, the breeding system can greatly affect the energy balance of lactating does, and influences both milk production and feed intake (Fig. 7.9).

Lactation and pregnancy completely overlap when a PP remating rhythm is adopted, thus causing a deterioration in the energy balance and the depletion of protein reserves in the does, as described above. Moreover, the lack of a rest period between lactations precludes the recovery of energy reserves and has negative repercussions on reproductive performance: low fertility, reduced litter sizes, and higher doe replacement rates. The female cannot become pregnant and only a consequently longer interval between lactations permits the restoration of body energy reserves.

An increase in milk production and an improvement in reproductive performance were observed when semi-intensive breeding rhythms were adopted (remating 9–15 days PP), while the diet was found to play a marginal role (Méndez *et al.*, 1986; Fraga *et al.*, 1989; Cervera *et al.*, 1993). All over the world, most breeders have by now abandoned the excessively intensive PP reproductive cycle in favour of a 10–12 days PP remating schedule (Fig. 7.10), which amounts to a theoretical 42-day interval between parturitions, or rather, remating every 6 weeks. This reproductive cycle, although still considered intensive, permits adequate recovery of the body energy loss of does and has proven to be the ideal compromise between economic convenience and the physiological and metabolic requirements of the does (Maertens, 1992; Xiccato, 1993).

Widespread opinion holds that, in addition to reducing reproductive performance, a more extensive remating interval (21–28 days PP or even post-weaning) requires the use of restricted feeding in order to prevent the excessive fattening of does and the negative consequences for fertility and prolificacy that usually result (Maertens, 1992). A technical choice such as this should be governed only by particular management necessities (e.g. the 'single-cycle' or 'bande unique' system) or the use of unimproved breeds and non-intensive production systems.

In contrast to this opinion, Parigi Bini *et al.* (1996) have recently demonstrated that primiparous rabbits remated 12 days PP do not recover

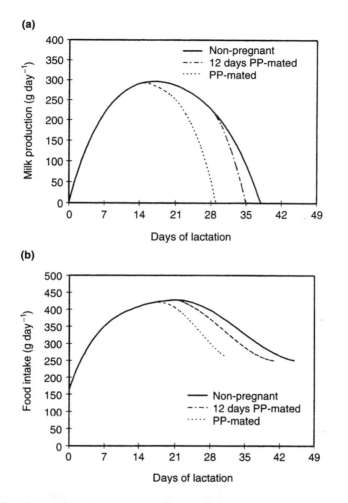

Fig. 7.9. Effect of breeding rhythm on milk production (a) and food intake (b) (Xiccato, 1993, 1996).

their lactation energy deficit, which remains rather high (−26% of the initial energy level) and not very much lower than the deficit recorded during the lactation period (−32%) (Fig. 7.11). More surprising, however, is the fact that not even a longer remating interval (28 days PP) permits the recovery of the energy deficit caused by lactation, which remains at −16%. While demonstrating the positive effect of more extensive remating intervals, this result is undoubtedly caused by the natural decrease in the food intake of the doe at the end of lactation, which makes return to the original levels of energy reserves even longer.

Another aspect to be considered in order to reduce the doe energy deficit is the correct preparation of the young female destined for reproductive activity. This animal has to withstand both the rapid and intense metabolic

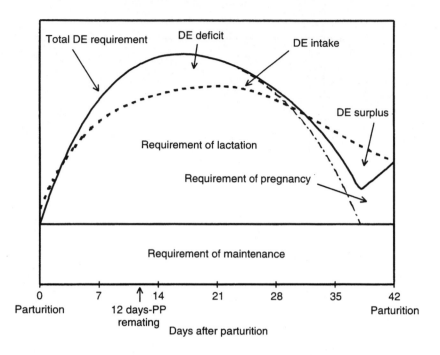

Fig. 7.10. Changes in energy balance and energy intake during lactation (Xiccato, 1993, 1996).

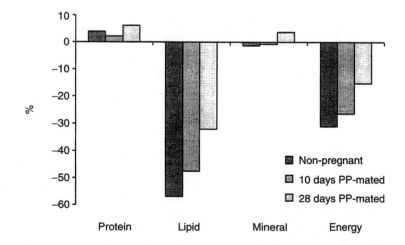

Fig. 7.11. Effect of different breeding rhythm on chemical and energy balance in rabbit does (Parigi Bini *et al.*, 1996).

Table 7.10. Influence of a low-energy diet given from weaning until first mating on the ensuing lactation and empty body balance (Parigi Bini *et al.*, 1995).

	Diet fed before first mating	
	Control diet	Low-energy diet
LW at parturition (g)	3846	3833
LW at the end of lactation (g)	3939	3848
Milk production (g day^{-1})	206	204
Food intake (g day^{-1})	331	340
DE intake (kJ day^{-1} kg^{-1} LW$^{0.75}$)	1203	1245
EB gain (g)	−233	−221
EB balance (% change)		
Water	10.8	6.9
Protein	6.9	6.4
Fat	−68.3	−59.4
Mineral	0	−3.2
Energy	−41.0	−36.1

processes of lactation and pregnancy and the pronounced modification in body composition following the depletion of body energy reserves. In contrast with the indications provided by Maertens (1992), which suggested reducing dietary energy intake in young females and giving them 35 g day^{-1} kg^{-1} LW of a lactation diet until 17 or 18 weeks of age, followed by a 4-day flushing before mating, Parigi Bini *et al.* (1995) proposed feeding them a non-restricted low-energy diet from weaning until mating, followed by an *ad libitum* 'lactation diet' for the first pregnancy onwards. Such a feeding programme stimulated voluntary feed intake without affecting milk production, therefore limiting the energy deficit (−36% vs. −41% of initial content) in the primiparous does (Table 7.10).

References

Battaglini, M. and Grandi, A. (1991) Effetto della fase fisiologica, della stagione e dell'ordine di parto sul comportamento alimentare della coniglia fattrice. In: *Proceedings IX Congresso Nazionale A.S.P.A., Roma*, Vol. I, pp. 465–475.

de Blas, J.C., Fraga, M. and Rodriguez, J. (1985) Units for feed evaluation and requirements for commercially grown rabbits. *JouÚäal of Animal Science* 60, 1021–1028.

de Blas, J.C., Taboada, E., Mateos, G., Nicodemus, N. and Mendez, J. (1995) Effect of substitution of starch for fibre and fat in isoenergetic diets on nutrient digestibility and reproductive performance of rabbits. *Journal of Animal Science* 73, 1131–1137.

Blaxter K. (1989) *Energy Metabolism in Animals and Man.* Cambridge University Press, Cambridge, UK.

Cambero, M.I., De La Hoz, L., Sanz, B. and Ordóñez, J.A. (1991) Lipid and fatty acid composition of rabbit meat. Part 2 – Phospholipids. *Meat Science* 29, 167–176.

Castellini, C. and Battaglini, M. (1991) Influenza della concentrazione energetica della razione e del ritmo riproduttivo sulle performance delle coniglie. In: *Proceedings IX Congresso Nazionale A.S.P.A.*, *Roma*, Vol. I, pp. 477–488.

Cervera, C., Fernandez-Carmona, J., Viudes, P. and Blas, E. (1993) Effect of remating interval and diet on the performance of female rabbits and their litters. *Animal Production* 56, 399–405.

Cheeke, P.R. (1987) *Rabbit Feeding and Nutrition*. Academic Press, Orlando, Florida.

Close, W.H. (1990) The evaluation of feeds through calorimetry studies. In: Wiseman, J. and Cole, D.J.A. (eds) *Feedstuffs Evaluation*. Butterworths, London, pp. 21–39.

De Rochambeau, H. (1990) Genetique du lapin domestique pour la production de poil et la production de viande: revue bibliographique 1984–1987. *Cuni-Sciences* 6(2), 17–48.

Fekete, S. (1992) The rabbit body composition: methods of measurement, significance of its knowledge and the obtained results. A critical review. *Journal of Applied Rabbit Research* 15, 72–85.

Fortun, L. (1994) Effets de la lactation sur la mortalité et la croissance fetale chez la lapine primipare. PhD thesis. Université de Rennes, France.

Fortun, L. and Lebas, F. (1994) Effets de l'origine et de la teneur en énergie de l'aliment sur les performances de reproduction de lapines primipares saillies post-partum. Premiers résultats. In: *Proceedings 6èmes Journées de la Recherche Cunicole en France*, *La Rochelle*, Vol. II, ITAVI, Paris, pp. 285–292.

Fortun, L., Prunier, A. and Lebas, F. (1993) Effect of lactation on fetal survival and development in rabbit does mated shortly after parturition. *Journal of Animal Science* 71, 1882–1886.

Fortun-Lamothe, L. and Bolet, G. (1995) Les effets de la lactation sur les performances de reproduction chez la lapine. *INRA Production Animale* 8, 49–56.

Fraga, M.J., Lorente, M., Carabaño, R.M. and de Blas, J.C. (1989) Effect of diet and remating interval on milk production and milk composition of the doe rabbit. *Animal Production* 48, 459–466.

Gidenne, T. (1996) Nutritional and ontogenic factors affecting rabbit caeco-colic digestive physiology. In: Lebas, F. (ed.) *Proceedings 6th World Rabbit Congress*, *Toulouse*, Vol. 1. Association Française de Cuniculture, Lempdes, pp. 13–28.

Hulot, F., Ouhayoun, J. and Dalle Zotte, A. (1992) The effects of recombinant porcine somatotropin on rabbit growth, feed efficiency and body composition. *Journal of Applied Rabbit Research* 15, 832–840.

Isar, O. (1981) Energy requirement of meat rabbits. *Lucrarile Stiintifice ale Institului de Cercetari pentru Nutritie Animale* 9/10, 253–264.

Jean-Blain, C. and Durix, A. (1985) Ketone body metabolism during pregnancy in the rabbit. *Reproduction, Nutrition and Development* 25, 545–554.

Jentsch, W., Schiemann, L., Hofmann, L. and Nehering, K. (1963) Die energetische Verwertung der Futterstoffe. 2. Die energetische Verwertung der Kraftfutterstoffe durch Kaninchen. *Archiv für Tierernährung* 13, 133–145.

Lebas, F. (1971) Composition minérale du lait de lapine. Variations en fonction du stade de lactation. *Annales de Zootechnie* 20, 487–495.

Lebas, F. (1972) Effet de la simultanéité de la lactation et de la gestation sur les performances laitières chez la lapine. *Annales de Zootechnie* 21, 129–131.

Lebas, F. (1989) Besoins nutritionnels des lapins. Revue bibliographique et perspectives. *Cuni-Sciences* 5(2), 1–28.

Lebas, F., Coudert, P., Rouvier, R. and De Rochambeau, H. (1984) *Le Lapin: Élevage et Pathologie*. FAO, Rome.

Lofgreen, G.P. and Garrett, W.N. (1968) A system for expressing net energy requirements and feed values for growing and finishing beef cattle. *Journal of Animal Science* 27, 793–806.

Maertens, L. (1992) Rabbit nutrition and feeding: a review of some recent developments. *Journal of Applied Rabbit Research* 15, 889–913.

Maertens, L. and De Groote, G. (1988) The influence of the dietary energy content on the performances of post-partum breeding does. In: *Proceedings of the 4th World Rabbit Congress, Budapest*, Vol. 3. Sandor Holdas, Hercegalom, pp. 42–52.

Maertens, L., Janssen, W.M.M., Steenland, E., Wolfers, D.F., Branje, H.E.B. and Jager, F. (1990) Tables de composition, de digestibilité et de valeur énergétique des matières premières pour lapins. In: *Proceedings 5èmes Journées de la Recherche Cunicole, Paris*, Vol. II, Communication 57, ITAVI, Paris, pp. 1–9.

Mendez, J., de Blas, J.C. and Fraga, M. (1986) The effects of diet and remating interval after parturition on the reproductive performance of the commercial doe rabbit. *Journal of Animal Science* 62, 1624–1634.

Milisits, G., Romvári, R., Dalle Zotte, A., Xiccato, G. and Szendrö, Z.S. (1996) Determination of body composition changes of pregnant does by X-ray computerised tomography. In: Lebas, F. (ed.) *Proceedings of the 6th World Rabbit Congress, Toulouse*, Vol. 1. Association Française de Cuniculture, Lempdes, pp. 207–212.

Nizza, A., Moniello, G. and Di Lella, T. (1995) Prestazioni produttive e metabolismo energetico di conigli in accrescimento in funzione della stagione e della fonte proteica alimentare. *Zootecnica Nutrizione Animale* 21, 173–183.

Noblet, J. and Close, V.H. (1980) The energy cost of pregnancy in the sow. In: Mount, L.E. (ed.) *Proceedings 8th Symposium on Energy Metabolism*. EEAP Publ. No. 26, Butterworth, London/Boston, pp. 335–339.

Ortiz, V., de Blas, C. and Sanz, E. (1989) Effect of dietary fibre and fat content on energy balance in fattening rabbits. *Journal of Applied Rabbit Research* 12, 159–162.

Parigi Bini, R. (1988) Recent developments and future goals in research on nutrition of intensively reared rabbits. In: *Proceedings of the 4th World Rabbit Congress, Budapest*, Vol. 3. Sandor Holdas, Hercegalom, pp. 1–29.

Parigi Bini, R. and Cesselli, P. (1977) Valutazione dell'escrezione urinaria di energia in conigli in accrescimento. *Rivista Zootecnica e Veterinaria* 5, 130–137.

Parigi Bini, R. and Dalle Rive, V. (1978) Utilizzazione dell'energia metabolizzabile per la sintesi di proteine e grassi nei conigli in accrescimento. *Rivista Zootecnica e Veterinaria* 6, 242–248.

Parigi Bini, R. and Xiccato, G. (1986) Utilizzazione dell'energia e della proteina digeribile nel coniglio in accrescimento. *Coniglicoltura* 23(4), 54–56.

Parigi Bini, R. and Xiccato, G. (1993) Recherches sur l'interaction entre alimentation, reproduction et lactation chez la lapine. Une revue. *World Rabbit Science* 1, 155–161.

Parigi Bini, R., Chiericato, G.M. and Lanari, D. (1974) I mangimi grassati nel coniglio in accrescimento. Digeribilità e utilizzazione energetica. *Rivista Zootecnica e Veterinaria* 2, 193–202.

Parigi Bini, R., Dalle Rive, V. and Mazzarella, M. (1978) Fabbisogni di energia netta dei conigli in accrescimento. *Rivista Zootecnica e Veterinaria* 6, 32–36.

Parigi Bini, R., Xiccato, G. and Cinetto, M. (1989) Influenza dell'intervallo parto-accoppiamento sulle prestazioni riproduttive delle coniglie fattrici. *Coniglicoltura* 26(7), 51–57.

Parigi Bini, R., Xiccato, G. and Cinetto, M. (1990a) Energy and protein retention and partition in pregnant and non-pregnant rabbit does during the first pregnancy. *Cuni-Sciences* 6(1), 19–29.

Parigi Bini, R., Xiccato, G. and Cinetto, M. (1990b) Repartition de l'énergie alimentaire chez la lapine non gestante pendant la première lactation. In: *Proceedings 5èmes Journées de la Recherche Cunicole en France, Paris*, Vol. II. Communication 47, ITAVI, Paris, pp. 1–8.

Parigi Bini, R., Xiccato, G. and Cinetto, M. (1991a) Utilization and partition of digestible energy in primiparous rabbit does in different physiological states. In: *Proceedings 12th International Symposium on Energy Metabolism, Zurich*. Institute for Animal Sciences, Swiss Federal Institute of Technology, Zurich, pp. 284–287.

Parigi Bini, R., Xiccato, G. and Cinetto, M. (1991b) Utilizzazione e ripartizione dell'energia e della proteina digeribile in coniglie non gravide durante la prima lattazione. *Zootecnica Nutrizione Animale* 17, 107–120.

Parigi Bini, R., Xiccato, G., Cinetto, M. and Dalle Zotte, A. (1992) Energy and protein utilization and partition in rabbit does concurrently pregnant and lactating. *Animal Production* 55, 153–162.

Parigi Bini, R., Battaglini, M., Xiccato, G., Castellini, C., Dalle Zotte, A. and Carazzolo, A. (1995) Effetto del piano alimentare post-svezzamento sulle prestazioni riproduttive delle giovani coniglie. In: *Proceedings XI Congresso Nazionale ASPA, Grado*. Dipartimento di Scienze della Produzione Animale, Udine, Italy, pp. 131–132.

Parigi Bini, R., Xiccato, G., Cinetto, M., Dalle Zotte, A., Carazzolo, A., Castellini, C. and Stradaioli, G. (1996) Effect of remating interval and diet on the performance and energy balance of rabbit does. In: Lebas, F. (ed.) *Proceedings 6th World Rabbit Congress, Toulouse*, Vol. 1. Association Française de Cuniculture, Lempdes, pp. 253–258.

Partridge, G.G. (1986) Meeting the protein and energy requirements of the commercial rabbit for growth and reproduction. In: *Proceedings 4th World Congress of Animal Feeding, Madrid*, Vol. IX. Editorial Garsi, Madrid, pp. 271–277.

Partridge, G.G., Fuller, M. and Pullar, J. (1983) Energy and nitrogen metabolism of lactating rabbits. *British Journal of Nutrition* 49, 507–516.

Partridge, G.G., Daniels, Y. and Fordyce, R.A. (1986a) The effects of energy intake during pregnancy in doe rabbits on pup birth weight, milk output and maternal body composition change in the ensuing lactation. *Journal of Agricultural Science, Cambridge* 107, 697–708.

Partridge, G.G., Lobley, G.E. and Fordyce, R.A. (1986b) Energy and nitrogen metabolism of rabbits during pregnancy, lactation, and concurrent pregnancy and lactation. *British Journal of Nutrition* 56, 199–207.

Partridge, G.G., Garthwaite, P.H. and Findlay, M. (1989) Protein and energy retention by growing rabbits offered diets with increasing proportions of fibre. *Journal of Agricultural Science, Cambridge* 112, 171–178.

Perez, J.M., Lebas, F., Gidenne, T., Maertens, L., Xiccato, G., Parigi Bini, R., Dalle Zotte, A., Cossu, M.E., Carazzolo, A., Villamide, M.J., Carabaño, R., Fraga, M.J., Ramos, M.A., Cervera, C., Blas, E., Fernandez, J., Falcao E Cunha, L. and Bengala Freire, J. (1995) European reference method for *in vivo* determination of diet digestibility in rabbits. *World Rabbit Science* 3, 41–43.

Rattray, P.V., Trigg, T.E. and Urlich, C.F. (1980) Energy exchanges in twin-pregnancy ewes. In: Mount, L.E. (ed.) *Proceedings 8th Symposium on Energy Metabolism.* EEAP Publ. No. 26, Butterworth, London/Boston, pp. 325–328.

Santomá, G., De Blas, J.C., Carabaño, R. and Fraga, M. (1989) Nutrition of rabbits. In: Haresign, W. and Lewis, D. (eds) *Recent Advances in Animal Nutrition.* Butterworths, London, pp. 97–138.

Scheele, C.W., Van Den Broek, A. and Hendricks, F.A. (1985) Maintenance requirements and energy utilisation of growing rabbits at different temperatures. In: Moe, P.W., Tyrrell, H.F. and Reynolds, P.J. (eds) *Proceedings 10th Energy Metabolism Symposium, Airlie.* EEAP Publication no. 32, Rowman and Littlefield, Totowa, New Jersey, pp. 202–205.

Simplicio, J.B., Fernandez-Carmona, J. and Cervera, C. (1988) The effect of the high ambient temperature on the reproductive response of the commercial doe rabbit. In: *Proceedings of the 4th World Rabbit Congress, Budapest,* Vol. 3. Sandor Holdas, Hercegalom, pp. 36–41.

Viudes-De Castro, P., Santacreu, M.A. and Vicente, J.S. (1991) Effet de la concentration énergétique de l'alimentation sur les pertes embryonnaires et foetales chez la lapine. *Reproduction, Nutrition and Development* 31, 529–534.

Walach-Janiak, M., Raj, S. and Fandrejewski, H. (1986) The effect of pregnancy on protein, water and fat deposition in the body of gilts. *Livestock Production Science* 15, 249–260.

Webster, A.J.F. (1989) Bioenergetics, bioengineering and growth. *Animal Production* 48, 249–269.

Xiccato, G. (1989) Quale sistema energetico per il coniglio. *Professioneallevatore* 16(2), 1–8.

Xiccato, G. (1993) Come alimentare il coniglio. *Professioneallevatore* 18(1), 27–40.

Xiccato, G. (1996) Nutrition of lactating does. In: Lebas, F. (ed.) *Proceedings of the 6th World Rabbit Congress, Toulouse,* Vol. 1. Association Française de Cuniculture, Lempdes, pp. 29–50.

Xiccato, G., Cinetto, M. and Dalle Zotte, A. (1992a) Effetto del livello nutritivo e della categoria di conigli sulla digeribilità degli alimenti e sul bilancio azotato. *Zootecnica Nutrizione Animale* 18, 35–43.

Xiccato, G., Parigi Bini, R., Cinetto, M. and Dalle Zotte, A. (1992b) The influence of feeding and protein levels on energy and protein utilization by rabbit does. *Journal of Applied Rabbit Research* 15, 965–972.

Xiccato, G., Parigi Bini, R., Dalle Zotte, A., Carazzolo, A. and Cossu, M.E. (1995) Effect of dietary energy level, addition of fat and physiological state on performance and energy balance of lactating and pregnant rabbit does. *Animal Science* 61, 387–398.

Young, M. (1979) Transfer of amino acids. In: Chamberlain, G.V.P. and Wilkinson, A.W. (eds) *Placental Transfer.* Pitman, London, pp. 142–158.

8.   Protein Requirements

M.J. Fraga

*Departamento de Producción Animal, ETS Ingenieros Agrónomos,
Universidad Politécnica, 28040 Madrid, Spain*

Introduction

To synthesize proteins (for example meat, milk and hair proteins) the rabbit
simultaneously requires all constituent amino acids. Some amino acids, those
that the animal does not synthesize, are defined as essential and have to be
supplied by the diet. The requirements of animals therefore are for amino
acids, rather than for protein.

Thus, the dietary protein level necessary to meet the requirements of the
rabbit vary according to: (i) its amino acid profile; (ii) the degree to which the
protein is digested; and (iii) the amount of feed ingested which, in turn, depends on the dietary digestible energy (DE) concentration. Consequently,
information about the digestible essential amino acid levels in relation to the
DE content of diets would be extremely valuable. However, there is only
limited information available on this in rabbits, such that it is not possible to
express requirements in such units.

Currently, information on the dietary digestible protein/digestible energy
(DCP/DE) ratio is valuable because it includes two of the most important
variable factors mentioned above. The utilization of digestible units to
express protein requirements is obviously more appropriate because of the
considerable differences in protein digestibility between feed ingredients, the
average values expressed as coefficients for protein concentrates, cereals,
forages and by-products being 0.79, 0.73, 0.61 and 0.55, respectively
(Villamide and Fraga, 1998).

Information on the optimal dietary DCP/DE ratio should be complemented with the amino acid requirements expressed, if possible, in terms of
digestible amino acids. Although recommended digestible levels of the more
limiting amino acids have recently been published (Taboada *et al.*, 1994,
1996; de Blas *et al.*, 1996), data on amino acid digestibility of different feedstuffs are still limited.

This chapter reviews the information that permits the calculation of daily
protein requirements (expressed in g DCP day^{-1}) of rabbits. The practical

recommendations presented in Chapter 13 should be used until more information about amino acid requirements becomes available.

Maintenance requirements

There are unavoidable (obligatory) losses of amino acids from the body that require replacement to maintain body conditions constant. These losses are important in tissues where there is significant sloughing of cells such as in skin, hair and intestinal mucosae. The requirements to support these activities are determined as the intercept of the regression equation which relates the amounts of an amino acid ingested to the quantities retained, and are called maintenance requirements. There are few data on maintenance requirements for amino acids in rabbits, but the values for crude protein (CP) are 2.9 and 3.7 g digestible CP kg $LW^{-0.75}$ day^{-1} in growing and doe rabbits, respectively (Box 8.1).

Box 8.1. Protein requirements for rabbits[a].

1. Maintenance
(i) Growing animals: 2.9 g DCP kg $LW^{-0.75}$ day^{-1}
(see text)
(ii) Pregnant and lactating does: 3.7–3.8 DCP kg $LW^{-0.75}$ day^{-1} (refs 1, 2).

2. Milk production
(i) Protein content of milk: 115 g CP kg^{-1} (360 g kg^{-1} on a DM basis) (refs 3, 4).
(ii) Efficiency of utilization of dietary digestible protein (DCP) in milk protein synthesis: 0.76–0.80 (refs 1, 2).
(iii) Efficiency of utilization of body protein for milk protein synthesis: 0.59–0.61 (refs 1, 2).

3. Pregnancy
(i) Efficiency of utilization of dietary DCP for fetal growth: 0.42–0.46 (refs 1, 3).

4. Growth
(i) Protein content in live weight gain: 180 g CP kg^{-1} (600 g kg^{-1} DM) (refs 6, 7, 8).
(ii) Efficiency of utilization of dietary DCP for body protein synthesis: 0.56 (see text).

[a]If the average live weight of the growing rabbit is known, the maintenance requirements in DCP day^{-1} can be calculated (1(i)). If the rate of growth, the protein content in weight gain (4(i)) and the efficiency of utilization of dietary DCP for body protein synthesis (4(ii)) are known, the growth requirements in DCP day^{-1} can be calculated. The sum is the total requirement in DCP day^{-1} of a growing rabbit. The same procedure may be used for the other physiological situations.
References: 1, Xiccato *et al.*, 1992; 2, Parigi Bini *et al.*, 1991; 3, Fraga *et al.*, 1989; 4, Xiccato *et al.*, 1995; 5, Parigi Bini *et al.*, 1992; 6, García *et al.*, 1992; 7, Motta Ferreira *et al.*, 1996; 8, Fernández and Fraga, 1996b.

Requirements for growth

Body composition

The amino acid composition of whole body protein of growing rabbits is presented in Table 8.1. There is a notable similarity with the amino acid pattern of other species such as pigs or rats; the main difference is related to the high cystine content (a high proportion of hair to rabbit whole body, and a high level of cystine in hair protein). On the other hand, the methionine content of the body of rabbits is low. There is a close agreement between the sum of the individual amino acid levels and the CP content.

When comparing rabbits of different breeds and strains at the same live weight, the differences are very small for body CP content. However, a higher content of body fat and energy and a lower content of water is observed in fast-growing rabbits.

In the same way, the influence of body weight and sex on the body CP content is small. The males had a slightly higher body CP content than females (1–2%; Fraga *et al.*, 1983; Fernández, 1993), but the differences for amino acids due to sex are small and generally not significant (Moughan *et al.*, 1988).

Table 8.1. Amino acid composition (mg g^{-1} N) of the whole growing body[a] and of the milk of doe[b] rabbits.

Amino acids	Whole body		Milk	
	Absolute value	Relative to lysine	Absolute value	Relative to lysine
Lysine	383	100	451	100
Alanine	365	74	228	50
Arginine	415	108	328	73
Aspartic acid	467	121	451	100
Histidine	193	50	159	35
Isoleucine	194	51	304	67
Leucine	429	112	567	125
Methionine	77	20	150	33
Cystine	158	41	175	39
Glutamic acid	788	205	1220	270
Glycine	466	121	106	23
Phenylalanine	249	65	281	62
Serine	283	74	228	50
Threonine	245	64	305	67
Tyrosine	192	50	332	73
Valine	239	62	382	85

[a]53-day-old New Zealand rabbits (Moughan *et al.*, 1988).
[b]New Zealand × Californian doe rabbits (M.J. Fraga, unpublished data).

The diet does not affect the body CP content, except when the rate of growth is modified. In any case, a variation of 10 g day^{-1} in the rate of growth (i.e. from 35 to 45 g day^{-1}) results in a variation in the body CP content of rabbits of only 2% (Fraga *et al.*, 1983).

Retained protein and efficiency of growth

As a consequence of the lack of variation in body CP content at slaughter, the CP content of live weight gain is generally constant (Box 8.1) and, therefore, daily protein retention (and daily requirements) depend on the rate of growth of the rabbit. Following from this, the efficiency of utilization of total digestible CP ingested increases with the rate of growth because the relative contribution of maintenance requirements of protein decreases.

Protein retention decreases linearly with an increase in slaughter weight (Table 8.2). In the later stages of the growing period the rate of growth is lower. As a consequence, the efficiency of DP utilization decreases regularly as slaughter weight increases. Males retain more protein, grow faster (8.7%, Partridge *et al.*, 1989; 4%, Fernández, 1993) and are slightly more efficient at DP utilization (Table 8.2) than females.

From the data obtained by Partridge *et al.* (1989), Motta Ferreira *et al.* (1996) and Fernández and Fraga (1996b), the following regression equation was derived relating retained protein and total digestible protein ingested:

$$\text{DCPI} = 2.88 + 1.78 \text{ CPR}, \qquad\qquad P < 0.0001, R^2 = 0.85, n = 17,$$

where DCPI and PR are digestible crude protein ingested and protein retained, respectively, as g kg LW$^{-0.75}$ day^{-1}. From this equation, the intercept (2.9 g kg LW$^{-0.75}$ day^{-1}) corresponds to the maintenance requirements of growing rabbits, and 0.56 (1/1.78) is the amount of dietary digestible CP used for growth that is retained in the body of rabbits (Box 8.1).

Table 8.2. Influence of slaughter weight and sex on retained protein and on overall efficiency of digestible crude protein utilization for growing (New Zealand × Californian) rabbits.

	Slaughter weight (kg)		Sex	
	2.0	2.5	Male	Female
Daily protein retention				
(g kg LW$^{-0.75}$ day^{-1})[a]	5.4	4.7	5.2	4.8
Overall DCP efficiency[a]	0.39	0.35	0.38	0.37
Daily protein retention				
(g kg LW$^{-0.75}$ day^{-1})[b]	4.6	3.9		
Overall DCP efficiency[b]	0.37	0.32		

[a] Fernández and Fraga (1996b).
[b] García *et al.* (1992).

Requirements of does

Milk production

Amino acids are necessary to synthesize milk protein in the mammary gland. Some amino acids may also be used for gluconeogenesis by does fed diets with low starch levels. In these diets the uptake of glucose from the gut may be insufficient to meet the requirements for milk lactose synthesis. De Blas *et al.* (1995) obtained a decrease of feed efficiency in doe rabbits fed diets with low levels of starch (the use of amino acids for gluconeogenesis implies a low efficiency in digestible energy utilization); the optimal value being obtained with a level of 190–210 g dietary starch kg⁻¹. During lactation, mobilization of body protein may occur to support synthesis of milk protein, mainly when lactation and pregnancy are simultaneous.

Milk production is high in hybrid does, the main factors contributing to its variation being the litter size and the remating interval. Total milk production can be predicted from easily measurable traits such as litter live weight at 21 days, which is when the pups began to consume solid feed. Using the data of Méndez *et al.* (1986), Fraga *et al.* (1989) and de Blas *et al.* (1996), the following regression equation was obtained (to a litter weight at 21 days from 2 to 3 kg):

$$MP = 1.77 + 1.39 \, LW21, \qquad\qquad P < 0.001, R^2 = 0.88, n = 13,$$

where MP is total milk production (kg) and LW21 is the litter weight (kg) at 21 days of lactation.

Daily milk production continuously increases from parturition to 17–21 days. The relatively low incremental requirements for milk yield in early lactation allows milk yield and live weight gain to occur simultaneously.

From 21 to 30 days, milk yield decreases but at different rates in pregnant and in non-pregnant does. Does re-mated 1 day after parturition showed a lower total milk yield than those re-mated 8 days later (approximately 0.5 kg, with 0.95 of the decrease occurring during the last week of lactation). Does that are not pregnant during lactation yield about 1 kg more than those re-mated at parturition (0.76 of the difference occurring during the last week). These amounts should be used to correct the values obtained by applying the equation.

The CP content of milk of does approaches 115 g kg⁻¹ (Box 8.1). However, a high CP content (about 140 g kg⁻¹; Partridge and Allan, 1982; Pascual *et al.*, 1996) was observed during the last week of lactation in lactating pregnant does. The efficiencies of utilization of dietary digestible protein and body protein for milk protein synthesis are shown in Table 8.1.

The type of diet will not substantially alter the CP content and the amino acid composition of milk. The more noticeable differences in the amino acid patterns of milk of does (Table 8.1), cows and pigs are the low contents of proline and serine in doe milk. The methionine/methionine + cystine ratio is also lower and can be related to the high requirements for the postnatal growth of pups, which is characterized by the rapid growth of skin and hair. With respect to lysine content, the threonine and sulphur amino acid levels of doe milk are higher than in other species.

During lactation, in non-pregnant does, there is a mobilization of fat deposits, and the empty body shows an increase in water. However, the body protein content remains virtually unchanged, or shows a positive protein balance (Box 8.2).

Pregnancy

During pregnancy, amino acids are retained in the conceptus, in the mammary glands and in the maternal body. The importance of protein changes in the body of the does during pregnancy is higher whenever the does are concurrently lactating.

First gestation

During the initial two thirds of pregnancy, body protein retention of does is higher than that of the conceptus (Box 8.2). However, to support the high growth of this in the last third of pregnancy, a portion of body protein is

---

Box 8.2. Protein balance in does.

(a) Partition of weight gain and retained protein during the first pregnancy of does (Parigi Bini *et al.*, 1990).

	0–21 days		21–30 days	
	Doe's empty body	Conceptus	Doe's empty body	Conceptus
Empty body gain (g)	180.2	192.3	−90.5	453.4
CP content (g kg^{-1} DM basis)	580	620	630	610
Protein retained (g)	60.3	18.3	−21.9	54.3

(b) Protein balance during the first lactation of does (Xiccato *et al.*, 1995).

	Physiological state	
	Lactating and pregnant	Lactating non-pregnant
Empty body gain (g)	−131	184
Retained protein (g)	−38	75
Protein balance (%)	−6	11

(c) Composition of litters (Xiccato *et al.*, 1995).
Water: 808 g kg^{-1}, CP: 123 g kg^{-1}, 615 g kg^{-1} DM, fat: 50 g kg^{-1}, 250 g kg^{-1} DM, energy: 4.56 MJ kg^{-1} (22.8 MJ kg^{-1} DM).

mobilized. From data on the body composition of pups at birth (Box 8.2), Partridge and Allan (1982) have determined that between 0.52 and 0.84 of retained protein is used to support fetal growth, with the remainder being for intrauterine deposition. The rapid turnover in fetal protein may explain why the efficiency of utilization of dietary digestible protein for pregnancy is lower than that for growth (Box 8.1).

Simultaneous pregnancy and lactation

The overlapping of pregnancy and lactation increases the protein requirements in response to the concurrent high demands for protein for fetal growth and milk yield. As a consequence, a negative balance of body protein was observed (Box 8.2) in pregnant lactating rabbits.

Digestible protein/digestible energy ratio of rabbit diets

The protein requirements, estimated as g DCP day^{-1} using the data in Box 8.1, can be converted to dietary concentrations for the purposes of feed formulation if the actual daily feed intake of rabbits is known. As the quantity of energy ingested daily in terms of DE tends to be constant, the feed intake of rabbits can be predicted from the energy concentration of the diet. As a consequence, it is advisable to express the protein requirements as a DCP/DE ratio.

According to de Blas *et al.* (1985), the requirements in terms of DCP/DE ratio for maintenance are close to 6.8 g MJ^{-1}, implying that maintenance represents a high energy cost in relation to protein; for average growth the requirement for protein is much higher. The young rabbit multiplies its birth weight sixfold during the first 3 weeks of life, in which milk (with a ratio of 13–14 g CP MJ^{-1} GE) is its only feed. From 21 days to weaning a progressive change is made from milk to doe feed. Later, the rate of growth in relation to live weight decreases (30–45 g day^{-1} up to 8 weeks). The best results in growing rabbits measured in productivity terms were obtained by de Blas *et al.* (1985) using Spanish Giant rabbits with diets containing about 10 g DCP MJ^{-1} DE (see Chapter 13). In countries where rabbits are raised up to 2.5–2.8 kg live weight, and using two or three different growing feeds, a decrease in dietary protein levels should be considered in the latter stages (see earlier, 'Retained protein and efficiency of growth').

However, the recommended value should be revised when using current fast-growing commercial strains. It is possible that the daily protein requirements do not vary, but the higher fat content in the live weight gain of fast-growing rabbits can decrease the recommended value of the DCP/DE ratio. However, more information is necessary on the effect of selection for growth on body composition.

Daily protein requirements for lactating does are relatively higher than for growing rabbits. Because of this, and taking into account the difficulty of obtaining a high feed intake during the lactation of highly productive does, the DCP/DE ratio recommended is higher than for growing rabbits, and varies from 11.0 to 12.5 g MJ^{-1} according to performance (see Chapter 13).

Amino acid requirements

The actual amino acid intake of rabbits is the sum of amino acids provided by the diet and from reingestion of soft faeces. In rabbits fed conventional diets the contribution of soft faeces to total intake of CP is around 0.17 (see Chapter 3) but the amino acid intake and its relation to dietary composition remains to be established, because of the limited information concerning the variation in amino acid composition of soft faeces. As a consequence, the amino acid requirements of growing and doe rabbits that have been determined to date, mainly by dose–response trials (see Chapter 13 for practical recommendations), should be used with caution.

The data obtained by Moughan *et al.* (1988) allow estimation of the amino acid requirements of growing rabbits according to their respective proportions of body protein. To apply this method it is necessary to know the optimal level, measured in productivity terms, for a single amino acid (for example lysine). From the data on amino acid composition of doe milk that are presented in Table 8.1, the amino acid requirements of lactating females may also be estimated according to their respective ratios in milk protein.

In fact, the values obtained for threonine and sulphur amino acids with respect to lysine content in milk (see Table 8.1) compare well with the optimal relative requirements of these amino acids (0.68 and 0.76 for threonine and sulphur amino acids, respectively) expressed in digestible units obtained by Taboada *et al.* (1966) and de Blas *et al.* (1996), respectively. In the same way, the optimal values of amino acids for growth obtained by the same authors were consistent with the data on body amino acid composition (Table 8.1).

However, the use of the pattern of amino acids in doe milk and in the whole body of growing rabbits to define their amino acid requirements assumes that the maintenance requirements are only a small proportion of the total amino acid requirements during lactation and growth. This proportion is low in pigs (approximately 0.05 in lactating sows; Pettigrew, 1995) but there are no data for rabbits. However, using the data in Box 8.1, the protein maintenance requirements of growing and doe rabbits are, respectively, about 0.30 and 0.25 of total protein requirements. The contribution of doe hair loss (approaching 300–500 mg day^{-1} in the 5 days before parturition; Sawin *et al.*, 1960) and its growth during the next lactation to maintenance requirements may partially explain these differences. In the same way, there is a mobilization of body protein stores in pregnant and lactating doe rabbits to support protein synthesis. However, in does fed diets with high CP contents, slightly positive balances can be obtained. More information is necessary to establish a method, similar to the one proposed by Pettigrew (1995) for pigs, where the amino acid requirements determined for different functions (replenishment of obligatory losses, mobilization of body protein) could be integrated with data on the amino acid composition of different products. Any improvement in the knowledge of amino acid requirements would allow a more appropriate use of synthetic amino acids with consequent reductions in the nitrogen content of diets and excreta, in the energy losses and in the risks of digestive disturbances.

References

de Blas, J.C., Fraga, M.J. and Rodríguez, J.M. (1985) Units for feed evaluation and requirements for commercially grown rabbits. *Journal of Animal Science* 60, 1021–1028.

de Blas, J.C., Taboada, E., Mateos, G., Nicodemus, N. and Méndez, J. (1995) Effect of substitution of starch for fiber and fat in isoenergetic diets on nutrient digestibility and reproductive performance of rabbits. *Journal of Animal Science* 73, 1131–1137.

de Blas, J.C., Taboada, E., Nicodemus, N., Campos, R., Piquer, J. and Méndez, J. (1996) The response of highly productive rabbits to dietary threonine content for reproduction and growth. In: Lebas, F. (ed.) *Proceedings of the 6th World Rabbit Congress, Toulouse*, Vol. 1. Association Française de Cuniculture, Lempdes, pp. 139–143.

Fernández, C. (1993) Efecto de la incorporación de grasa en piensos fibrosos sobre la utilización digestiva de la dieta, crecimiento y calidad de la canal de conejos en cebo. PhD Thesis, Universidad Politécnica de Madrid, Spain.

Fernández, C. and Fraga, M.J. (1996a) The effect of fat inclusion on growth performance, carcass characteristics, and chemical composition of rabbits. *Journal of Animal Science* 74, 2088–2094.

Fernández, C. and Fraga, M.J. (1996b) Effect of fat inclusion in diets for rabbits on the efficiency of digestible energy and protein utilization. *World Rabbit Science* 4, 19–23.

Fraga, M.J., de Blas, J.C, Pérez, E., Rodríguez, J.M., Pérez, C.J. and Gálvez, J.F. (1983) Effect of diet on chemical composition of rabbits slaughtered at fixed body weights. *Journal of Animal Science* 56, 1097–1104.

Fraga, M.J., Lorente, M., Carabaño, R.M. and de Blas, J.C. (1989) Effect of diet and remating interval on milk production and milk composition of the doe rabbit. *Animal Production* 48, 459–466.

García, G., Gálvez, J.F. and de Blas, J.C. (1992) Substitution of barley grain by sugar-beet pulp in diets for finishing rabbits. 1. Effect on energy and nitrogen balance. *Journal of Applied Rabbit Research* 15, 1008–1016.

García, G., Gálvez, J.F. and de Blas, J.C. (1993) Effect of substitution of sugarbeet pulp for barley in diets for finishing rabbits on growth perfomance and on energy

142   *M.J. Fraga*

and nitrogen efficiency. *Journal of Animal Science* 71, 1823–1830.

Méndez, J., de Blas, J.C. and Fraga, M.J. (1986) The effects of diet and remating interval after parturition on the reproductive perfomance of the commercial doe rabbit. *Journal of Animal Science* 62, 1624–1634.

Motta Ferreira, W., Fraga, M.J. and Carabaño, R. (1996) Inclusion of grape pomace, in substitution for lucerne hay, in diets for growing rabbits. *Animal Science* 63, 167–174.

Moughan, P.J., Schultze, W.H. and Smith, W.C. (1988) Amino acid requirements of the growing meat rabbit. 1. The amino acid composition of rabbit whole-body tissue – a theoretical estimate of ideal amino acid balance. *Animal Production* 47, 297–301.

Parigi Bini, R., Xiccato, G. and Cinetto, M. (1990) Energy qnd protein retention and partition in rabbit does during the first pregnancy. *Cuni-Science* 6, 12–29.

Parigi Bini, R., Xiccato, G. and Cinetto, M. (1991) Utilizzazione e ripartizione dell'energia e della proteina digeribile in coniglie non gravide durante la prima lattazione. *Zootecnica e Nutrizione Animale* 17, 107–120.

Parigi Bini, R., Xiccato, G., Cinetto, M. and Dalle Zotte, A. (1992) Energy and protein utilization and partition in rabbit does concurrently pregnant and lactating. *Animal Production* 55, 153–162.

Partridge, G.G. and Allan, S.J. (1982) The effects of different intakes of crude protein on nitrogen utilization in the pregnant and lactating rabbit. *Animal Production* 35, 145–155.

Partridge, G.G., Garthwaite, P.H. and Findlay, M. (1989) Protein and energy retention by growing rabbits offered diets with increasing proportions of fibre. *Journal of Agricultural Science* 112, 171–178.

Pascual, J.J., Cervera, C., Blas, E. and Fernández-Carmona, J. (1996) Milk yield and composition in breeding rabbit does using high fat diets. In: Lebas, F. (ed.) *Proceedings of the 6th World Rabbit Congress, Toulouse*, Vol. 1. Association Française de Cuniculture, Lempdes, pp. 259–262.

Pettigrew, J.E. (1995) Amino acid requirements of breeding pigs. In: Haresign, W. (ed.) *Recent Advances in Animal Nutrition.* Butterworths, London, pp. 241–256.

Sawin, P.B., Deneberg, V.H., Ross, S., Hafter, E. and Zarrow, M.X. (1960) Maternal behavior in the rabbit: hair loosening during gestation. *American Journal of Physiology* 198, 1099–1102.

Taboada, E., Méndez, J., Mateos, G.G. and de Blas, J.C. (1994) The response of highly productive rabbits to dietary lysine content. *Livestock Production Science* 40, 329–337.

Taboada, E., Méndez, J. and de Blas, J.C. (1996) The response of highly productive rabbits to dietary sulphur amino acid content for reproduction and growth. *Reproduction, Nutrition and Development* 36, 191–203.

Villamide , M.J. and Fraga, M.J. (1998) Prediction of the digestible crude protein and protein digestibility of feed ingredients for rabbits from chemical analysis. *Animal Feed Science and Technology* 70, 211–224.

Xiccato, G., Parigi Bini, R., Cinetto, M. and Dalle Zotte, A. (1992) The influence of feeding and protein levels on energy and protein utilization by rabbit does. *Journal of Applied Rabbit Research* 15, 965–972.

Xiccato, G., Parigi Bini, R., Dalle Zotte, A., Carazzolo, A. and Cossu, M.E. (1995)

Effect of dietary energy level, addition of fat and physiological state on perfomance and energy balance of lactating and pregnant rabbit does. *Animal Science* 61, 387–398.

9. Minerals, Vitamins and Additives

G.G. Mateos and C. de Blas

*Departamento de Producción Animal, Universidad Politécnica de Madrid,
Ciudad Universitaria, 28040 Madrid, Spain*

Mineral requirements of rabbits

The amount of scientific information on the requirements of highly
productive rabbits for minerals published in the last 10 years is very limited
for some macroelements and nil for the majority of the oligo-elements.
Therefore, most of the information gathered is based on old data and on
practical figures obtained under commercial conditions.

Rabbit meat is rich in protein and low in energy but its ash content is
similar to or greater than that of other domestic species (Ouhayoun and
Lebas, 1987). According to Rao *et al.* (1979) rabbit meat is comparatively
poor in potassium (K) and sodium (Na) and rich in calcium (Ca) and
phosphorus (P). Mean values found (in ppm) were 393 for Na, 2052 for K,
129 for Ca, 145 for magnesium (Mg), 54 for zinc (Zn), 29 for iron (Fe) and
1.6 for manganese (Mn).

Compared with other mammals, rabbit milk is higher in ash, especially in
Ca, P and Na. This is not surprising since bones of animals that are immature
at birth need extensive mineralization (Widdowson, 1974). Table 9.1 presents

Table 9.1. Macromineral composition of rabbit milk compared with that of other mammals (g kg^{-1}).

	Cow[a]	Sheep[a]	Sow[b]	Rabbit doe[c]
Na	0.45	0.45	0.5	1.11
K	1.50	1.25	0.84	1.71
Ca	1.20	1.9	2.2	4.78
Mg	0.12	0.15	—	0.37
P	0.90	1.5	1.6	2.57
Cl	1.10	1.2	—	0.64

[a]Gueguen *et al.* (1988).
[b]Partridge and Gill (1993).
[c]El Sayiad *et al.* (1994).

the approximate macromineral composition of rabbit milk compared with that of other mammals.

Macrominerals

Macrominerals are defined as elements required in grams per day and whose requirements are expressed as concentrations in the diet. The definition includes Ca, P, Mg, Na, K, chloride (Cl) and sulphur (S). Currently, only Ca, P and Na are taken into account in practical formulation of rabbit diets.

Calcium

Ca is the main component of the skeleton. Over 0.98 of the total calcium is in bones and teeth. In addition, Ca plays a key role in numerous organic processes such as heart function, muscle contraction, coagulation and electrolyte equilibrium of blood. In addition, doe milk is rich in Ca. Therefore, the requirements for this element in the diet are expected to be greater for fast-growing young animals and does in late gestation and at the peak of milk production than for fattening or maintenance.

The metabolism of Ca in the rabbit has distinct differences compared with other domestic species:

1. Ca is absorbed in direct proportion to its concentration in the diet, regardless of metabolic needs. Blood levels of Ca rise with increasing intakes (Chapin and Smith, 1967).
2. Urine is the main route through which excess is eliminated.

The regulation of Ca absorption is not precisely controlled in the rabbit. The role of the parathyroid hormone (PTH) and of 1,25-OH D$_3$ in the process of absorption of Ca is not well understood. Ca is absorbed and passed into the bloodstream of rabbits with greater efficiency than in other species such as rats, pigs or ruminants (Cheeke and Amberg, 1973). Excess of blood Ca over the kidney threshold is excreted as a white thick, creamy, urine precipitate which will be deposited beneath the cages. Swick *et al.* (1981) proposed that the filtration of Ca and the crystals formed in the process might damage the structure of the kidneys, producing red pigmentation of the urine. As in other species, prolonged feeding of an excess of Ca (>40 g kg^{-1}) may result in calcification of soft tissues, particularly with elevated vitamin D intakes (Löliger and Vogt, 1980; Cheeke, 1987).

High milk-producing does might express a syndrome similar to that of milk fever in dairy cows. Late in gestation and early in lactation, does may suffer a drop in plasma Ca levels (from 14 to below 7 mg 100 ml^{-1}) and other minerals (P and Mg) with loss of appetite, tetany, muscle tremours, ear flapping, lying on their sides and death (Barlet, 1980). Injection of calcium gluconate induced a rapid recovery within 2 h. Whether or not a modification of the electrolyte equilibrium of the diet towards a negative balance (acidotic diet) will benefit rabbits, as it does in dairy cattle, remains unknown.

Phosphorus

P is a major constituent of bones. It also plays an important role in energy metabolism. In most mammalian species, inorganic P is absorbed at the duodenal and jejunum level and the mechanism is modulated both by endocrine (calcitriol, triodothyronine) and nutritional factors (Barlet *et al.*, 1995). The role of 1,25-OH D_3 and other metabolites on phosphorus absorption in the rabbit is unknown. Borowitz and Granrud (1993) have demonstrated in 3-month-old rabbits an active mechanism for P transport at the duodenum and proximal jejunum. Absorption is more efficient in very young animals. Because horses and rabbits have a similar digestive tract, Cheeke (1987) proposed that similar mineral absorption values should be used for both species. Recently, there has been an increasing interest in reducing the excretion of P through feed manipulation to control environmental pollution. Unfortunately, no studies in this area have been conducted with rabbits. Recent studies with fryers (Steenland, 1991) and breeding does (Lebas and Jouglar, 1990) have demonstrated that rabbit performance was not reduced when phosphorus levels were decreased below former recommendations. In fact, these authors found that P levels below 5 g kg^{-1} were adequate for all types of production.

A major factor influencing phosphorus availability to non-ruminant animals is the presence of phytates and phytases in plant materials. Phytates are not degraded by animal enzymes. Natural phytases, such as those existing in wheat, or addition of exogenous commercial sources, will improve utilization of P. In the rabbit, phytate P is well utilized because of phytase production by the microorganisms of the caecum. Swick *et al.* (1981) found an apparent P digestibility in a maize–soybean meal diet of 0.75 which indicates utilization close to that of dicalcium phosphate. However, the utilization of P contained in lucerne seems rather low, compared with that of pigs (Cheeke *et al.*, 1985; Cromwell, 1992). This is surprising, since most of the P in lucerne is found in the leaves rather than in the stems. Therefore, most of it will be recycled through soft faeces and coprophagy and an almost complete utilization of phytate P should be expected.

In most species, Ca and P requirements are closely interrelated. A dietary relationship of Ca to available P of 2:1 to 1.5:1 is widely accepted (Vandelli, 1995). In fact, the milk of rabbits maintains a constant 2:1 relationship between both minerals throughout the lactation period (El-Sayiad *et al.*, 1994). An excess of Ca is probably more detrimental at marginal P levels. Ca might decrease the absorption of P and therefore create an artificial deficiency of P when low dietary levels of this mineral are present. The need to maintain this relationship closely in rabbits is not evident and at least in fatteners does not seem to be critical. Diets for growing rabbits with a Ca:P ratio of 12:1 did not show any detrimental effect in terms of performance (Chapin and Smith, 1967). However, Assane *et al.* (1993) have observed an increase of P and Mg in the blood at the end of the gestation period when the Ca:P ratio in feed was 1:1 compared with 2:1. Narrowing the Ca:P ratio in the feed improved phosphataemia and magnesaemia in this period (Table 9.2). This information needs further substantiation.

Practical recommendations on dietary levels of Ca and P vary according to age, breed, productivity and composition of the diet. For growing-fattening rabbits, values in the literature vary from 4 to 10 g for Ca and from 2.2 to 6 g for P (NRC, 1977; AEC, 1987; Schlolaut, 1987; Mateos, 1989; Lebas, 1990; Burgi, 1993; Mateos and Piquer, 1994; Vandelli, 1995; Maertens, 1996; Xiccato, 1996) (Table 9.3a).

Lactating does have higher requirements for Ca and P than growing rabbits or non-lactating does, as rabbit milk is particularly rich in Ca and P. Average contents are 4.5–6.5 g Ca kg^{-1} and 3.5–4.4 g P kg^{-1}, which are approximately three to five times higher than those of cow milk (Burgi, 1993; El-Sayiad *et al.*, 1994). A doe can excrete up to 2 g of Ca, at maximal milk production. Practical recommendations in doe feeds vary from 7.5 to 13.5 g for Ca and from 5 to 8 g for P, according to the same authors (Table 9.3b).

Based on a literature review and practical experience, the recommended concentrations of Ca and P in complete diets are presented in Table 9.4.

Dietary levels below requirements of Ca and P will lead to rickets

Table 9.2. Influence of the Ca:P ratio of the feed on Ca, P and Mg levels in serum of gestating rabbit does (Assane *et al.*, 1993).

Ca:P ratio	Premating	3rd week of gestation	Preparturition	SEM
Calcaemia, mg l^{-1}				
1:1[a]	123*	120†	111‡	3.7
2:1[b]	145*	125‡	114§	6.2
Phosphataemia, mg l^{-1}				
1:1	38*	39*	41*	6.3
2:1	43*	38*	25†	5.9
Magnesaemia, mg l^{-1}				
1:1	28‡	27*	27*	3.7
2:1	31*	23‡	19‡	4.7

[a] 5.2 g Ca and 5.1 g kg^{-1} P.
[b] 8.3 g Ca and 3.9 g kg^{-1} P.
Means in the same row with different superscripts differ ($P < 0.05$).

Table 9.3. Macromineral recommendations (g kg⁻¹ diet as-fed) for intensively reared rabbits.

Author	Ca	P	Na	Cl	K
a. *Growing-fattening*					
NRC (1977)	4.0	2.2	2.0	3.0	6.0
AEC (1987)	8.0	5.0	3.0	—	—
Schlolaut (1987)[a]	10.0	5.0	—	—	10.0
Lebas (1990)	8.0	5.0	2.0	3.5	6.0
Burgi (1993)	5.0	3.0	—	—	—
Mateos and Piquer (1994)	5.5	3.5	2.5	—	—
Vandelli (1995)	4.0–8.0	3.0–5.0	—	—	—
Maertens (1996)	8.0	5.0	—	3.0	—
Xiccato (1996)[b]	8.0–9.0	5.0–6.0	2.0	3.0	–
b. *Lactating does*					
NRC (1977)	7.5	5.0	2.0	3.0	6.0
AEC (1987)	11.0	8.0	3.0	—	—
Schlolaut (1987)[a]	10.0	5.0	—	—	10.0
Lebas (1990)	12.0	7.0	2.0	3.5	9.0
Mateos and Piquer (1994)	11.5	7.0	—	—	—
Vandelli (1995)	11.0–13.5	6.0–8.0	—	—	—
Maertens (1996)	12.0	5.5	—	3.0	—
Xiccato (1996)[b]	13.0–13.5	6.0–6.5	2.5	3.5	—

[a]Angora rabbits.
[b]Young does.

(young), osteomalacia (adults), lack of fertility and abnormal behaviour. An excess of Ca (over 15 g kg⁻¹) increases the calcification of soft tissues (Kamphues, 1991) and reduces Zn and P absorption. An excess of P (over 9 g kg⁻¹) may depress feed intake and impair prolificacy in does (Chapin and Smith, 1967; Lebas and Jouglar, 1984, 1990). In all cases, an excess has a negative impact on the environment.

Table 9.4. Requirements of Ca and P for rabbits (g kg⁻¹ as-fed basis).

	Ca	P
Breeding does		
Recommendation	12.0	6.0
Commercial range	10.0–15.0	4.5–7.5
Fattening rabbits (1–2 months of age)		
Recommendation	6.0	4.0
Commercial range	4.0–10.0	3.5–7.0
Finishing rabbits (> 2 months)		
Recommendation	4.5	3.2
Commercial range	3.0–8.0	3.0–6.0

Other macrominerals

Mg is a major bone component (0.7 of total Mg is in the skeleton) and also acts as a cofactor in many energy metabolism reactions. A deficiency is indicated by poor growth, alopecia, hyperexcitability, convulsions, lack of fur texture and fur chewing. The metabolism of Mg is not well known but, as in the case of Ca, the excess is eliminated through the urine. The requirements in Mg for growing rabbits vary from 0.3 (NRC, 1977) to 3 g kg^{-1} diet (Lebas, 1990). Evans *et al.* (1983a, b) found that 3.4 g Mg kg^{-1} diet fulfilled requirements but that 1.7 g was insufficient. The content and apparent digestibility of Mg in most raw materials is high and the need to add extra Mg to commercial rabbit diets has not been established.

K plays a key role in the regulation of the acid–base balance and is a cofactor for numerous enzymes. Symptoms of deficiency include muscle weakness, paralysis and respiratory distress. The problems might be exacerbated with diarrhoea (Licois *et al.*, 1978). Current estimates indicate that 6 g kg^{-1} diet avoids symptoms of deficiency. Because most feed ingredients used in diets are rich in K (soybean meal, lucerne, molasses), a deficiency is difficult to envisage. Several authors have indicated the need to avoid a diet with K content over 10 g kg^{-1}. This level can be reached when a high proportion of heavily fertilized, early mature lucerne is used. Surdeau *et al.* (1976) observed an increased incidence of nephritis at 8 g kg^{-1} diet. Evans *et al.* (1983a) reported a reduction of intake when levels of K over 10 g kg^{-1} were used. In addition, an excess of K antagonizes Mg absorption, although the importance of the problem has not been studied in the rabbit. Practical recommendations range between 6.5 and 10 g kg^{-1}.

Na is involved in the regulation of pH and osmotic pressure. In contrast to K, Na concentrates in the plasma, outside the cells. The requirements for Na have not been studied in the rabbit. Under practical conditions, levels of 2.0–2.3 and 2.2–2.5 g kg^{-1} of the diet are used for fryers and lactating does, respectively. Excess of Na in the form of salt (NaCl above 15 g kg^{-1}) is detrimental to growth (Harris *et al.*, 1984b).

Cl is also involved in acid–base regulation. In addition, this ion is concentrated by gastric cells. It is secreted as HCl and is involved in the process of protein digestion. Cl requirements have been estimated as being between 1.7 and 3.2 g kg^{-1}, but an excess (4.7 g) does not impair performance (Colin, 1977).

Practical diets for high-producing rabbits are unlikely to be deficient in Cl. Sodium chloride and lysine hydrochloride are routinely used in diets as a source of Na and lysine respectively and indirectly serve as a supplement for Cl. Practical levels vary between 2.8 and 4.8 g kg^{-1}.

It is well known that the relationship between Na$^+$, K$^+$ and Cl$^-$ affects animal performance. In addition to influencing resistance to thermal stress, leg score, kidney function and incidence of milk fever, a large negative value might decrease feed intake while a positive value might increase problems around farrowing. No information on the effects of variation of this ratio on productivity is available for the rabbit. However, this species is particularly vulnerable to acid loads. In fact its urine is more alkaline than that of rats or other mammals fed the same type of diet. Therefore, care should be taken to avoid imbalances that might promote nephritis, reproductive problems or decreased feed intake.

S is one of the more abundant elements in nature. It forms part of chondroitin sulphate, a major component of cartilage, tendons, the wall of blood vessels and bones. In addition, S is a constituent of numerous organic substances such as haemoglobin, glutathione, coenzyme A, and the amino acids methionine and cystine. Practical diets include over 2.0 g S kg^{-1} but no supplemental sources are used. There are no reports in the literature indicating any benefit of S supplementation on rabbit performance, although inorganic S can be incorporated into microbial protein in the hindgut and be used for protein accretion in the rabbit.

Microminerals

Microminerals are defined as those elements required in mg per day and whose requirements are expressed as ppm of the diet. The definition includes Fe, copper (Cu), Mn, Zn, selenium (Se), iodine (I) and cobalt (Co). Other trace elements that are required by the rabbit but are not supplemented under practical conditions are molybdenum (Mo), fluorine (F) and chromium (Cr). All those mentioned in the first group are routinely added to rabbit diet as salts through an oligo-element premix.

In addition to having other functions, iron is a major constituent of the pigments and enzymes involved in oxygen transport and metabolism. Therefore, a deficiency will produce anaemia because of impaired haemo-globin formation. Mammalian mechanisms for transporting Fe to milk are very poor but rabbit does are capable of passing reasonable amounts of Fe through the placenta. Rabbits are born with large iron reserves, provided that the doe receives a diet properly supplemented. Therefore, rabbits are not as dependent as piglets on an exogenous supply of Fe for survival. Even though milk is poor in Fe, no deficiency is expected in the young rabbit. Furthermore, young rabbits start eating doe feed at around 14 days. Since most ingredients used in feeds are rich in iron (soil-contaminated lucerne, macromineral sources and micromineral premix) an Fe deficiency is not expected to develop early in life. Recent data obtained by El-Masry and Nasr (1996) indicate that a benefit can be obtained from adding 80 ppm of Fe to diets for does that contained a total of 129 ppm of Fe. Does fed the Fe-supplemented diet produced more milk and had greater litter size and weight than controls. However, the low productivity and high mortality of the animals on trial, as well as the small number of replicates and the composition of the premix used, do not allow the results of this trial to be adopted with confidence.

The Fe recommendations reported in the literature vary from 30 to 100 ppm with greater levels recommended for does and fur-producing animals (Table 9.5). Under commercial conditions, most commercial premixes supply extra Fe as ferrous sulphate heptahydrate of the order of 30–50 ppm. At these levels, the requirements of Fe for all types of production are easily met if calcium carbonate and dicalcium phosphate are used as sources of Ca and P, respectively.

Cu is a major component of metalloenzymes involved in energy and iron metabolism, and collagen and hair formation. A deficiency will produce symptoms such as retarded growth, grey hair, bone abnormalities and anaemia. Published recommendations for Cu intake vary between 5 and 20 ppm, with higher levels for fur production and breeding does. Practical levels used in Spain vary from 4 to 30 ppm. Even at lower levels, no deficiency symptoms are expected because of the high content of Cu in most raw materials used in rabbit feeds and the potential for accumulation in the liver. As a precaution, the use of forages high in S and Mo should be avoided. The competition between Cu and Mo for absorption is well known, as is the fact that S exacerbates this antagonism.

Table 9.5. Micromineral requirements of rabbits (ppm).

	NRC (1977)	Schlolaut (1987)[a]	Lebas (1990)	Mateos and Piquer (1994)[b]	Xiccato (1996)[c]	Maertens (1995)
a. *Growing-fattening*						
Cu	3	20	15	5	10	10
I	0.2	–	0.2	1.1	0.2	0.2
Fe	+	100	50	35	50	50
Mn	8.5	30	8.5	25	5	8.5
Zn	+	40	25	60	25	25
Co	0	–	0.1	0.25	0.1	0.1
Se	0	–	–	0.01	0.15	–
b. *Lactating does*						
Cu	5	10	15	5	10	10
I	1	–	0.2	1.1	0.2	0.2
Fe	30	50	100	35	100	100
Mn	15	30	2.5	25	5	2.5
Zn	30	40	50	60	50	50
Co	1	–	0.1	0.25	0.1	0.1
Se	0.08	–	0	0.01	0.15	0

[a]Angora rabbits.
[b]Common feed for does and fryers.
[c]Young does.

In addition to its role as an essential nutrient, Cu is used worldwide as a growth promoter in poultry and pigs. Indeed, several reports (Bassuny, 1991; Ayyat *et al.*, 1995; Abbo El-Ezz *et al.*, 1996) have indicated that CuSO₄ at 100–400 ppm improves performance in growing-fattening rabbits. The influence seems more positive in young animals, poor sanitation status, and presence of diseases such as enteritis and enterotoxaemia (Patton *et al.*, 1982). There are, however, conflicting results among authors. In fact, King (1975) and Harris *et al.* (1984a) did not find any benefit when CuSO₄ was used. The use of CuSO₄ as a growth promoter is not allowed in the European Union which precludes its utilization at high levels in commercial feeds.

Mn acts as a coenzyme in amino acid metabolism and formation of the matrix of cartilages. A deficiency produces bones of poor consistency, which may result in brittle bones and crooked legs. In most domestic species, an Mn deficiency results in reproductive failure but information on rabbit does is not available. Hidiroglou *et al.* (1978) observed that, when fed in excess, Mn concentrates in the kidney, liver and spleen of rabbits but not in the reproductive tract. In fact, even in the absence of supplementary Mn, its concentration in uterine tissues remained stable.

Published values on Mn requirements for rabbits vary between 2.5 and 30 ppm (Table 9.5). Commercial mineral premixes include levels between 10 and 75 ppm. Based on field information and cost, it is advisable to use Mn levels of around 8–15 ppm.

Zn, a component of numerous enzymes, is involved in the biosynthesis of nucleic acids, therefore it is especially important in cell division processes. Higher levels are recommended for reproduction and fur and hair production than for maintenance or meat production. Because of enzyme production by hindgut microorganisms, phytates are less detrimental for Zn absorption in rabbits than in other non-ruminant species. Therefore, a lower requirement must be expected for rabbits than for pigs or poultry (Swick *et al.*, 1981).

Zn requirements published in the literature vary between 30 and 60 ppm, with higher values proposed for breeders. Practical commercial diets contain a similar range. These values are considered sufficient for rabbits fed practical diets, but no trials have been conducted to quantify the requirements.

Until recently, selenium was considered a toxic element. In 1957, the essentiality of this nutrient was demonstrated. Diseases such as muscle and liver degeneration, exudative diathesis and impaired reproduction and immunity have been associated with an Se deficiency. In most species the role of Se is closely linked to vitamin E. Selenium is a constituent of the enzyme glutathione peroxidase which plays a role in the detoxification of peroxides formed during metabolic processes. However, rabbit tissues are less dependent on Se for disposal of peroxides than tissues from other mammals. Lee *et al.* (1979) observed that, in the rabbit, most of the existing GSH does not use Se as a cofactor. Therefore, the rabbit is more dependent on vitamin E and less on Se to reduce oxidation load on tissues than other mammals

(Jenkins *et al.*, 1970). Struklec *et al.* (1994) observed improved fetal and birth weight when does received 0.1 ppm of supplemental Se but not with 0.3 ppm. Commercial premixes without supplemental Se have been marketed for years in Europe without evidence of impaired productivity both in does and in growing-fattening rabbits (Mateos, 1989; Lebas, 1990; Mateos and Piquer, 1994). Since no detailed experiments have been conducted to date on this subject, it is wise to include a small amount of supplemental Se (0.05 ppm) in the feeds to avoid potential problems in long-term rabbit production.

I is a component of the thyroid hormones which regulate energy metabolism. Lack of I results in goitre. The incidence of this disease increases when goitrogens are included in the diet. Brassica species, such as cabbage, turnips and rape seeds are rich in goitrogens and therefore their use will increase I requirements.

No requirements for I have ever been established for any type of production. Does are probably more sensitive to an I deficiency than growing-fatteners. Literature requirements vary from nil to 1.1 ppm (Table 9.5). Practical premixes in Spain include between 0.4 and 2 ppm. At these levels of inclusion no goitre or other classical symptoms of deficiency have ever been detected. If marine salt is used as a source of I, the requirements are fully satisfied and a supplemental source of the mineral is no longer recommended.

Co requirements are often overestimated for non-ruminants. The only metabolic role currently accepted for Co is as a component of Vitamin B_{12}. Therefore, symptoms of Co or vitamin B_{12} deficiency are similar. Since animals do not have the enzymes required to attach Co to the molecule to form vitamin B_{12}, the supply of this mineral is ineffective in non-ruminants without access to faeces. Rabbits, however, depend on Co to produce vitamin B_{12}. In fact, the bacteria of rabbit hindguts are more efficient than bacteria of the rumen in utilizing Co (Simnett and Spray, 1965a). Also, absorption of vitamin B_{12} is more efficient in the rabbit than in most mammals (Simnett and Spray, 1965b). Furthermore, requirements for Co are greater for ruminants than for rabbits (Underwood, 1977). Rumen microorganisms use large amounts of Co to synthesize non-active compounds. In addition, ruminants require extra amounts of vitamin B_{12}, and therefore of Co, for propionic acid metabolism. This volatile fatty acid is a major source of energy in these species. In the case of a Co deficiency, the rate of propionate clearance from blood is depressed and, as a consequence, feed intake and productivity are decreased (McDowell, 1992).

Literature requirements for Co vary between 0 and 0.25 ppm, although AEC (1987) recommends 1.0 ppm. In fact, up to 0.8 of commercial Spanish premixes for rabbits include more than 0.4 ppm of Co (Mateos and Piquer, 1994). However, a deficiency of Co, even in unsupplemented vitamin B_{12} diets, is unlikely to occur in the rabbit. Based on the above data supplementation to rabbit diets of 0.25 ppm of Co is recommended.

Vitamin requirements of rabbits

Vitamins are defined as a group of complex organic compounds present in minute amounts in natural feeds that are essential for metabolism. A deficiency always causes a decrease in performance and often pathological symptoms. Vitamins and trace minerals differ in their nature: the former are organic and the latter are inorganic.

Rao *et al.* (1979) found that rabbit meat contains on average the following amounts of vitamins per 100 g of dried product: 0.11 mg of thiamine, 0.37 mg of riboflavin, 21.2 mg of niacin, 0.27 mg of pyridoxine, 0.1 mg of pantothenic acid, 14.8 μg cobalamine, 40.6 μg of folic acid and 2.8 μg of biotin. Vitamins A and E were only found in trace amounts, while vitamin C was essentially absent. Except for riboflavin and biotin, these values are similar to those found for other meats (Ouhayoun and Lebas, 1987).

Except for choline, vitamins are required in minute amounts and requirements are expressed as micrograms or milligrams per day. All vitamins perform essential functions: most act as metabolic catalysts of organic processes. Not all vitamins are essential in a strict sense. Some can be derived from other substances obtained through metabolic changes. For example, choline and vitamin C can be synthesized by several species; niacin can be obtained from tryptophan; most B vitamins are synthesized by microorganisms of the gut and recycled into the body. Finally, vitamin D can be obtained from precursors by the action of ultraviolet light on the skin. Therefore, many vitamins do not fit the classic vitamin definition. They are metabolically but not nutritionally essential.

Vitamins are classified on the basis of their solubility. Vitamins A, D, E and K are soluble in fat while all the others (B complex, vitamin C) are soluble in water. Fat-soluble vitamins are absorbed with dietary lipids, probably by a similar mechanism, and in general are stored in the body (predominantly liver and fat) in appreciable amounts. Water-soluble vitamins are not stored but rapidly excreted, the exception being vitamin B_{12}. In addition, both groups differ in the excretion pattern: fat-soluble vitamins are mainly excreted through bile and faeces while water-soluble vitamins are excreted via urine. A continuous supply is therefore more important for water than for fat-soluble vitamins. Since rabbits have a functional hindgut, their need for supplementation is much higher for fat- than for water-soluble vitamins. In fact, the benefits of adding B complex vitamins to commercial rabbit feeds have not been experimentally demonstrated.

Fat-soluble vitamins

Vitamin A

Vitamin A is only found in feeds of animal origin or synthetic supplements. Plants do contain a series of precursors, the carotenes, with variable vitamin

A-like activity. In the rabbit, β-carotene, the most important precursor, is converted into vitamin A in the intestinal mucosa. In most domestic species, the process is not very efficient. However, estimates of Bondi and Sklan (1984) in the rabbit indicate a conversion efficiency of around 1700 IU of vitamin A by 1 mg of β-carotene, similar to that of chickens. The process, which requires the presence of a Cu-dependent enzyme, is more efficient with low β-carotene intake. For this reason, vitamin A toxicity is more likely to occur when the vitamin itself rather than the precursor is supplemented in the diet (Deeb *et al.*, 1992).

Vitamin A levels in the plasma of rabbits are around 150 μg 100 ml^{-1}, somewhat higher than that for most domestic species (Cheeke, 1987). This value is very variable, since vitamin A is stored and released from the liver as needed. Vitamin A participates in numerous metabolic reactions and it is involved in the mechanism of vision, bone development, maintenance of epithelial integrity, reproduction and immunological response. A particular problem that responds to vitamin A supplementation and affects young rabbits is hydrocephaly. It results from defective bone growth with stenosis of the cerebral aqueduct and elevated cerebrospinal fluid pressure, which may affect nerve function. Also, a deficiency impairs fertility in both sexes. Vitamin A-deficient rabbits show increased abortion rates, resorption of fetuses and diminution of milk production (Cheeke *et al.*, 1984; Moghaddam *et al.*, 1987; Deeb *et al.*, 1992).

The liver can store large quantities of vitamin A. When excessive amounts are supplied, the organ becomes overloaded and toxicity symptons may appear. Rabbit does are particularly sensitive to vitamin A excesses, showing symptoms of toxicity similar to those observed in case of deficiencies (Cheeke *et al.*, 1984; Grobner *et al.*, 1985). The NRC (1987) recommends the addition of no more than 16,000 IU to rabbit diets as an upper safe level. Therefore, high doses of vitamin A supplied continuously through feed or water to combat stress or other field conditions should be avoided.

Vitamin A requirements for growth and reproduction have not been experimentally determined. Values in the literature vary from 600 to 10,000 IU (Table 9.6). In practice, feeding levels of 6000 IU for growing-fattening and of 10,000 IU for breeders are recommended.

In the last 20 years, several authors have observed the benefit of supplementing β-carotene to sows and dairy cattle irrespective of vitamin A status (Byers *et al.*, 1956; Czarnecki *et al.*, 1992; Chew, 1994a, b). According to these data, cows receiving extra β-carotene show more intense oestrus, increased conception rates and reduced incidence of follicular cysts. The suggestion is that β-carotene has a specific function in reproduction, independent of its role as a precursor of vitamin A. The mechanism is not known. In the cow, β-carotene is absorbed intact through the intestinal wall and concentrates in the ovarian follicles where it could exert a beneficial action on reproduction. However, other authors have not found any benefit of

Table 9.6. Vitamin requirements of rabbits.

	NRC (1977)	Schlolaut (1987)[a]	Lebas (1990)	Mateos and Piquer (1994)[b]	Maertens (1996)	Xiccato (1996)[c]
a. *Growing-fattening*						
Vitamin A (mIU)	0.580	8	6	10	6	6
Vitamin D (mIU)	+	1	1	1	1	0.8
Vitamin E (ppm)	40	40	50	20	30	30
Vitamin K_3 (ppm)	?	1	0	1	0	2
Niacin (ppm)	180	50	50	31	50	50
Pyridoxine (ppm)	39	400	2	0.5	2	2
Thiamine (ppm)	—	—	2	0.8	2	2
Riboflavin (ppm)	—	—	6	3	6	6
Folic acid (ppm)	—	—	5	0.1	5	5
Pantothenic acid	—	—	20	10	20	20
Cianocobalamin (ppb)	—	—	10	10	10	10
Choline (mg)	1200	1500	0	300	50[d]	50[d]
Biotin (ppb)	—	—	200	10	200	200
b. *Lactating does*						
Vitamin A (mIU)	10	8	10	10	10	10
Vitamin D (mIU)	1	0.8	1	1	1	1
Vitamin E (ppm)	30	40	50	20[e]	50	50
Vitamin K_3 (ppm)	1	2	2	1	2	2
Niacin (ppm)	50	50	—	31	50	—
Pyridoxine (ppm)	2	300	—	0.5	2	—
Thiamine (ppm)	1	—	—	0.8	2	—
Riboflavin (ppm)	3.5	—	—	3	6	—
Folic acid (ppm)	0.3	—	—	0.1	5	—
Pantothenic acid (ppm)	10	—	—	10	20	—
Cianocobalamin (ppb)	10	—	—	10	10	—
Choline (mg)	1000	1500	—	300	100[d]	100[d]
Biotin (ppb)	—	—	—	10	200	—

[a]Angora rabbits; [b]common feed for does and fryers; [c]young does; [d]as choline chloride; [e]increase to 50 ppm for high-producing does.

β-carotene supplementation other than as a source of vitamin A (Wang *et al.*, 1982, 1988). In the rabbit, this theory has been tested by several authors (Parigi Bini *et al.*, 1983; Elmarimi *et al.*, 1989; Kormann *et al.*, 1989; Besenfelder *et al.*, 1996) with conficting results. In some cases, the injection or the addition through feed of 30–40 ppm of β-carotene improved doe conception and young rabbit survival rates. In others, no benefits were noted. Hoffmann-La Roche (1995) recommend supplementation of doe feeds with

10–20 ppm of β-carotene in order to improve prolificacy. In contrast with cattle, horses and poultry, rabbits are 'white fat' animals which means they are not capable of storing carotenoids. Therefore, it is surprising that supplementation of β-carotene to diets rich in vitamin A through the feed can improve fertility. The β-carotene molecule will be split at the intestinal mucosa by a 15,15′-dioxygenase and converted into a molecule of vitamin A. In fact, the ovarian follicles of rabbits fed β-carotene showed no detectable levels of β-carotene or other carotenoids (Kormann *et al.*, 1989). Therefore, the only explanation could be that the mechanism by which β-carotene exerts its benefits in rabbits is different from that of the cow. Kormann *et al.* (1989) speculated that cleavage of β-carotene may yield a biologically active metabolite of an as yet unknown nature. Based on the lack of agreement among authors on the influence of β-carotene on reproduction and the cost of supplementation, caution is needed. Therefore, it is advisable not to make any recommendation until additional research has confirmed the benefits of supplementation.

Vitamin D

Vitamin D is synthesized by the animal when exposed to sunlight. The two major natural sources are cholecalciferol (vitamin D_3; animal origin) and ergocalciferol (vitamin D_2; predominantly plant origin). Vitamin D_3 is preferred to vitamin D_2 by rabbit tissues.

Vitamin D, after dihydroxylation in the liver and kidney, acts as a hormone and plays a central role in the metabolism of Ca and P, influencing bone mineralization and mobilization. The classical symptoms of a deficiency are rickets in growing animals and osteomalacia in adults.

As previously discussed, rabbits are very efficient in absorbing Ca and the process seems to be quite independent of vitamin D status. Levels of supplementation as low as 2300–3000 IU kg^{-1} have been shown to be detrimental to rabbits (Ringler and Abrams, 1970; Lebas, 1987). Symptoms of toxicity include fetal mortality, depressed appetite, diarrhoea, ataxia, paralysis and death (Kubota *et al.*, 1982; Zimmerman *et al.*, 1990). In addition, excess of the vitamin causes resorption of bones and calcification of soft tissues such as arteries, liver and kidneys (Löliger and Vogt, 1980; Kamphues, 1991). The incidence of problems is more acute when calcium is fed in excess of requirements.

Levels of vitamin D_3 recommended for rabbits are very low and should not exceed 1000–1300 IU (Table 9.6). However, 12 out of 16 of the commercial premixes for rabbits studied in the survey of Mateos and Piquer (1994) used levels of vitamin D in excess of this recommendation. According to Cheeke (1987) and Lebas (1990), an excess rather than a deficiency is more likely to be a problem in practical conditions.

Vitamin E

Vitamin E activity is found in a series of compounds of plant origin. Eight forms of the vitamin are found in nature, of which α-tocopherol is the more important.

As discussed before, the functionality of vitamin E is closely related to that of Se for most species. Both microelements function in the prevention of oxidative damage to cells. In addition, other major functions of vitamin E are synthesis of prostaglandins, blood clotting, stability of membrane structure and maintenance of immunity. As a result, a deficiency of vitamin E results in muscular dystrophy, myocardial damage, exudative diathesis, hepatosis, oedema, ulcerations, increased incidence of mastitis, mammitis and agalaxia (MMA) and reproductive failure.

The main sign of vitamin E deficiency is muscular dystrophy in the growing rabbit and poor reproductive performance, with infertility, abortions and stillbirths in the pregnant doe (Yamini and Stein, 1989). As mentioned earlier, Se does not seem to have a sparing effect on vitamin E requirements. No recent experiments have been conducted to determine the vitamin E requirement of growing-fattening and breeding rabbits under intensive production patterns. Recommended levels in the literature and practical supplementation values are based either on old data (Ringler and Abrams, 1971) or on extrapolation from other species. Until more information is available, it seems advisable to recommend 15 and 50 ppm of vitamin E for fatteners and doe rabbits, respectively. These amounts should be increased in cases of impaired immunity or coccidiosis infection (Diehl and Kristler, 1961).

Recent research conducted in cattle, pigs, poultry and other species has shown the benefits of including high doses of vitamin E (>200 ppm) on the maintenance of the quality of the meat after slaughter. Meat cuts produced by animals that received diets supplemented with vitamin E had greater stability, better colour, less dripping losses and longer shelf-life than cuts from control animals. Bernardini *et al.* (1996) have obtained similar data with meat from rabbits that were fed 200 ppm of vitamin E.

As for other fat-soluble vitamins, the term vitamin K is used to describe a group of compounds that have as a common characteristic their antihaemorrhagic effects. These compounds can be of vegetable (phyloquinone or K_1), and/or microbial, or animal (menaquinones or K_2) origin.

Vitamin K is mainly involved in the mechanisms of blood coagulation. It is required for the synthesis of prothrombin and other plasma-clotting factors. A deficiency may result in haemorrhagic conditions and lameness in growing rabbits and in placental haemorrhage and abortion of kits in pregnant does (NRC, 1977). Recently, other vitamin K-dependent enzymes have been discovered, indicating that this vitamin does have more roles than previously thought. For example, osteocalcin, a metabolite involved in the mineralization and formation of bone is a vitamin K-dependent protein (McDowell, 1989).

A considerable number of microorganisms present in the rumen and hindgut synthesize large amounts of vitamin K. Animal faeces contain substantial amounts of the vitamin even if none is present in the feed. Therefore, the requirement of rabbits for this vitamin is partly satisfied through coprophagy.

Most ingredients used in feeds are poor sources of vitamin K. The exception is lucerne meal, which may contain up to 20–25 ppm of vitamin K. The requirement of rabbits for this vitamin is difficult to evaluate. In fact, no studies have been conducted in this respect. Most commercial feeds include levels of vitamin K between 1 and 2 ppm (Mateos and Piquer, 1994). These quantities should suffice in most situations. However, in cases of subclinical coccidiosis, use of sulpha drugs and other medication, or inclusion of antimetabolites in the feed (mouldy ingredients, amprolium), an increase in vitamin K supplementation is recommended, especially in pregnant does.

Water-soluble vitamins

Vitamin C

Vitamin C, or ascorbic acid, plays an important role in many biochemical reactions in which oxygen is incorporated into the substrate. It is involved in the biosynthesis of collagen (hydroxylation of lysine and proline) and carnitine, and stimulates phagocytic activity of leukocytes. It concentrates in the seminal fluid, but whether or not it protects sperm from oxidation is not known.

Vitamin C is synthesized from D-glucose in the liver by most mammals, including the rabbit. Therefore, it is not strictly considered as a vitamin for these species (Jennes *et al.*, 1978). In other species, supplementation with vitamin C has been shown to be beneficial in reducing the effects of stress. Under adverse conditions, such as hot summers, intensive production, high stocking density, transport, weaning and subclinical diseases, the synthesis of ascorbic acid from glucose is inadequate and plasma concentration of vitamin C is reduced. Under these conditions a supplemental amount of ascorbic acid might be useful (Mahan *et al.*, 1994; Zakaria and Al-Anezi, 1996). No confirmation of these effects has been obtained in the rabbit. However, rabbits under heat stress showed reduced vitamin C concentration in plasma (Verde and Piquer, 1986). Ismail *et al.* (1992a,b) found an improved reproductive response in rabbits fed vitamins C and E when subjected to high ambient temperature. In stressful situations, Xiccato (1996) recommended a supplement to rabbit feed of 50–100 ppm of vitamin C. Any supplement of this vitamin must be added to the premix in a protected form, since ascorbic acid is easily destroyed by oxidation, especially under moist conditions and when exposed to oxygen, Cu, Fe and other minerals.

B vitamins

Microflora of the hindgut synthesize appreciable amounts of water-soluble vitamins which are utilized through caecotrophy. By this mechanism, the requirements of rabbits for maintenance and average levels of production are fulfilled (NRC, 1977; Harris *et al.*, 1983). However, fast-growing fryers and high-producing does may respond to additional supplementation of B vitamins, namely thiamine (B_1), pyridoxine (B_6), riboflavin (B_2) and niacin (PP) (Lebas, 1987; Maertens, 1996).

Few detailed studies have been conducted recently on the requirements of rabbits for B vitamins. In fact it would appear that none have been reported on the requirements of high-producing rabbits. Therefore, most of the recommendations for intensive production have been extrapolated from other species or are based on field observations. Dietary ingredients used in rabbit diets, such as lucerne meal, wheat middlings and soybean meal, are excellent sources of most B vitamins (Cheeke, 1987). Even with semipurified diets, classical symptoms of deficiency are seldom demonstrated.

Choline is utilized by the organism both as a building unit and as an essential component of other molecules and is involved in the regulation of certain metabolic processes. Unlike the remaining vitamins of the B group, choline is synthesized in the liver and its role in the metabolism is more as a structural constituent than as a coenzyme. Choline is essential for: (i) building and maintenance of cell structure as a component of phospholipids; (ii) fat metabolism in liver, preventing abnormal lipid accumulation; (iii) acetylcholine formation to allow the transmission of nerve impulses; and (iv) donation of labile methyl groups for the formation of methionine, betaine and other metabolites.

In the rabbit, a choline deficiency results in retarded growth, fatty liver and necrosis of the kidney tubules (McDowell, 1992). Also, a progressive muscular dystrophy has been reported in fryers (NRC, 1977).

No report has been published on the requirements of rabbits for choline. Recommendations for supplementation vary from 0 to 1500 ppm (Table 9.6). Under practical conditions, the survey of Mateos and Piquer (1994) observed that choline was supplemented through the premix at levels between 0 and 800 ppm. Based on this, and data from other species, a choline supplement of 200 ppm should suffice for most situations.

Folic acid is necessary for the transfer of single-carbon units, a role analogous to that of pantothenic acid in the transfer of two-carbon units. Therefore, it is important for the biosynthesis of nucleic acids and is required for cell division. Folic acid has attracted attention on the basis of studies conducted with gestating sows which showed an improvement in number of piglets born alive when it was supplemented to the diet (Lindemann, 1993; Matte and Girard, 1996). The response in litter size seems to be a consequence of improved embryo or fetal survival.

No information is available on the requirements of folic acid for reproduction in does. In fact, the NRC (1977) does not even consider this vitamin in its recommendations for rabbits. The study of El-Masry and Nasr (1996) indicated that additional supplementation of doe diets with 5 ppm of folic acid may improve performance and prolificacy. However, as mentioned before, the experimental conditions used were inadequate (poor productivity, high mortality at weaning, small number of replicates used, variable data, inappropriate composition of the premix of the control diet).

Values recommended in the literature vary from 0 to 5 ppm (Table 9.6). This variability indicates the lack of information on the vitamin requirement for rabbits. Commercial premixes used in Spain have a folic acid content varying from 0 to 1.5 ppm without any evident sign of deficiency even at the lowest level (Mateos and Piquer, 1994). Based on available information, 0.1 ppm is recommended for growing-fattening rabbits and 1.5 ppm for does until further information becomes available.

Biotin, or vitamin H, is involved in many metabolic reactions, including the interconversions of protein to carbohydrate and of carbohydrate to fat. It plays a role in maintaining normal blood glucose when carbohydrate intake is low. A deficiency is detected by abnormal function of the thyroid and the adrenal glands, the reproductive tract and the nervous system (McDowell, 1992). The more obvious clinical signs are dermatitis, horn cracks and secondary lameness.

In the rabbit, no deficiency signs have been observed even in the absence of supplementary biotin. Only when raw egg white was fed, which contains avidin (an antivitamin H), was loss of hair and dermatitis noted (NRC, 1977). There is, however, a lack of information on the requirements for reproduction and growth under intensive rearing conditions. Recommended values in the literature vary from 0 to 200 ppb, with greater values for young rabbits. Commercial premixes used in Spain vary in vitamin H content from 0 to 100 ppm although most of them do not include any supplement at all (Mateos and Piquer, 1994). Data available for pigs indicate a benefit from biotin supplementation on horn cracks, growth and reproduction (Kornegay, 1986; Lewis *et al.*, 1991). Therefore, until more information is available, 10 ppb is recommended for fatteners and 80 ppb for does and rearing kits.

Thiamine, or vitamin B_1, is a coenzyme in certain reactions of the citric acid cycle. The classic symptoms of a deficiency are neurological disorders, cardiovascular damage and lack of appetite. In the rabbit, a mild ataxia and flaccid paralysis have been reported when extremely low thiamine diets were used (NRC, 1977). Recommended values in the literature are presented in Table 9.6. Commercial premixes include between 0 and 2 ppm. Until more information is available a thiamine supplementation of 0.6 to 0.8 ppm is recommended.

Riboflavin, or vitamin B_2, is required as a coenzyme in many metabolic processes. Most flavoproteins (flavin adenine dinucleotide, flavin mononucleotide) contain vitamin B_2. Therefore, it is involved in the release of food energy and the assimilation of nutrients. Typical clinical signs often involve the eye, skin and nervous system. Milk is rich in vitamin B_2 and therefore a deficiency should not be expected in suckling animals.

Diets deficient in riboflavin have more effect on early embryonic mortality than on fertility. Riboflavin concentrates in the uterus in early pregnancy, and a massive increase in dietary riboflavin improves embryo survival and litter size (Bazer and Zavy, 1988). Pettigrew *et al.* (1996) observed an increase in farrowings in early pregnant sows but no increases in litter size.

Values recommended in the literature vary between 0 and 6 ppm (Table 9.6); Spanish premixes contain between 0 and 5 ppm (Mateos and Piquer, 1994). Based on the few available data, 3 ppm is recommended for fryers and 5 ppm for rabbit does.

Niacin is involved in many metabolic reactions such as electron transport, which yields energy to the animal. It plays a role in tissue integrity, especially for the skin, gastrointestinal tract and nervous systems. A deficiency is characterized by hair loss, dermatitis, diarrhoea, lack of appetite and ulcerative lesions. Therefore, in cases of deficiency, bacterial infection and enteric conditions are likely to develop.

In the rabbit, substantial amounts of niacin are synthesized by the hindgut microorganisms. In addition, the vitamin can be derived from tryptophan, although the process is very inefficient. However, rabbits respond significantly to extra niacin supplementation (NRC, 1977). Recommended values in the literature vary between 0 (Lebas, 1990, for lactating does) and 180 ppm (NRC, 1977, for growing fattening rabbits). These large discrepancies are inexplicable, even in the absence of experimental information.

Pyridoxine, or vitamin B_6, refers to a group of three related compounds: pyridoxine, pyridoxal and pyridoxamine. Their activities are equivalent in mammals. This vitamin plays a role in the Krebs cycle and amino acid, carbohydrate and fatty acid metabolism. Synthesis of niacin from tryptophan, conversion of linoleic to arachidonic acid, formation of adrenalin from phenylananine or tyrosine, incorporation of iron into haemoglobin and antibody formation are some of the reactions in which pyridoxine is involved. A deficiency produces retarded growth, dermatitis, convulsions, anaemia, scaly skin, alopecia, diarrhoea and fatty livers, among other symptoms.

In the rabbit, vitamin B_6 deficiency causes inflammation around the eyes and nose, scaly thickening of the skin around the ears, alopecia on forelegs and skin desquamation (Bräunlich, 1974). Supplemental values recommended in the literature vary from 0 for lactating does (Lebas, 1990) to 39 ppm for growing-fattening (NRC, 1977) up to 400 ppm for fattening Angora rabbits (Schlolaut, 1987). Again, this extreme range of recommendations is equivocal and is partly due to lack of experimental data. Spanish premixes use between 0 and 2 ppm of vitamin B_6. More than 0.4 of feed manufacturers do not use any supplementation at all and no clinical symptoms have been detected in the field (Mateos and Piquer, 1994). Based on current information, obtained from the field, 0.5 and 1 ppm are recommended for fatteners and does, respectively.

Pantothenic acid is a constituent of coenzyme A and acyl carrier protein, key substances in tissue metabolism. A deficiency reduces growth, and provokes symptoms such as skin lesions, nervous disorders, gastrointestinal disturbances, impairment of adrenal function and decreased resistance to infection (McDowell, 1992).

No deficiency has ever been described in the rabbit (McDowell, 1992). Kulwich *et al.* (1953) observed that caecotrophs had up to six times more pantothenic acid than hard faeces. Literature requirements vary from 0 to 20 ppm (Table 9.6). Contents of Spanish premixes vary also from 0 to 20 ppm. Until new data are available, 8 and 10 ppm for growers' and does' diets, respectively, are recommended.

Vitamin B_{12} was the last but the most potent vitamin to be discovered. It is synthesized in nature only by microorganisms and is not found in feeds of plant origin. As mentioned before, Co is part of this molecule (being the only role known for this micromineral). Vitamin B_{12} is involved as a coenzyme in reactions such as formation of one-carbon units (methyl group synthesis). Therefore, vitamin B_{12} is metabolically related to choline, methionine and folacin, among other essential nutrients. Symptoms of deficiency include anaemia, loss of appetite, rough skin, diarrhoea and reduced litter size. Rabbits are capable of producing substantial amounts of vitamin B_{12} provided that Co is available, and no deficiency symptoms have ever been described with diets currently in use.

Values recommended in the literature vary from 0 to 10 ppb. Commercial premixes contain between 0 and 15 ppb. On the basis of the current situation 9 to 10 ppb for growers and does is recommended.

Additives

A number of feed additives are used in animal feeding to improve certain characteristics of the feed or to enhance animal performance. The list is very broad and includes anticoccidial drugs, growth promoters, preservatives, enzymes, feed flavours, oligosaccharides, probiotics, acidifiers and pellet binders. Some of them are legally permitted but not in all countries. A selection of those additives authorized by European Union legislation and more likely to be used in conventional rabbit feeds will be discussed.

Anticoccidial drugs

Coccidiosis is one of the most important diseases affecting rabbitries. Intestinal coccidiosis may cause diarrhoea and mortality. In most commercial situations the disease occurs subclinically, with growth retardation and impairment of the feed conversion. The use of wire cages and preventive medication reduces the incidence of the disease. Anticoccidial drugs are commonly used in intensive production as a prophylactic therapy to reduce losses caused by intestinal and hepatic coccidiosis.

Several products are available to control the disease. Those used in the European Union include methylclorpindol, robenidine (Licois and Coudert, 1980) and salinomycin (Gaca-Lagodzinska *et al.*, 1994; Paefgen *et al.*, 1996) (Table 9.7). Other products that have been shown to be effective but not registered are diclazuril (Vanparijs *et al.*, 1989; Van Meirhaeghe *et al.*, 1996) and the combination of methylclorpindol and methylbenzoquate. In addition, sulpha drugs and certain ionophores (monensin, narasin, etc.) have been shown to be effective.

All the products available have some benefits but also some drawbacks, especially when the recommended doses are not followed. For example, robenidine can taint the liver and it is not as effective as others in controlling hepatic coccidiosis. Furthermore, if cross-contamination occurs, a taint may appear in yolks. Salinomycin in excess of recommended levels decreases feed intake, which is quite common for most ionophore drugs (Okerman and Moermans, 1980; Peeters *et al.*, 1980; Morisse *et al.*, 1989). In addition, ionophore cross-contamination may be responsible for toxicity in animals other than target species. Horses, turkeys, guinea fowls and broiler breeders are the species more affected by toxicity of ionophores when cross-contamination occurs.

Table 9.7. Anticoccidial drugs in rabbits.

Name	Dose (ppm)	EU registration status	Withdrawal period (day)
Methylclorpindol	125–200	Annex I All types	5
Robenidine	50–66	Annex I All types	5
Salinomycin	20–25	Annex II Growing-fattening	5
Diclazuril	1	Requested	?
Methylclorpindol + methylbenzoquate	220[a]	Withdrawn from Annex II	5

[a]Proportion of 100:8.35 in the mixture.

Antibiotics and growth promoters

Numerous antibiotics have side effects on rabbit performance (Licois, 1980; Thilsted *et al.*, 1981). For example, ampicillin, lincomycin and numerous ionophores disturb the normal microflora of the rabbit intestine, increasing mortality and depressing growth (Mateos, 1989; Morisse *et al.*, 1989; Lafargue-Hauret *et al.*, 1994; Licois, 1996).

Several antibiotics, usually effective against Gram-positive micro-organisms, have been shown to improve effectively rabbit growth and efficiency. Flavophospholipol (2–4 ppm), avoparcin (10–20 ppm), virginiamycin (10–20 ppm) and zinc-bacitracin (20–100 ppm) are growth promoters effective in improving performance (Escribano *et al.*, 1982; Mateos, 1989; Maertens *et al.*, 1992). However, only flavophospholipol is registered in the EU for all types of rabbits; the use of the others is not authorized.

Probiotics

Because, of the uncertainty of the use of antibiotics and growth promoters in animal feeds, new lines of products, more acceptable to the consumer, are appearing in the market to combat subclinical enteric diseases or to improve digestibility.

Probiotics, such as yeasts (Maertens and De Groote, 1992), *Bacillus* (McCartney, 1994) and oligosaccharides (Morisse *et al.*, 1992; Lebas, 1993), are finding a place in rabbit feeding.

Probiotics are supplements that contain beneficial live or reviable microorganisms. It is expected that these microorganisms are capable of colonizing the gut, contributing to the maintenance of the flora equilibrium (Maertens and De Groote, 1992). Their objective is to create a gut barrier against pathogens. Some of them have been shown to benefit rabbit performance (de Blas *et al.*, 1991). Others, however, did not (Aoun *et al.*, 1994; Maertens *et al.*, 1994).

Supplementation of rabbit feeds with oligosaccharides increases the caecal levels of volatile fatty acids in weanling rabbits and decreases caecal ammonia. These changes may help in the prevention of colibacillosis (Peeters *et al.*, 1992). Several authors (Aguilar *et al.*, 1996; Lebas, 1996) have observed a decrease in mortality and an improvement in performance when oligosaccharides were added to rabbit feeds. The results, however, are not always beneficial and more research and information are needed prior to recommendations on their inclusion in diets.

Enzymes

Extensive research conducted in poultry throughout the world has clearly demonstrated that adding exogenous enzymes to diets rich in cereals such as wheat and barley improves bird performance (Ward, 1995; Bedford and Morgan, 1996). Results obtained with pigs and rabbits may not be as promising (Officer, 1995). The mode of action of enzymes has not been elucidated. Non-starch polysaccharides may coat the nutrients contained in the grain. The addition of cell wall-degrading enzymes such as xylanases and β-glucanases may release nutrients and favour their digestion (Classen, 1996; Cowan *et al.*, 1996). It is well known that enzymes decrease the viscosity of the digestive contents (Bedford, 1995). This may allow a better contact of nutrients with endogenous enzymes and absorptive mucosae cells and therefore a better use of the diet. Marquardt *et al.* (1996) have observed a decrease in the water content of faeces which will benefit management, productivity and quality of the end product. In addition, enzyme supplementation increases rate of passage which may improve feed intake (Brenes *et al.*, 1996) and decreases multiplication of anaerobes of the genus *Clostridium* (Ward, 1995). Other enzymes (pectinases, proteases, oligosaccharidases) have also been utilized with limited success in domestic species (Annison *et al.*, 1996; Simbaya *et al.*, 1996).

In the rabbit, the use of enzymes as feed additives has not been extensively studied. Except for fibre, rabbits are very efficient in the utilization of nutrients. Makkar and Singh (1987) found that activity of proteases and amylases was higher in the caecum of rabbits than in the rumen. Reports by Tor-Aghidye *et al.* (1992), Remois *et al.* (1996) and Fernández *et al.* (1996) did not find any benefit from the inclusion of different enzymes in rabbit diets. The authors claim that hydrolases needed for nutrient utilization are available in sufficient amounts in the rabbit digestive tract and that dietary β-glucan does not have any adverse effect in rabbits. In fact, Bolis *et al.* (1996) found a negative effect on nutrient digestibility when a commercial protease or a glycosidase was added to the diet. Others, however, did find a benefit when an enzyme cocktail was added to diets for rabbits under low intense productivity patterns (Bhatt *et al.*, 1996).

Therefore, at present, a recommendation on the use of exogenous enzymes in rabbit diets is not possible.

References

Abbo El-Ezz, Z.R., Salem, M.H., Hassan, G.A., El-Komy, A.G. and Adb El Moula, E. (1996) Effect of different levels of copper sulphate supplementation on some physical traits of rabbits. In: Lebas, F. (ed.) *Proceedings of the 6th World Rabbit Congress, Toulouse*. Association Française de Cuniculture, Lempdes, pp. 59–64.

AEC (1987) *Tables AEC. Recommendations for Animal Nutrition*, 5th Edn. Rhône-Poulenc, Commentry, France, 86 pp.

Aguilar, J.C., Roca, T. and Sanz, E. (1996) Fructo-oligosaccharides in rabbit diets. Study of efficiency in suckling and fattening periods. In: Lebas, F. (ed.) *Proceedings of the 6th World Rabbit Congress, Toulouse*. Association Française de Cuniculture, Lempdes, pp. 73–77.

Annison, G., Hughes, R.J. and Choct, N. (1996) Effects of enzyme supplementation on the nutritive value of dehulled lupins. *British Poultry Science* 37, 157–172.

Aoun, M., Grenet, L., Mousset, J.L. and Robart, P. (1994) Effet d'une supplementation avec de l'oxytetracycline et des levures vivantes sur les performances d'engraissement. In: *6èmes Journées de la Recherche Cunicole*. INRA-ITAVI, La Rochelle, pp. 277–284.

Assane, M., Gongnet, G.P., Coulibaly, A. and Sere, A. (1993) Influence du rapport calcium/phosphore de la ration sur la calcemie, la phosphatemie et la magnesiemie de la lapine en gestation. *Reproduction, Nutrition and Development* 33, 223–228.

Ayyat, M.S., Marai, I.F.M. and Alazab, A.M. (1995) Copper protein nutrition of New Zealand White rabbits under Egyptian conditions. *World Rabbit Science* 3, 113–118.

Barlet, J.P. (1980) Plasma calcium, inorganic phosphorus and magnesium levels in pregnant and lactating rabbits. *Reproduction, Nutrition and Development* 20, 647–651.

Barlet, J.P., Davicco, M.J. and Coxam, V. (1995) Physiologie de l'absorption intestinale du phosphore chez l'animal. *Reproduction, Nutrition and Development* 35, 475–489.

Bassuny, S.M. (1991) The effect of copper sulfate supplement on rabbit performance under Egyptian conditions. *Journal of Applied Rabbit Research* 14, 93–97.

Bazer, F.W. and Zavy, M.T. (1988) Supplemental riboflavin and reproductive performance of gilts. *Journal of Animal Science* 66, 324 (abstract).

Bedford, M.R. (1995) Mechanism of action and potential environmental benefits from the use of feed enzymes. *Animal Feed Science and Technology* 53, 145–155.

Bedford, M.R. and Morgan, A.J. (1996) The use of enzymes in poultry diets. *World's Poultry Science Journal* 52, 61–68.

Bernardini, M., Dal Bosco, A., Castellini, C. and Miggiano, G. (1996) Dietary vitamin E supplementation in rabbit: antioxidant capacity and meat quality. In: Lebas, F. (ed.) *Proceedings of the 6th World Rabbit Congress, Toulouse.* Association Française de Cuniculture, Lempdes, pp. 137–140.

Besenfelder, U., Solti, L., Seregi, J., Muller, M. and Brem, G. (1996) Different roles of β-carotene and vitamin A in the reproduction of rabbits. *Theriogenology* 45, 1583–1591.

Bhatt, R.S., Bhasin, V. and Bhatia, D.R. (1996) Effect of kemzyme on the performance of German angora weaners. In: Lebas, F. (ed.) *Proceedings of the 6th World Rabbit Congress, Toulouse.* Association Française de Cuniculture, Lempdes, pp. 97–99.

Bolis, S., Castrovilli, C., Rigoni, M., Tedesco, D. and Luzi, F. (1996) Effect of enzymes addition in diet on protein and energy utilization in rabbit. In: Lebas, E. (ed.) *Proceedings of the 6th World Rabbit Congress, Toulouse.* Association Française de Cuniculture, Lempdes, pp. 111–115.

Bondi, A. and Sklan, D. (1984) Vitamin A and carotene in animal nutrition. *Progress in Food Nutrition Science* 8, 165–191.

Borowitz, S.M., and Granrud, G.S. (1993) Ontogeny of intestinal phosphate absorption in rabbits. *American Journal of Physiology* 262, 847–853.

Bräunlich, K. (1974) *Vitamin B₆*. Report No. 1451, Hoffmann-La Roche, Basel.

Brenes, A., Lázaro, R., García, M. and Mateos, G.G. (1996) Utilización práctica de complejos enzimáticos en avicultura. In: Rebollar, P.G., Mateos, G.G. and de Blas, C. (eds) *XII Curso de Especialización FEDNA*. FEDNA, Madrid, pp. 135–157.

Burgi, A.R. (1993) Nutriçao mineral de coelhos. *Zootecnia* 31, 89–95.

Byers, J.H., Jones, I.R. and Bone, J.F. (1956) Carotene in the ration of dairy cattle. II. The influence of suboptimal levels of carotene intake upon the microscopic aspect of selected organs. *Journal of Dairy Science* 39, 1556–1564.

Chapin, R.E. and Smith, S.E. (1967) Calcium requirement of growing rabbits. *Journal of Animal Science* 26, 67–71.

Cheeke, P.R. (1987) *Rabbit Feeding and Nutrition*, 1st Edn. Academic Press, London, 376 pp.

Cheeke, P.R. and Amberg, J.W. (1973) Comparative calcium excretion by rats and rabbits. *Journal of Animal Science* 37, 450–454.

Cheeke, P.R., Patton, N.M., Diwyanto, K., Lasmini, A., Nurhadi, A., Prawirodigdo, S. and Sudaryanto, B. (1984) The effect of high dietary Vitamin A levels on reproductive performance of female rabbits. *Journal of Applied Rabbit Research* 7, 135–137.

Cheeke, P.R., Bronson, J., Robinson, K.L. and Patton, N.M. (1985) Availability of calcium, phosphorus and magnesium in rabbit feeds and mineral supplements. *Journal of Applied Rabbit Research* 8, 72–74.

Chew, B.P. (1994a) Beta-carotene appears to improve reproductive performance. *Feedstuffs* 66, 14–15.

Chew, B.P. (1994b) Beta-carotene, other carotenoids push inmunity defense. *Feedstuffs* 66, 17–51.

Classen, H.L. (1996) Cereal grain starch and exogenous enzymes in poultry diets. *Animal Feed Science and Technology* 62, 21–27.

Colin, M. (1977) Effet d'une variation du taux de chlore dans l'alimentation du lapin en croissance. *Annales de Zootechnie* 26, 99–103.

Cowan, W.D., Korsbak, A., Hastrup, T. and Rasmussen, P.B. (1996) Influence of added microbial enzymes on energy and protein availability of selected feed ingredients. *Animal Feed Science and Technology* 60, 311–319.

Cromwell, G.L. (1992) The biological availability of phosphorus in feedstuffs for pigs. *Pig News and Information* 13, 75N–78N.

Czarnecki, R., Iwanska, S., Falkowska, A., Delikator, B., Karmelita, M. and Pycio, Z. (1992) Effects of β-carotene containing caromix on reproductive performance of primiparous sows. *World Review of Animal Production* 27, 3–30.

de Blas, C., García, J. and Alday, S. (1991) Effects of dietary inclusion of a probiotic (Paciflor) on performance of growing rabbits. *Journal of Applied Rabbit Research* 14, 148–150.

Deeb, B.J., Digiacomo, R. and Anderson, R.J. (1992) Reproductive abnormalities in rabbits due to Vitamin A toxicity. *Journal of Applied Rabbit Research* 15, 973–984.

Diehl, J.F. and Kristler, B.G. (1961) Vitamin E saturation test in coccidiosis infected rabbits. *Journal of Nutrition* 74, 495–499.

Elmarimi, A.A., Ven, E. and Bardos, L. (1989) Preliminary study on the effects of Vitamin A and β-carotene on growth and reproduction of female rabbits. *Journal of Applied Rabbit Research* 12, 163–168.

El-Masry, K.A. and Nasr, A.S. (1996) The role of folic acid and iron in reproductive performance of New Zealand White does and their kits. *World Rabbit Science* 4, 127–131.

El-Sayiad, G.A., Habbeb, A.A. and Maghawry, A.M. (1994) A note on the effects of breed, stage of lactation and pregnancy status on milk composition of rabbits. *Animal Production* 58, 153–157.

Escribano, F., Garcia Alfonso, J., Lozon, C. and Mateos, G.G. (1982) Avotan avoparcina, un nuevo estimulante de crecimiento en conejos en crecimiento-cebo. In: *VII Symposium de Cunicultura, Santiago de Compostela*. ASESCU, pp. 241–249.

Evans, E., Jebelian, V. and Rycquart, W.C. (1983a) Effects of potassium and magnesium levels upon performance of fryer rabbits. *Journal of Applied Rabbit Research* 6, 49–50.

Evans, E., Jabelian, V. and Rycquart, W.C. (1983b) Further evaluation of the magnesium requirements of fryer rabbits. *Journal of Applied Rabbit Research* 6, 130–131.

Fernández, C., Merino, M.J. and Carabaño, R. (1996) Effect of enzyme complex supplementation on diet digestibility and growth performance in growing rabbits. In: Lebas, F. (ed.) *Proceedings of the 6th World Rabbit Congress, Toulouse*. Association Française de Cuniculture, Lempdes, pp. 163–166.

Gaca-Lagodzinska, K., Provot, F. and Coudert, P. (1994) Tolerance de la salinomycine et efficacite contre trois coccidioses du lapin. In: *6èmes Journées de la Recherche Cunicole en France*. INRA-ITAVI, La Rochelle, pp. 67–72.

Grobner, M.A., Cheeke, P.R. and Patton, N.M. (1985) A note on the effect of a high dietary vitamin level on the growth of fryer rabbits. *Journal of Applied Rabbit Research* 8, 6.

Gueguen, L., Lamand, M. and Meschy, F. (1988) Nutrition mineral. In: Jarrige, R. (ed.) *Alimentation des Bovines, Ovins and Caprines*. Institut National de la Recherche Agronomique, Paris, pp. 97–111.

Harris, D.J., Cheeke, P.R. and Patton, N.M. (1983) Effect of suupplemental vitamins on fryer rabbit performance. *Journal of Applied Rabbit Research* 6, 29–31.

Harris, D.J., Cheeke, P.R., and Patton, N.M. (1984a) Effect of supplemental copper on postweaning performance of rabbits. *Journal of Applied Rabbit Research* 7, 10–12.

Harris, D.J., Cheeke, P.R., and Patton, N.M. (1984b) Effect of feeding various levels of salt on growth performance, mortality and feed preferences of fryer rabbits. *Journal of Applied Rabbit Research* 7, 117–119.

Hidiroglou, M., Ho, S.K., Ivan, M. and Shearer, D.A. (1978) Manganese status of pasturing ewes, of pregnant ewes and doe rabbits on low manganese diets and of dairy cows with cystic ovaries. *Canadian Journal of Comparative Medicine* 42, 100–107.

Hoffmann-La Roche (1995) Recommended vitamin supplementation levels for domestic animals. *Feed Milling Yearbook*, Hoffman-La Roche, pp. 124–126.

Ismail, A., Shalash, S., Kotby, E., Cheeke, P.R. and Patton, N.M. (1992a) Hypervitaminosis A in rabbits. 3. Reproductive effects and interactions with vitamins E and C and ethoxyquin. *Journal of Applied Rabbit Research* 15, 1206–1218.

Ismail, A., Shalash, S., Kotby, E. and Cheeke, P.R. (1992b) Effects of vitamin A, C and E on the reproductive performance of heat stressed female rabbits in Egypt. *Journal of Applied Rabbit Research* 15, 1291–1300.

Jenkins, K.J., Hidiroglow, M., Mackay, R.R. and Proulx, J.G. (1970) Influence of selenium and linoleic acid on the development of nutritional muscular dystrophy in beef calves, lambs and rabbits. *Canadian Journal of Animal Science* 50, 137–146.

Jenness, R., Birney, E.C. and Ayaz, K.L. (1978) Ascorbic acid and L-gulonolactone oxidase in lagomorph. *Comparative Biochemistry and Physiology* 61, 395–399.

Kamphues, J. (1991) Calcium metabolism of rabbits as an etiological factor for urolithiasis. *Journal of Nutrition* 121, 595–596.

King, J.O.L. (1975) The feeding of copper sulphate to growing rabbits. *British Veterinary Journal* 131, 70–75.

Kormann, A.W., Riss, G. and Weiser, H. (1989) Improved reproductive performance of rabbit does supplemented with dietary β-carotene. *Journal of Applied Rabbit Research* 12, 15–21.

Kornegay, E.T. (1986) Biotin in swine production: a review. *Livestock Production Science* 14, 65–89.

Kubota, M., Ohno, K., Shiina, Y. and Suda, T. (1982) Vitamin D metabolism in pregnant rabbits: differences between the maternal and fetal response to administration of large amounts of Vitamin D_3. *Endocrinology* 110, 1950–1956.

Kulwich, R., Struglia, L. and Pearson, P.B. (1953) The effect of coprophagy on the excretion of B vitamins by the rabbit. *Journal of Nutrition* 49, 639–645.

Lafargue-Hauret, P., Javrin, D., Ricca, V. and Rouillere, H. (1994) Toxicite de l'amoxicilline chez le lapin. In: *6èmes Journées de la Recherche Cunicole*. INRA-ITAVI, La Rochelle, pp. 81–84.

Lebas, F. (1987) Feeding conditions for top performance in the rabbit. In: Auxilia, T. (ed.) *Report EUR 10983 EN*. Commission of the European Community, Brussels, pp. 27–39.

Lebas, F. (1990) Recherche et alimentation des lapines. *Cuniculture* 17, 12–15.

Lebas, F. (1993) Incidence du Profeed sur l'efficacité alimentaire chez le lapin en croissance. *Cuniculture* 20, 169–172.

Lebas, F. (1996) Effects of fructo-oligosaccharides origin on rabbit's growth performance in 2 seasons. In: Lebas, F. (ed.) *Proceedings of the 6th World Rabbit Congress, Toulouse*. Association Française de Cuniculture, Lempdes, pp. 211–215.

Lebas, F. and Jouglar, J. (1984) Apports alimentaires de calcium et de phosphore chez la lapine reproductrice. In: Finzi, A. (ed.) *Proceedings of the 3rd World Rabbit Congress, Rome*. WRSA, pp. 461–466.

Lebas, F. and Jouglar, J.Y. (1990) Influence du taux de phosphore alimentaire sur les performances de lapines reproductrices. In: *5èmes Journées Recherche Cunicole*. Communication 48, INRA, ITAVI, Paris.

Lee, Y.H., Layman, D.K. and Bell, R.R. (1979) Selenium dependent and non selenium dependent glutathione peroxidase activity in rabbit tissue. *Nutrition Reports International* 20, 573–578.

Lewis, A.J., Cromwell, G.L. and Pettigrew, J.F. (1991) Effects of supplemental biotin during gestation and lactation on reproductive performance of sows: a cooperative study. *Journal of Animal Science* 69, 207–214.

Licois, D. (1980) Action toxique de certains antibiotiques chez le lapin. *Recueil Médecine Vétérinaire* 156, 915–919.

Licois, D. (1996) Risques associes a l'utilization des antibiotiques chez le lapin: une mini revue. *World Rabbit Science* 4, 63–68.

Licois, D. and Coudert, P. (1980) Action de la robenidine sur l'excretion des oocystes de differentes especes de coccidies du lapin. In: Camps, J. (ed.) *Proceedings of the 2nd World Rabbit Congress, Barcelona*, pp. 285–289.

Licois, D., Coudert, P. and Mongin, P. (1978) Changes in hydromineral metabolism in diarrhoeic rabbits. 2. Study of the modifications of electrolyte metabolism. *Annales Recherches Vétérinaires* 9, 1–10.

Lindemann, M.D. (1993) Supplemental folic acid: a requirement for optimizing swine reproduction. *Journal of Animal Science* 71, 239–246.

Löliger, H.C. and Vogt, H. (1980) Calcinosis of kidneys and vessels in rabbits. In: Camps, J. (ed.) *Proceedings of the 2nd World Rabbit Congress, Barcelona*, p. 284 (abstract).

McCartney, E. (1994) *Bacillus toyoi*: Aplicación en conejos de carne y en conejas reproductoras. *Boletin de Cunicultura* 17, 1–6.

McDowell, L.R. (1989) *Vitamins in Animal Nutrition*. Academic Press, San Diego, 486 pp.

McDowell, L.R. (1992) *Minerals in Animal and Human Nutrition*. Academic Press, San Diego, 524 pp.

Maertens, L. (1992) Rabbit nutrition and feeding: a review of some recent developments. *Journal of Applied Rabbit Research* 15, 889–913.

Maertens, L. (1996) Nutrition du lapin. Connaissances actuelles et acquisitions recentes. *Cuniculture* 23, 33–35.

Maertens, L. and De Groote, G. (1992) Effect of a dietary supplementation of live yeast on the zootechnical performance of does and weanling rabbits. *Journal of Applied Rabbit Research* 15, 1079–1086.

Maertens, L., Moermans, R. and De Groote, G. (1992) Flavomycin for early weaned rabbits: response of dose on zootechnical performance. Effect on nutrient digestibility. *Journal of Applied Rabbit Research* 15, 1270–1277.

Maertens, L., Van Renterghem, R. and De Groote, G. (1994) Effects of dietary inclusion of Paciflor (Bacillus CIP 5832) on the milk composition and performance of does on caecal and growth parameters of their weanlings. *World Rabbit Science* 2, 67–73.

Mahan, D.C., Lepine, A.J. and Dabrowski, K. (1994) Efficacy of magnesium-L-ascorbyl 2-phosphate as a Vitamin C source for weanling and growing-finishing swine. *Journal of Animal Science* 72, 2354–2361.

Makkar, H.P.S. and Singh, B. (1987) Comparative enzymatic profiles of rabbit cecum and bovine rumen content. *Journal of Applied Rabbit Research* 10, 172–174.

Marquardt, R.R., Brenes, A., Zhang, Z. and Boros, D. (1996) Use of enzymes to improve nutrient availability in poultry feedstuffs. *Animal Feed Science and Technology* 60, 321–330.

Mateos, G.G. (1989) Minerales, vitaminas, antibióticos, anticcodiósicos y otros aditivos en la alimentación del conejo. In: de Blas, C. (ed.) *Alimentación del Conejo*, 2nd Edn. Mundi Prensa, Madrid, 177 pp.

Mateos, G. and Piquer, J. (1994) Diseño de programas alimenticios para conejos: aspectos teóricos y formulación práctica. *Boletín de Cunicultura* 17, 16–31.

Matte, J.J. and Girard, C.L. (1996) Le besoin en acide folique chez la truie gravide. *Journées Recherche Porcine en France* 28, 365–370.

Moghaddam, M.F., Cheeke, P.R. and Patton, N.M. (1987) Toxic effects of Vitamin A on reproduction in female rabbits. *Journal of Applied Rabbit Research* 10, 65–67.

Morisse, J.P., Le Gall, G., Boilletot, E. and Maurice, R. (1989) Intoxication alimentaire chez le lapin par des residus d'antibiotiques. *Cuniculture* 16, 288–290.

Morisse, J.P., Maurice, R., Boilletot, E. and Cotte, J.P. (1992) Effect of a fructo-oligo-saccharides compound in rabbit experimentally infected with *E. coli* 0103. *Journal of Applied Rabbit Research* 15, 1137–1143.

NRC (1977) *Nutrient Requirements of Rabbits*. National Academy of Science, National Research Council, Washington, DC, 30 pp.

NRC (1987) *Vitamin Tolerance of Animals*. National Academy of Sciences, National Research Council, Washington, DC, 96 pp.

Officer, D.I. (1995) Effect of multi-enzyme supplements on the growth performance of piglets during the pre- and postweaning periods. *Animal Feed Science and Technology* 56, 55–65.

Okerman, F. and Moermans, R.J. (1980) L'influence du coccidiostatique salinomycin en tant qu'additif des aliment composés sur les resultats de production des lapins de chair. *Revue de l'Agriculture* 33, 1311–1322.

Ouhayoun, J. and Lebas, F. (1987) Composition chimique de la viande de lapin. *Cuniculture* 14, 33–45.

Paefgen, D., Scheuermann, S.E. and Raether, W. (1996) The anticoccidial activity of sacox in fattening rabbits. In: Lebas, F. (ed.) *Proceedings of the 6th World Rabbit Congress. Toulouse.* Association Française de Cuniculture, Lempdes, pp. 103–106.

Parigi Bini, R., Cinetto, M. and Carotta, N. (1983) The effect of β-carotene on the reproductive performance of female rabbits. In: *Proceedings of the 5th World Conference on Animal Production*, Vol. 2, pp. 231–232.

Partridge, G.G. and Gill, B.P. (1993) New approaches with weaner diets. In: Garnsworthy, P.C. and Cole, D.J.A. (eds) *Recent Advances in Animal Nutrition*. Nottingham University Press, Nottingham, UK, pp. 221–248.

Patton, N.M., Harris, D.J., Grobner, M.A., Swick, R.A. and Cheeke, P.R. (1982) The effect of dietary copper sulfate on enteritis in fryer rabbits. *Journal of Applied Rabbit Research* 5, 78–82.

Peeters, J.E., Janssens-Geroms, R. and Halen, P.H. (1980) Activité anticoccidienne de la narasine chez le lapin: essais de laboratorie. In: Camps, J. (ed.) *Proceedings of the 2nd World Rabbit Congress*, Barcelona, pp. 325–334.

Peeters, J.E., Maertens, L. and Geeroms, R. (1992) Influence of galacto-oligosaccharides on zootechnical performance, cecal biochemistry and experimental colibacilosis 0103/8 + in weanling rabbits. *Journal of Applied Rabbit and Research* 15, 1129–1136.

Pettigrew, J.F., El-Kandelgy, S.M., Johnston, L.J. and Shurson, G.C. (1996) Riboflavin nutrition of sows. *Journal of Animal Science* 74, 2226–2230.

Rao, D.R., Chawan, C.B., Chen, C.P. and Sunki, G.R. (1979) Nutritive value of rabbit meat. In: Cheeke, P.R. (ed.) *The Domestic Rabbit: Potentials, Problems and Current Research*. Oregon State University Press, Corvallis, Oregon, pp. 53–59.

Remois, G., Lafargue-Hauret, P. and Rouillere, H. (1996) Effect of amylases supplementation in rabbit feed on growth performance. In: Lebas, F. (ed.) *Proceedings of the 6th World Rabbit Congress, Toulouse*. Association Française de Cuniculture, Lempdes, pp. 289–292.

Ringler, D.H. and Abrams, G.D. (1970) Nutritional muscular distrophy and neonatal mortality in a rabbit breeding colony. *Journal of the American Veterinary Medicine Association* 157, 1928–1934.

Ringler, D.H. and Abrams, G.D. (1971) Laboratory diagnosis of Vitamin E deficiency in rabbits fed a faulty commercial ration. *Laboratory Animal Science* 21, 383–388.

Schlolaut, W. (1987) Nutritional needs and feeding of German Angora Rabbits. *Journal of Applied Rabbit Research* 10, 111–121.

Simbaya, J., Slominski, B.A., Guenter, W., Morgan, A. and Campbell, L.D. (1996) The effects of protease and carbohydrase supplementation on the nutritive value of canola meal for poultry: *in vitro* and *in vivo* studies. *Animal Feed Science and Technology* 61, 219–234.

Simnett, K.E. and Spray, G.H. (1965a) The effect of a low-Cobalt diet in rabbits. *British Journal of Nutrition* 19, 119–123.

Simnett, K.E. and Spray, G.H. (1965b) The absorption and excretion of ^{58}Co cyanocobalamin by rabbits. *British Journal of Nutrition* 19, 593–598.

Steenland, E. (1991) Effekt van verschillende Lichtschema's, bezettingsdichtheid en forforgehalte van het voeder op de gezondheid en produktieresultaten van slachtkonijnen. *NOK-blad* 9, 60–68.

Strucklec, M., Dermelj, M., Stibilij, V. and Rajh, I. (1994) The effect of selenium added to feedstuffs on its content in tissues and on growth of rabbits. *Krmiva* 36, 117–123.

Surdeau, P., Henaff, R. and Perrier, G. (1976) Apport et equilibre alimentaire du sodium, du potasium et du chlore chez le lapin croissance. In: Lebas, F. (ed.) *Proceedings of the 1st World Rabbit Congress, Dijon*. Communication No. 21, WRSA, Dijon.

Swick, R.A., Cheeke, P.R. and Patton, N.M. (1981) The effect of soybean meal and supplementary zinc and copper on mineral balance in rabbits. *Journal of Applied Rabbit Research* 4, 57–65.

Thilsted, J.P., Newton, W.M. and Crandell, R.A. (1981) Fatal diarrhea in rabbits resulting from the feeding of antibiotic contaminated feed. *Journal of the American Veterinary Medicine Association* 179, 360–362.

Tor-Aghidye, Y., Cheeke, P.R., Nakaue, H.S., Froseth, J.A. and Patton, N.M. (1992) Effects of β-glucanase (Allzyme B6) on comparative performance of growing rabbits and broiler chicks fed rye, triticale and high- and low-glucan barley. *Journal of Applied Rabbit Research* 15, 1144–1152.

Underwood, E.J. (1977) *Trace Elements in Human and Animal Nutrition*, 4th Edn. Academic Press, New York, 545 pp.

Vandelli, A. (1995) Attenti a calcio e fosforo. *Rivista di Coniglicoltura* 12, 36–37.

Van Meirhaeghe, P., Rochette, F. and Homedes, J. (1996) Anticoccidial efficacy of diclazuril in fattening rabbits: comparative field trials in Spain and Belgium. In: Lebas, F. (ed.) *Proceedings of the 6th World Rabbit Congress, Toulouse*. Association Française de Cuniculture, Lempdes, pp. 119–124.

Vanparijs, O., Hermans, L., Van der Flaes, L. and Marsboom, R. (1989) Efficacy of diclazuril in the prevention and cure of intestinal and hepatic coccidiosis in rabbits. *Veterinary Parasitology* 32, 109–117.

Verde, M.T. and Piquer, J.G. (1986) Effect of stress on the corticosterone and ascorbic acid (Vitamin C) content of the blood plasma of rabbits. *Journal of Applied Rabbit Research* 9, 181–185.

Wang, J.Y., Larson, L.L. and Owen, F.G. (1982) Effect of β-carotene supplementation on reproductive performance of dairy heifers. *Theriogenology* 18, 461–467.

Wang, J.Y., Owen, F.G. and Larson, L.L. (1988) Effect of β-carotene supplementation on reproductive performance of lactating Holstein cows. *Journal of Dairy Science* 71, 181–186.

Ward, N.E. (1995) Enzyme use in viscous inducing cereal diets examined. *Feedstuffs* 67(49), 1–4.

Widdowson, E.M. (1974) Feeding the newborn: comparative problems in man and animals. *Proceedings of the Nutrition Society* 33, 97–102.

Xiccato, G. (1996) Nutrition of lactating does. In: Lebas, F. (ed.) *Proceedings of the 6th World Rabbit Congress, Toulouse.* Association Française de Cuniculture, Lempdes, pp. 29–47.

Yamini, B. and Stein, S. (1989) Abortion, stillbirth, neonatal death and nutritional myodegeneration in a rabbit breeding colony. *Journal of the American Veterinary Medicine Association* 194, 561–562.

Zakaria, A.H. and Al-Anezi, M.A. (1996) Effect of ascorbic acid and cooling during egg incubation on hatchability, culling, mortality, and the body weights of broiler chickens. *Poultry Science* 75, 1204–1209.

Zimmerman, T., Giddens, W., Di Giacomo, R. and Ladiges, W. (1990) Soft tissue mineralization in rabbits fed a diet containing excess Vitamin D. *Laboratory Animal Science* 40, 212–215.

10. Influence of the Diet on Rabbit Meat Quality

J. Ouhayoun

*Station de Recherches Cunicoles, INRA Centre de Toulouse, BP 27-31326,
Castanet-Tolosan Cedex, France*

Introduction

The objective of meat production, and especially rabbit meat production, is the
transformation of proteins which are not usually consumed by humans into a
product better adapted to their nutritional requirements and tastes. From this
point of view, the rabbit is a suitable animal, especially when compared with
other herbivores. Nevertheless, because of its reliance on fibre-rich feed and
labour-intensive systems, rabbit meat production is more expensive than pork
or poultry (broiler and turkey) production. The qualitative properties of rabbit
meat, which depend largely on its nutrition, place it high among meats.

Definition and determination of meat quality

Meat quality depends on the practices applied throughout production. The
feed manufacturer plays a major role in the formulation and preparation of a
balanced feed mixture in selecting the raw ingredients as a function of their
nutritional properties, availability and price, and determines constituent
proportions in order to meet the requirements of the various animal categories
and to facilitate the processing of pelleted feeds. The *producer* manages the
feed (choice of origin, storage of pellets) and determines whether feeding
should be *ad libitum* or restricted, and thus influences quality by controlling
growth rate, food conversion ratio and animal viability. Processing conditions
(transport, slaughtering, chilling and storage of carcasses, cutting and
conditioning) determine the yield which is considered as the first criterion of
quality by the *processor*, as well as the level of bacterial contamination, the
aspect and the rheological properties of meat. Finally, the *distributor*
considers that the main criterion of meat quality is its attractiveness to
consumers which depends on the shape of the carcasses or the cuts, the level
of fat, the colour and the absence of exudate in the packs. The storage ability
of meat is also essential. Meat should contain as few bacteria as possible and,

to this end, should have an acid pH which is not favourable
to their development. To meet *consumer* requirements, meat should be
nutritional and safe, i.e. contain the appropriate proportion of protein, lipids
and other components necessary for the growth of children and to maintain
the weight of adults. For example, rabbit meat has a low cholesterol level
(50 mg 100 g^{-1}) and a high protein/energy ratio (24 g MJ^{-1} for the whole
carcass and 32 g MJ^{-1} for the hind part). Moreover, fat in rabbit meat is
relatively rich in essential fatty acids.

A series of regulations govern the *hygienic properties* of meat, which
should contain no toxic residues or bacterial growth likely to cause a health
risk. The *serviceable quality* of meat is more subjective: it depends mainly on
the time available for cooking. Since this is partly undertaken by the agrifood
industry (cutting, cooking) the time has decreased from 3 h to 20 min day^{-1}
in the last 40 years. *Organoleptic properties* are crucial. They include the
aspect (presentation and conditioning of the carcass or the cuts, colour,
consistency of raw meat), easy separation of cooked meat from the bone,
texture (tenderness and juiciness) and flavour (taste, smell and aroma).
Flavour, and to a lesser extent juiciness, depend on the meat fat content and
its fatty acid composition. Texture is determined mainly by the water-
holding capacity of meat during cooking and by the balance and the
structure of muscle proteins.

The properties of rabbit meat in relation to the anatomical criteria of the
carcass and the physico-chemical and sensory criteria of the edible fraction
are influenced by specific rules determining the relative growth of organs and
tissues, patterns of muscle differentiation and dietary factors.

General growth rules

In the rabbit, the genetic variability between breeds and crosses plays a major
role, insofar as adults weigh between 1 and 8 kg. Under practical conditions,
however, the most commonly used breed types for meat production have an
adult mean weight of at least 3.5 kg and usually range between 4 and 5 kg.
Meat and carcass characteristics depend on the age of the animal and its
weight at a given age. Diet influences meat properties, i.e. the composition of
the carcass tissues and the chemical components of the muscle and adipose
tissues, mainly through variation in growth.

Animal body weight increases as a function of age. Body weight gain is a
function of the growth of the different body components, which all develop at
different rates. In order to remain functional it appears that the relative size of
individual body components in the growing rabbit cannot remain constant.
Thus the allometry of growth of the organs and tissues is considerable, which
leads to major changes in the morphology and body composition of rabbits
from birth onwards.

The relative growth of body components can be examined using the allometric relationship which compares the relative growth rate of each organ *y* to that of the whole body *x*:

dy/y = a.dx/x

Three situations are possible:

- $a = 1$: *x* is related to *y* by an isometric relationship; the organ *y* has the same relative growth rate as the whole body *x* and represents a constant proportion of the body.
- $a > 1$: *y* is related to *x* by a positive allometry; the organ *y* grows relatively more rapidly than the whole body *x* and represents an increasing proportion of the body.
- $a < 1$: *y* is related to *x* by a negative allometry; the organ *y* grows relatively more slowly than the whole body *x* and represents a decreasing proportion of the body.

Ranking allometric coefficients in increasing order shows that the growth of individual tissues and organs occurs at different times (Table 10.1). The growth of brain is the earliest, whilst that of adipose tissue is the latest.

However, allometric coefficients usually vary during growth. With the exception of the adipose tissue and the skin, they usually decrease. Table 10.2 shows a few representative examples of the allometric changes in a medium-sized rabbit.

The allometry of growth of blood, digestive tract and the skin is negative, whilst that of the carcass is positive. This explains the increase in carcass yield as a function of weight.

The variation in carcass composition results mainly from the allometry of growth of bone, muscle and adipose tissues. The relative growth rate of the bone tissue decreases when the empty body weight reaches about 1000 g, whereas that of the muscle tissue decreases at an empty body weight of about

Table 10.1. Mean allometric coefficients between 9 and 182 days of the main tissues and organs, ranked in increasing order (male rabbits, reference variable: empty body weight) (according to Cantier *et al.*, 1969).

Organs	Allometric coefficients
Brain	0.27
Kidney	0.70
Skin	0.79
Digestive tract	0.79
Skeleton	0.81
Liver	0.94
Blood	0.94
Muscle tissue	1.15
Adipose tissue	1.31

Table 10.2. Values of the allometric coefficients of the main tissues and organs and of empty body weights (g) at which allometry changes (according to Cantier *et al.*, 1969).

Blood							0.94		
Digestive tract	1.13	**650**					0.46		
Skin	0.44		**850**				0.86		
Adipose tissue	0.82			**950**			1.87	**2100**	3.21
Skeleton	0.91				**1000**		0.55		
Liver	1.25					**1675**	0.47		
Muscle tissue	1.20							**2450**	0.50

2450 g. Between those two weights, the muscle/bone ratio of the carcass increases very rapidly and reaches its maximum value. Beyond 2450 g it tends to decrease. The relative growth rate of the adipose tissue is slow before reaching an empty body weight of 950 g, then rapid up to 2100 g and very rapid thereafter. The increase in the cost of feeding associated with adipogenesis is therefore very high before the muscle/bone ratio has reached its maximum value. The liver proportion increases up to an empty body weight of 1500 g and then rapidly decreases.

The effects of age and weight

Body composition varies as a function of body weight and growth rate. In order to interpret the effects of the diet, it is necessary to distinguish between experiments where animals end up being slaughtered at the same weight or those at the same age.

At the same age, whatever their weight, rabbits have approximately the same slaughter yield (Table 10.3). This results from the opposite relationships of the skin ($r = + 0.25$) and of the digestive tract ($r = - 0.37$) with body weight. Rabbits also have similar muscle/bone ratios. Body weight at slaughter has no significant effect on these characteristics, probably as a result of both the variability in adult body weight and in early growth rate of rabbits. The heaviest rabbits are, however, characterized by a perirenal adiposity and a muscle lipid content higher than those of lighter rabbits. This means that they are more mature. This maturity is associated with increased glycolytic energy metabolism and to low ultimate pH and water-holding capacity of the meat.

When rabbits are slaughtered at a given weight, their body characteristics depend on how quickly they have reached this weight. Rapid growth disadvantages early-maturing tissues; conversely, it enhances late-maturing tissues. Rapidly growing rabbits are thus characterized by a smaller relative development of the skin, the digestive tract and the bones as well as by a better relative development of the muscles and especially the adipose tissue. They consequently have a slaughter yield, muscle/bone ratio, total adiposity and muscle lipid content higher than those of slower growing rabbits (Table 10.4).

Table 10.3. Correlations between body weight and body composition in 11-week-old rabbits (according to Ouhayoun, 1978).

Slaughter yield	Digestive tract weight	Skin weight	Meat/bone ratio	Perirenal adiposity proportion	Muscular lipids proportion	Muscular water proportion	Carcass compacity (weight/length ratio)
Non-significant	−0.37	+0.25	Non-significant	+0.28	+0.33	−0.23	+0.88

Table 10.4. Carcass and meat characteristics of slow- and rapid-growing rabbits slaughtered at the same carcass weight (according to Prud'hon *et al.*, 1970).

Age	74 days	106 days
Chilled carcass weight (g)	1522	1543
Slaughter yield (proportion of live weight)	0.6259	0.6125
Bone (proportion of carcass)	0.118	0.140
Meat (proportion of carcass)	0.691	0.685
Meat/bone ratio	5.85	4.91
Total dissectable fat (proportion of carcass)	0.075	0.036
Perirenal fat (proportion of carcass)	0.023	0.008

However, rabbits which are characterized by both rapid growth rate and high weight at weaning have a slaughter yield and a muscle/bone ratio lower than slow-growing rabbits which are lighter at weaning (Table 10.5). They are thus characterized by late maturity which corresponds to rabbits with a high potential adult weight.

Muscle development and growth

Muscle development occurs in two phases: a cell proliferation phase followed by a morphological, functional and metabolic differentiation phase. Premyo-

Table 10.5. Carcass and meat characteristics of rapid- and slow-growing rabbits slaughtered at 2.45 kg (according to Roiron-Cabanes and Ouhayoun, 1994).

Traits	Fast growth rate	Slow growth rate
Slaughter age (days)	62	73
Weaning weight (35 days) (g)	988	816
Average daily gain from weaning to slaughter (g day^{-1})	52.8	42.0
Slaughter yield	0.532	0.561
Hindleg meat/bone ratio	5.0	5.3

blasts actively divide during the early stages of development, then lose their ability to undergo mitosis, but acquire the ability to fuse. The fusion of myoblasts produces myotubes, i.e. elongated multinucleate cells whose nuclei, which are initially located in a central position, are gradually displaced towards peripheral areas by specific proteins. This *morphological differentiation* is followed by *functional differentiation*, which continues after birth. The first generation of myotubes leads to slow muscle fibres. The second generation of smaller myotubes form the fast fibres of fast muscles and part of the slow fibres of slow muscles. *Metabolic differentiation* occurs later than the differentiation into contraction types.

During the prenatal period, fibres are moderately oxidative and very weakly glycolytic. All fibres are of the oxidative type at birth (αR or βR) and differences appear only later. αR fibres are then converted into αW fibres. The rate and degree of this conversion will depend on the extent to which relevant muscles in the adult are glycolytic. However, this conversion is reversible: exercise, for instance, will increase the number of mitochondria in αW fibres and hence turn them into αR fibres. In red muscles (which are oxidative in the adult) oxidative enzyme activity increases after birth, reaches a maximum 30 days later and then decreases, whilst glycolytic activity remains very low up to the adult stage. In white muscles (which are glycolytic in the adult), the glycolytic activity increases as the animal grows and this continues beyond day 30, whilst the oxidative activity decreases after birth, especially before day 30. The ratios between the different nitrogenous fractions, including myofibrillar proteins (myosin, actin, troponin and tropomyosin), sarcoplasmic proteins (enzymes involved in glycolysis, creatine kinase and myoglobin), stroma proteins (collagen, reticulin and elastin) and non-protein components change after birth at different rates according to the muscles in question, but always rapidly enough for a permanent equilibrium to be reached on the 11th day after birth.

In the rabbit, muscle fibres continue to multiply after birth. This process lasts longer (up to 17 days) in the muscles whose relative growth is slow. Muscle weight thereafter increases as a result of the lengthening of the fibres, which follows bone growth, and an increase in diameter which continues for much longer. Muscle fibres gradually acquire the functional and metabolic properties which make them suitable for specific movements and postures, i.e. oxidative or glycolytic energy metabolism, slow or fast contraction, which enables either a moderate and prolonged or short and intense muscular activity (Ouhayoun and Dalle Zotte, 1993).

Influence of dietary factors on meat quality

Over the past 20 years, various experiments have been carried out worldwide, and especially in France, which have established reliable recommendations so that the nutritional requirements of growing rabbits are met through feed formulation (INRA, 1989; Lebas, 1991). Some of these studies take into account growth and feed efficiency as well as anatomical and biochemical criteria in relation to carcass and meat properties.

For maximum meat production, the values recommended for a diet fed *ad libitum* are as follows: 10.45 MJ digestible energy (DE) kg^{-1}, 150–160 g kg^{-1} crude protein or preferably 11–11.5 g digestible protein MJ^{-1} of DE, 130–140 g crude fibre kg^{-1} and a fat content of about 30 g kg^{-1}. Any substantial variation in these recommendations may modify the anatomical equilibrium of the carcass and the composition of the edible fraction, i.e. the amount of meat produced and its quality.

Influence of feeding level

Studies on the quantitative decrease in the daily feed supply were generally carried out to limit the frequency of digestive disorders (limited results), improve the feed efficiency or modify body composition. Once rabbits consume less than 0.85 of the *ad libitum* intake, growth rate slows down and feed efficiency is decreased. The slaughter yield is frequently affected, whatever the length and time of restriction. This decrease is due to an increase in the relative weight of the digestive tract resulting from two mechanisms: (i) feed restriction results in rabbits consuming bigger meals, which increases the retention time of feed in the digestive tract and the relative weight of the full digestive tract; (ii) the decrease in growth rate enhances the relative growth of early-maturing tissues, in particular those of the digestive tract.

Modifying the allometric relationships by increasing the time required to reach the same body weight also leads to an increase in skeletal weight and a decrease in adiposity. Water, mineral and protein contents of the carcass show an overall increase whilst the lipid content shows a corresponding decrease (Table 10.6).

The effect of feed restriction depends on how it is implemented. Thus, for the same overall degree of feed restriction (0.80 of *ad libitum*) a period of

Table 10.6. Influence of feed restriction on slaughter yield and composition of the carcass in rabbits slaughtered at 3.2 kg (according to Schlolaut *et al.*, 1978).

Feeding	*Ad libitum*	0.8 *ad libitum*	0.6 *ad libitum*
Age (days)	73.4	91.9	132.9
Slaughter yield	0.591	0.563	0.555
Carcass composition (proportion)			
Water	0.609	0.658	0.674
Fat	0.166	0.098	0.055
Proteins	0.186	0.194	0.200
Minerals	0.036	0.039	0.046

feed restriction followed by *ad libitum* feeding is more favourable to growth and feed efficiency than the reverse. Compensatory growth which results from *ad libitum* feeding following feed restriction can be observed for body composition and muscle energy metabolism. Such a process favours late-maturing tissues, which improves the muscle/bone ratio and the adiposity of the carcass. It stimulates the glycolytic pathway of muscle energy metabolism, which leads to a decrease in the ultimate pH of meat (Table 10.7). By increasing general anabolism, it increases the relative growth of the liver.

Given that the liver plays a major role in growth anabolism and that it develops relatively early, its proportion varies considerably in relation to rationing conditions. For example, this proportion is high in rabbits slaughtered at a weight of 3.2 kg after feeding at 0.60 of *ad libitum* from the age of 4 weeks (Schlolaut *et al.*, 1978) or in rabbits slaughtered at the age of 8 or 11 weeks after feeding at 0.71 of *ad libitum* for 3 weeks before slaughter (Lebas and Laplace, 1982). The proportion is low in rabbits subjected to severe feed restriction (0.54 of *ad libitum*) from the age of 11 weeks and slaughtered at a weight of 2.7 kg (Ouhayoun *et al.*, 1986).

Influence of dietary fibre content

A significant increase in dietary fibre content does not always lead to a modification in the growth rate and thus in the slaughter yield (Table 10.8).

Diets with a high fibre content invariably decrease growth rate. The dressing proportion is then reduced. For example, Machin *et al.* (1980) obtained growth rates ranging from 33.1 to 20.7 g day^{-1} and slaughter yields ranging from 0.591 to 0.575 in rabbits slaughtered at a weight of 2.1 kg with diets containing from 87 to 265 g crude fibre kg^{-1}. The decrease in the

Table 10.7. Influence of feeding patterns[a] on growth, feed efficiency, carcass and meat quality (fitted values for a live weight at weaning of 1046 g) (according to Perrier and Ouhayoun, 1996).

Traits	Control	LH	HL	Significance
Live weight at 77 days (g)	2302	2402	2217	**
Average daily gain between 35 and 77 days (g day^{-1})	29.88	32.26	27.87	**
Food consumption between 35 and 77 days (g day^{-1})	122.7	123.7	121.1	ns
Consumption index between 35 and 77 days (g day^{-1})	4.13	3.87	4.40	**
Slaughter yield (proportion of live weight)	0.603	0.590	0.601	**
Scapular and perirenal fat (proportion of carcass)	0.027	0.030	0.025	ns
Hindleg meat / bone ratio	6.28	6.32	5.98	*
Average ultimate pH of the thigh	5.77	5.73	5.77	*

[a]Control: 0.80 of the *ad libitum* diet supplied between days 35 and 77; LH: supply of the same amount of feed but with a more severe restriction (0.70) between days 35 and 56 than between days 56 and 77 (0.90); HL: supply of the same amount of feed but with a less severe restriction (0.90) between days 35 and 56 than between days 56 and 77 (0.70).

Table 10.8. Influence of the crude fibre content of the diet on the slaughter yield in the absence of effects on the growth rate between 4 and 11 weeks.

Reference	Crude fibre (g kg^{-1})	Average daily gain (g day^{-1})	Slaughter yield
Auxilia *et al.* (1979)	159/202	32.2/32.7	0.609/0.604
Lebas *et al.* (1982)	124/269	36.7/37.2	0.601/0.596
Masoero *et al.* (1984)	139/169	36.1/36.8	0.594/0.594

slaughter yield is due to the increase in the weight of the digestive tract (293 vs. 204 g) and the increase in the proportion of the digestive tract which results from the slowing down of growth rate. The decrease in the slaughter yield is accompanied by major modifications in the proportions of the other body components: the adiposity of the carcass is reduced, the skeleton is more developed, the water and protein contents of the carcass are higher and the lipid content is lower (Table 10.9).

Since the feed intake capacity of rabbits is a limiting factor, the increase in the dietary fibre content as well as feed restriction leads to an energy deficiency. The dilution of the digestible energy and feed restriction thus have common effects on the overall body growth rate as well as on the relative growth of tissues and organs and body composition. The dietary fibre content does not seem to have a direct negative effect on slaughter yield.

Influence of protein quality

For rabbits to achieve maximum growth, dietary protein should have an optimum composition of constituent amino acids. In some experiments designed to determine the requirements for essential amino acids (EAA) of

Table 10.9. Influence of partial substitution of a pelleted diet with grass on the slaughter yield and the carcass composition of rabbits slaughtered at 3.0 kg (according to Schlolaut *et al.*, 1984).

Type of feeding	Concentrate	Concentrate + grass
Age of slaughter (days)	91	166
As a proportion of live weight		
Carcass	0.613	0.555
Full digestive tract	0.182	0.260
Skin	0.154	0.151
As a proportion of carcass weight		
Water	0.630	0.730
Lipid	0.170	0.040
Protein	0.180	0.200

the growing rabbit, growth rate (which is a fundamental criterion), as well as the anatomical and chemical composition of carcasses have been studied.

An inadequate or excessive supply of sulphur amino acids (SAA) in comparison with requirements (0.037 of protein) affects growth rate and sometimes feed efficiency. A deficiency in SAA does not modify the slaughter yield or the composition of the carcass and meat in rabbits slaughtered at 90 days (Czajkowska *et al.*, 1980). In younger rabbits (77 days) the same holds true for the composition of the carcass, but the effects on slaughter yield are erratic (Berchiche 1985; Berchiche and Lebas, 1994). Conversely, an excessive supply of SAA (4.3 times the recommended level by supplying D, L-methionine) decreases the slaughter yield (0.540 vs. 0.553) and the adiposity of the carcass (Schlolaut and Lange, 1973). These variations probably result from an increase in the relative growth rate of the digestive tract and from the decrease in that of the adipose tissue, which are both due to a decrease in the overall growth rate.

When lysine requirements are not met, growth rate is reduced but the slaughter yield (Colin and Allain, 1978) and carcass and meat values (Czajkowska *et al.*, 1980) remain unchanged. A small decrease in the threonine supply (0.80 of requirement estimated at 0.034 of protein) leads to a decrease in the growth rate without concomitant effects on the slaughter yield and the composition of the carcass (Berchiche, 1985).

In most experiments, the variations in the growth rate which are attributable to composition of the EAA are generally small (less than 10% in comparison with the control). It is consequently not surprising that the parameters of the relative growth of the body components (and thus the slaughter yield), the muscle/bone ratio and the adiposity of the carcass, which is highly variable, are not significantly affected.

Influence of the protein/energy ratio

It is difficult to draw general conclusions from experiments on the influence of the protein/energy ratio because authors have either evaluated different protein contents (while theoretically maintaining the energy level constant), or different energy levels, which sometimes leads to significant modifications to diet formulations. Moreover, energy values are too often calculated from unreliable data (Perez and Lebas, 1992) instead of being measured *in vivo*.

For a given dietary energy level the slaughter yield is not modified when the variation in the protein level has no significant effect on the growth rate. The lowest adiposity of the carcass is found in rabbits fed diets with the highest protein content (Table 10.10).

The optimum level of protein with a balanced EAA composition increases with the energy level of the diet (Lebas, 1983). The effect of the protein content on growth and body composition depends on the energy concentration of the diet. For example, two isoenergetic diets (10.04 MJ DE

Table 10.10. Influence of the protein content of the diet, with a constant energy level (net energy, NE, and DE), on carcass and meat quantity.

References	Raimondi *et al.* (1973)		Ouhayoun and Cheriet (1983)	
Energy level (MJ kg⁻¹)	6.69 (calculated NE)		10.04 (determined DE)	
Slaughter age (days)	91	91	77	77
Protein content (g kg⁻¹)	200	176	172	138
Slaughter weight (kg)	2.96	3.03	2.27	2.27
Slaughter yield	0.563	0.567	0.572	0.580

kg⁻¹) containing 160 and 180g protein kg⁻¹ gave equivalent growth rates and slaughter yield in rabbits slaughtered at the age of 90 days. Conversely, with a higher energy level (10.67 MJ DE kg⁻¹) the diet containing only 160 g crude protein kg⁻¹, i.e. 15 g MJ⁻¹ DE, resulted in lower performance (Table 10.11).

If the digestible protein/DE ratio is satisfactory (11–11.5 g of digestible protein MJ⁻¹ DE) the feed intake of rabbits increases while the energy level of the diet decreases (Lebas, 1992). Thus, for a given protein content, if the energy level of the diet increases, the intake of the protein-poor diet decreases. This leads to a decrease in growth rate. Surprisingly, this decrease can lead to an increase in the lipid level of the carcass and to a decrease of the perirenal adiposity (Table 10.12). Slowing down the growth rate delays the development of the perirenal adipose tissue, whereas the relatively high energy level of the diet enhances muscle fat which represents a relatively constant proportion of total body fat.

Insofar as the protein content of the diet greatly influences growth rate during fattening, consequences for the biochemistry of muscles, especially for the protein fraction and energy metabolism, are expected. Ouhayoun and Delmas (1983) compared the productive performance and biochemistry of muscles in 11-week-old rabbits fed *ad libitum* with weaning diets differing in their protein/digestible energy (P/E) ratio (10–17 g MJ⁻¹ DE). The growth rate increased with the P/E ratio of the feed and was accompanied by a better relative development of the muscles, which had a higher nitrogen content. The sarcoplasmic protein content and the glycolytic enzyme activity

Table 10.11. Interaction of the protein content and the digestible energy level of the diet on the performance of rabbits slaughtered at a given age (according to Martina *et al.*, 1974).

Energy level (MJ DE kg⁻¹)	10.04		10.67	
Protein content (g kg⁻¹)	160	180	160	180
Protein/DE ratio (g MJ⁻¹)	15.9	17.9	15.0	16.8
Slaughter weight (kg)	2.12	2.15	1.83	2.39
Slaughter yield	0.550	0.544	0.527	0.566

Table 10.12. Influence of a protein deficiency on the carcass and meat quality of rabbits slaughtered at 77 days (diet containing 10.04 MJ DE kg⁻¹) (according to Ouhayoun and Cheriet, 1983).

Protein content (g kg⁻¹)	138	104
Slaughter weight (kg)	2.27	1.90
Slaughter yield	0.580	0.570
As a proportion of carcass weight		
Lipids	0.102	0.109
Perirenal fat	0.024	0.021
Protein	0.212	0.206

increased (Table 10.13). That these two were correlated was due to the fact that 0.70 of sarcoplasmic proteins were involved in the glycolytic process. Two hypotheses could account for these effects on muscle biochemistry: (i) the feed protein content acts directly on the overall protein anabolism and on each of the protein fractions; or, more probably, (ii) the protein content acts indirectly by modifying the weight at which rabbits are slaughtered. It should be noted that rabbits both selected and not selected for growth rate have similar responses to these diets. The decrease in growth rate which results from the protein-poor diet enhances meat quality by limiting the development of the muscle glycolytic metabolism. This leads to an increase in the intracellular lipid content, restricted fall of the pH during the processing of the muscle into meat and better water-holding capacity of this meat.

In most cases, variations in meat quality resulting from a change in the growth rate induced by the diet can be interpreted using general rules on the relative growth of tissues and organs (Fig. 10.1).

When nutrients are diluted by undigestible fibre, and provided that the feed intake capacity of young rabbits is not a limiting factor, growth rate is not

Table 10.13. Influence of the protein content of isoenergetic diets on the muscle biochemistry of the longissimus dorsi muscle of rabbits selected for growth rate and slaughtered at 77 days, according to Ouhayoun and Delmas (1983).

Protein content of diet (g kg⁻¹)	Slaughter weight (kg)	Total muscle nitrogen (g kg⁻¹)	Sarcoplasmic nitrogen (proportion of total muscle N)	Aldolase activity[a] (UI g⁻¹)
172	2.65	37.6	0.244	854
138	2.58	37.0	0.243	803
104	2.28	35.4	0.228	711

[a]Fructose 1,6-diphosphate aldolase, a typical enzyme of the glycolytic pathway involved in muscle energy metabolism.

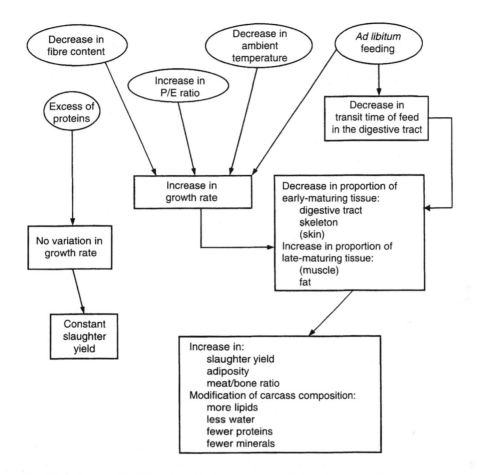

Fig. 10.1. Influence of deviations from the dietary recommendations (content and balance of nutrients and energy) on carcass and meat values (from Ouhayoun, 1991). The organs in parentheses are only slightly affected by the treatments.

slowed down. This is also the case when the protein content is excessive in comparison with the energy level of the diet. In both cases, the slaughter yield remains unchanged and only the excess of dietary proteins reduces the adiposity of the carcass.

If the growth rate is increased by *ad libitum* feeding, a decrease in room temperature which stimulates the feed intake, a major decrease in the undigestible fibre content or an increase in the P/E ratio, then carcass quality is modified. The slaughter yield, muscle/bone ratio (or the protein/mineral ratio of the carcass) and adiposity (or lipid content of the carcass) are increased. The reason may be that reducing the time required to reach a given body weight favours late-maturing tissues (muscles and especially fat) at the expense of early-maturing tissues (digestive tract, bones).

Some factors have specific effects. For example, *ad libitum* feeding diminishes the relative weight of the digestive tract by limiting the retention time of feed. This contributes to an increase in dressing proportion. The acceleration of growth rate resulting from an increase in the protein content of the diet with a constant energy level (protein requirements are better met) is accompanied by an increase in the nitrogen content of the muscle, in particular in the sarcoplasm fraction, and by an increase in the glycolytic activity of muscle energy metabolism. The consequences are *a priori* unfavourable to meat quality.

Influence of growth factors

Although their use is prohibited in meat production, β-adrenergic agonists have been studied in the main animal species. These substances exert an action through their affinity for cell β-adrenergic receptors. Ruminants are particularly sensitive to these substances (Lafontan *et al.*, 1988). Growth rate and feed efficiency are increased; carcasses contain more muscle and adiposity is reduced. The increase in muscle mass is due to a decrease in protein catabolism, while proteogenesis remains constant. The reduction in the adipose mass results from an antilipogenic effect as well as a lipolytic effect. In the rabbit, *cimaterol* improves the growth rate and feed efficiency. Optimal muscle development is due to a decrease in proteolysis, while the activity of the CDP (calcium-dependent proteinase) is halved (Forsberg *et al.*, 1989). This lack of proteolytic activity may limit the maturation rate of meat, as in beef (Fiems *et al.*, 1990) and mutton (Kretchmar *et al.*, 1990). *Clenbuterol* also improves the productive performance of the rabbit: nitrogen retention is increased, but energy retention is not modified. The muscle water and protein contents are higher, while the lipid content is lower (Parigi Bini *et al.*, 1990). Moreover, *clenbuterol* leads to an increase in the ultimate pH of muscle (Hulot *et al.*, 1996). This increase is explained by a relative increase in the oxidative metabolic pathway represented by the *fructose-1, 6-diP aldolase/isocitrate dehydrogenase* ratio and by a decrease in the carbohydrate content (total glycosyl residues) of the muscle. It is probable that *clenbuterol* causes a change in meat quality due to its effects on muscle biology.

Influence of the quality of dietary fat

In the rabbit, the most common method of varying the energy content of a diet consists of substituting a portion of highly digestible carbohydrates, such as starch and sugars, by slightly or non-digestible cell wall constituents, such as cellulose, hemicellulose or lignin. The influence of this substitution is discussed in the chapter on fibre. Another method involves substituting a portion of the carbohydrates by well-digested lipids which contain 2.3 times more energy. This makes it possible to increase the energy level of the diet without over-reducing the fibre content.

Growth performance (growth rate, feed efficiency, slaughter yield) is not affected by the type of lipids added, i.e. whether of animal or plant origin, insofar as the substitution of lipids for digestible carbohydrates does not modify the energy level of the feed estimated *in vivo*.

Highly polyunsaturated fat in the rabbit is often wrongly considered as being of endogenous origin. According to studies of Ouhayoun *et al.* (1985, 1987), the endogenous fatty acids of the rabbit which are synthesized from carbohydrates are mainly palmitic (C16:0), oleic (C18:1, *n*-9) and stearic (C18:0) acids, as in beef. The highly polyunsaturated lipids of the rabbit result from the influence of exogenous lipids. Cereals and oil seed meal provide linoleic acid (18:2, *n*-6). The galactolipids of lucerne provide linolenic acid (C18:3, *n*-3). These polyunsaturated fatty acids are not synthesized by higher animals. Supplementing a basal diet, in which most of the energy is provided by carbohydrates, with lipids prevents their endogenous production. The composition of storage lipids is all the more modified as the profile of dietary fatty acids is different from that of endogenous fatty acids and the added fat content is high. The direct effect of the composition of the fat ingested by the rabbit results from its high digestibility in the small intestine before being transformed by the digestive flora, as happens in ruminants.

The composition of dietary fat does not modify the total lipid content in the muscle tissue but can alter its fatty acid composition considerably (Raimondi *et al.*, 1975). The saturated fatty acid to unsaturated fatty acid (S/U) ratio is 0.54 when the diet is supplemented with beef tallow with a high palmitic and stearic acid content, and 0.38 when the diet is supplemented with groundnut oil which contains mainly unsaturated fatty acids (oleic and linoleic acids).

The composition of storage fat considerably influences carcass quality and the organoleptic properties of meat (Ouhayoun *et al.*, 1987). The melting point of fatty acids is known to decrease as the carbon chain length shortens (C18:0 = 70°C; C12:0 = 44°C). This melting point also falls when the number of double bonds increases (C18:1 = 13.5°C; C18:3 = −11°C). For example, if rabbits are fed a diet containing 50 g olive oil kg^{-1}, 0.93 of the lipids extracted from the adipose tissue melt at a temperature of +20°C or less. The carcass looks shiny and the perirenal adipose tissue is soft and translucent. In this case, 0.70 of the fatty acids which mainly (0.72) have a long chain (C18) are mono- or polyunsaturated. Conversely, if rabbits are fed on a diet containing 50 g copra oil (coconut) kg^{-1}, only 0.48 of the lipids extracted from the adipose tissue melt at a temperature of +20°C or less. The carcass looks dry, the perirenal fat is hard and white and comparable to beef tallow. Carcass lipids contain 0.63 saturated fatty acids (C12:0, C14:0, C18:0 and especially C16:0). Dietary lipids have an effect on the physical properties of fat as well as the organoleptic properties of meat (Table 10.14).

Table 10.14. Substitution of lipids for digestible carbohydrates in rabbit diets. Consequences for the consistency of fat and meat quality (from Ouhayoun, 1991).

Added fat (g kg⁻¹)	Principal fatty acid added	Organoleptic quality of meat	Physical quality of fat
Copra (50)	Lauric	Unfavourable (soapy taste)	Firm fat
Cocoa (50)	Stearic	Favourable	
Beef tallow (50)	Stearic	ns	
Olive oil (50)	Oleic	ns[b]	Liquid fat
Soybean oil (30)	Linoleic	Unfavourable	ns
Linseed oil (50)	Linolenic	Unfavourable (rancid)	ns
Rapeseed oil (40)[a]	Oleic + linoleic	ns[c]	Soft fat

[a]Full fat rapeseed.
ns, not significantly different from the control without fat.
[b]ns, but reasonably good.
[c]ns when fresh but unpleasant taste after freezing.

The use of copra oil leads to carcasses with a pleasant appearance. The meat, however, has a soapy taste, which makes it unfit for consumption. This defect is due to the high level of lauric acid (C12:0) in the oil. Conversely, when olive oil is incorporated into the diet of rabbits, their carcasses have an unappealing, slimy appearance, although the meat is acceptable. Linseed oil and soybean oil both to a certain degree give an unpleasant flavour to meat, which results from the high content of polyunsaturated fatty acids in the adipose tissue which are prone to oxidation. Only cocoa butter produces rabbit meat with a high organoleptic quality. Substituting more saturated fats for oils in the diet of rabbits has varying degrees of success and therefore should only be undertaken with caution by feed formulators.

Influence of dietary flavours

Flavours are added to feeds in order to stimulate and regulate the feed consumption of fattening young rabbits. The flavours used most frequently are terpenic hydrocarbons. As they can dissolve in adipose tissue, they can theoretically influence the organoleptic quality of meat. The incorporation of a flavour containing thymol into the finishing feed (2 weeks) of rabbits slaughtered at the age of 11 weeks was studied by Delmas et al. (1991). Neither the concentration of 200 ppm, which is the amount usually used in rabbit feeds, nor double that concentration had an effect on the growth performance or the meat and carcass quality. Although the adipose tissue of the rabbits fed the diet with the highest aroma content contained 10,000 times less thymol than the lipid of the feed, the flavour of the meat was considered to have an unpleasant acrid taste as well as a strong odour. Given the low transfer of this flavour to adipose tissue and its unpleasant consequences for flavour, it should only be incorporated into the feed of growing rabbits at a concentration which promotes normal levels of feed consumption.

Conclusions

For a given breed type, carcass and meat quality depend mainly on the age of the animal. By playing a role in the expression of growth potential, the diet has a marked effect on the relative development of organs and tissues and thus on anatomical carcass composition, as well as on the biochemical balance of muscle. Moreover, in the rabbit, which is a non-ruminant animal, the effect of exogenous fatty acids on the chemical composition and the physical and sensorial properties of body lipids demonstrates that supplementing feeds with fat should be undertaken with caution.

References

Auxilia, M.T., Masoero, G. and Terramoccia, S. (1979) lmpiego di mais disidratato integrale nelle diete per conigli in accrescimento. *Annali dell'Istituto Sperimentale per la Zootecnia* 12, 43.

Berchiche, M. (1985) Valorisation des protéines de la féverole par le lapin en croissance. Thèse de Doctorat, Université de Toulouse.

Berchiche, M. and Lebas F. (1994) Supplémentation en méthionine d'un aliment à base de féverole: effet sur la croissance, le rendement à l'abattage et la composition de la carcasse chez le lapin. *World Rabbit Science* 2, 135.

Cantier, J., Vezinhet, A., Rouvier, R. and Dauzier, L. (1969) Allométrie de croissance chez le lapin (*Oryctolagus cuniculus*). 1 – Principaux organes et tissus. *Annales de Biologie Animale Biochimie, Biophysique* 9, 539.

Colin, M. and Allain, D. (1978) Etude du besoin en lysine du lapin en croissance en relation avec la concentration énergétique de l'aliment. *Annales de Zootechnie* 27, 17.

Czajkowska, J., Jedryka, J., Kawinska, J., Niedzwiadek, S. and Ryba, Z. (1980) Obnizenie posiomu bialka w tuczu krolikow przy zastosowaniu aminokwasow syntetycznych. *Roczniki Naukowe Zootechniki* 7, 289.

Delmas, D., Lapanouse, A. and Ouhayoun, J. (1991) Qualité de la viande de lapin. 2-Aromatisation de l'aliment de finition. Compte-rendu d'une Étude Financée par le GIE Midi-Pyrénées.

Fiems, L.O., Buts, B., Boucqué, Ch. V., Demeyer, D.I. and Cottyn, B.G. (1990) Effect of a ß agonist on meat quality and myofibrillar protein fragmentation in bulls. *Meat Science* 27, 29.

Forsberg, N.E., Ilian, M.A., Ali-Bar, A., Cheeke, P.R. and Wehr, N.B. (1989) Effects of cimaterol on rabbit growth and myofibrillar protein degradation and on calcium-dependent proteinase and calpastatin activities in skeletal muscle. *Journal of Animal Science* 67, 3313.

Hulot, F., Ouhayoun, J. and Manoucheri, M. (1996) Effect of clenbuterol on productive performance, body composition and muscle biochemistry in the rabbit. *Meat Science* 42, 457.

INRA (1989) *L'Alimentation des Animaux Monogastriques: Porc, Lapin, Volailles,* 2nd Edn. INRA, Paris, 282 pp.

Kretchmar, D.H., Hathaway, M.R., Epley, R.J. and Dayton W.R. (1990) Alterations in post mortem degradation of myofibrillar proteins in muscle of lambs fed a β adrenergic agonist. *Journal of Animal Science* 68, 1760.

Lafontan, M., Berlan, M. and Prud'hon, M. (1988) Les agonistes β-adrénergiques. Mécanismes d'action: lipomobilisation et anabolisme. *Reproduction, Nutrition and Development* 28, 61.

Lebas, F. (1983) Bases physiologiques du besoin protéique des lapins. Analyse critique des recommandations. *Cuni-Sciences* 1, 16.

Lebas, F. (1991) Alimentation pratique des lapins en engraissement. *Cuniculture* 18, 273.

Lebas, F. (1992) Alimentation pratique des lapins en engraissement. 2ème partie. *Cuniculture* 19, 83.

Lebas, F. and Laplace, J.P. (1982) Mensurations viscérales chez le lapin. 4 – Effets de divers modes de restriction alimentaire sur la croissance corporelle et viscérale. *Annales de Zootechnie* 31, 391.

Lebas, F., Laplace, J.P. and Droumenq, P. (1982) Effet de la teneur en énergie de l'aliment chez le lapin. Variations en fonction de l'âge des animaux et de la séquence alimentaire. *Annales de Zootechnie* 31, 233.

Machin, D.H., Butcher, C., Owen, E., Bryant, M. and Owen, J.E. (1980) The effects of dietary metabolizable energy concentration and physical form of the diet on the performance of growing rabbits. In: Camps, J. (ed.) *Proceedings of the 2nd World Rabbit Congress, Barcelona,* Vol. 2, p. 65.

Martina, C., Damian, C., Palamaru, E. (1974) Retete de nutreturi combinate-granulate cu diferite nivele energo-proteice pentru cresterea si ingrasarea tineretului cunicul. *Lucrarile stiintifice ale Institutului de Cercetari pentru Nutritia Animalia* 2, 313.

Masoero G., Chicco R., Ferrero A., Rabino I. (1984) Paglie di riso e di frumento, trattate o non con soda, in diete per conogli in accrescimento. In: Finzi, A. (ed.) *Proceedings of the 3th World Rabbit Congress, Rome,* Vol. 1, p. 335.

Ouhayoun, J. (1978) Etude comparative de races de lapins différant par le poids adulte. Incidence du format paternel sur les composantes de la croissance des lapereaux issus de croisement terminal. Thèse de Doctorat, Université de Montpellier.

Ouhayoun, J. (1991) La viande de lapin. Caractéristiques et variabilité qualitative. *Cuni-Sciences* 7, 1.

Ouhayoun, J. and Cheriet, S. (1983) Valorisation comparée d'aliments à niveaux protéiques différents par des lapins sélectionnés sur la vitesse de croissance et par des lapins issus d'élevages traditionnels. 1 – Etude des performances de croissance et de la composition du gain de poids. *Annales de Zootechnie* 32, 257.

Ouhayoun, J. and Dalle Zotte, A. (1993) Muscular energy metabolism and related traits in rabbit. A review. *World Rabbit Science* 1, 97–108.

Ouhayoun, J. and Delmas, D. (1983) Valorisation comparée d'aliments à niveaux protéiques différents par des lapins sélectionnés sur la vitesse de croissance et par des lapins issus d'élevages traditionnels. 2 – Etude de la composition azotée et du métabolisme énergétique des muscles *longissimus dorsi* et *biceps femoris*. *Annales de Zootechnie* 2, 277.

Ouhayoun, J., Gidenne, T. and Demarne, Y. (1985) Evolution postnatale de la composition en acides gras des lipides du tissu adipeux et du tissu musculaire chez le lapin en régime hypolipidique. *Reproduction, Nutrition and Development* 25, 505.

Ouhayoun, J., Poujardieu, B. and Delmas, D. (1986) Etude de la croissance et de la composition corporelle des lapins au-delà de l'âge de 11 semaines. 2 – Composition corporelle. *Proceedings 4èmes Journées de la Recherche Cunicole, Paris.*

Ouhayoun, J., Kopp, J., Bonnet, M., Demarne, Y. and Delmas D. (1987) Influence de la composition des graisses alimentaires sur les propriétés des lipides périrénaux et la qualité de la viande de lapin. *Sciences des Aliments* 7, 521.

Parigi Bini, R., Xiccato, G. and Cinetto, M. (1990) Utilisation des protéines et de l'énergie alimentaire par le lapin en croissance traité au clenbutérol. In: *Proceedings 5èmes Journées de la Recherche Cunicole, Paris.*

Perez, J.M. and Lebas, F (1992) Peut-on estimer la valeur énergétique des aliments destinés aux lapins? *Cuniculture* 19, 271.

Perrier, G. and Ouhayoun, J. (1996) Growth and carcass traits of the rabbit. A comparative study of three modes of feed rationing during fattening. In: Lebas, F. (ed.) *Proceedings of the 6th World Rabbit Congress, Toulouse*, Vol. 3. Association Française de Cuniculture, Lempdes, p. 225.

Prud'hon, M., Vezinhet, A. and Cantier J. (1970) Croissance, qualités bouchères et coût de production des lapins de chair. *Bulletin Technique d'Information* 248, 203.

Raimondi, R., Auxilia, M.T., De Maria, C. and Masoero, G. (1973) Effeto comparativo di diete a diverso contenulo energetico e proteico sull'accrescimento, il consumo alimentare, la resa alla macelazione e le caratteritiche delle carni di coniglio. Atti Convegno Internazionale di Coniglicoltura, Erba.

Raimondi, R., De Maria, C., Auxilia, M-T-. and Masoero, G. (1975) Effet della grassatura dei mangimi sulla produzione della carne di conigli. 3 – Contenuto in acidi grassi delle carni e del grasso perirenale. *Annali dell'Istituto Sperimentale per la Zootecnia* 8, 167.

Roiron-Cabanes, A. and Ouhayoun, J. (1994) Précocité de croissance. Influence de l'âge à l'abattage sur la valeur bouchère et les caractéristiques de la viande de lapins abattus au même poids vif. In: *Proceedings 6èmes Journées de Recherche Cunicole, La Rochelle.*

Schlolaut, W. and Lange, K. (1973) Der Einfluss von Methionin auf die Mastleitung und den Woolertrag von Kaninchen. *Archiv für Geflügelkunde* 37, 208.

Schlolaut, W., Lange, K. and Schulter, H. (1978) Der Einfluss der Fütterungintensität auf die Mastleitung und Schlachtkörper Qualität beim Jungmastkaninchen. *Züchtungskunde* 50, 401.

Schlolaut, W., Walter, A. and Lange, K. (1984) Fattening performance and carcass quality in the rabbit in dependence on the final fattening weight and fattening method. In: Finzi, A. (ed.) *Proceedings of the 3th World Rabbit Congress, Rome*, Vol. 1, p. 445.

11.   Nutrition and Pathology

F. Lebas,[1] T. Gidenne,[1] J.M. Perez[1] and D. Licois[2]

[1]Station de Recherches Cunicoles, INRA Centre de Toulouse, BP 27, 31326 Castanet-Tolosan Cedex, France; [2]INRA Station de Pathologie Aviaire et de Parasitologie, Laboratoire de Pathologie du Lapin, 37380 Nouzilly, France

Nutrition can be related to impaired rabbit health through two main routes. The first is associated with the presence of toxic compounds in the feed or utilization of unbalanced diets. The second is the presence of pathogenic agents (viruses, bacteria, parasites) in the feed or in drinking water. This latter topic will not be discussed in this chapter since it is not directly linked to nutrition, but is only a question of feeding management and hygiene. Similarly, the presence of undesirable pesticides in feed ingredients can impair rabbit health. Very few specific data on this topic are available for rabbits under production conditions and the subject will not be considered further. There are specialist texts devoted to this aspect of animal feeding.

In this chapter, it is assumed that some efforts were made to provide the daily minimum requirements of the rabbit for the main individual dietary components such as energy, protein and amino acids, minerals and vitamins as recommended in previous chapters. Nevertheless it is generally difficult to provide all of these exactly at the optimum and, as a consequence of the composition of available raw materials, it is necessary to accept an excess or an imbalance of some components to be certain that the minimum level of others is met.

By itself, an imbalance is only responsible for low performance, but not for health problems, if the general production conditions are acceptable. For example, under highly controlled conditions, a diet containing only 60 g fibre (acid detergent fibre (ADF) kg⁻¹ dry matter) may induce non-specific digestive disorders (Davidson and Spreadbury, 1975). A similar situation was observed with diets containing up to 280–300 g crude protein kg⁻¹ (Lebas, 1973). Such imbalances provoke a greater susceptibility of rabbits to problems (mainly digestive disorders) which is why the extreme levels quoted must never be recommended for practical feeding (Lebas, 1989). One of the objects of this chapter is to indicate those rules, when known, which are able to minimize the risk of disorders and to present some ideas on 'acceptable' imbalances in everyday feeding practice.

In addition to the problems of imbalance, an absolute excess of components, such as some minerals (phosphorus) or vitamins (vitamin D), can be toxic independently of the health status of rabbits. The only relevant question is 'At which point does a nutrient supplied above the recommended minimum or optimum become toxic?'.

The present chapter will, accordingly, consider health problems linked to the balance of dietary components and to the presence of an excess of different nutrients, mainly in relation to the initial composition of feed ingredients. One section is devoted to the health consequences of non-nutritional components frequently associated with feed ingredients such as mycotoxins. The chapter will conclude with a discussion of water quality, since water is also able to induce nutritional disorders when some soluble components are present in too concentrated amounts.

Problems related to an imbalance of nutrients and energy

Among the various health problems related to feeding, intestinal pathology along with respiratory diseases are the predominant causes of morbidity and mortality in commercial rabbit husbandry. The former occurs mainly in young rabbits after weaning (4–10 weeks of age), while the latter preferentially affect adults. Enteritis in growing rabbits will induce a mortality rate of 0.11–0.12 (Koehl, 1997) but may frequently exceed 0.15 and even reach up to 0.50. Moreover, digestive disorders are responsible for significant morbidity as characterized by growth depression and poor feed conversion. These economic losses, whilst being less apparent than mortality, are often underestimated by rabbit breeders.

Diagnosis of intestinal diseases is difficult because, whatever the cause (nutritional problems or specific illness), symptoms and lesions are generally similar. The difficulty of recognizing the aetiology for an intestinal disorder in rabbits is reinforced by the fact that, as for most diseases in man or animals, several factors are involved in the development of enteritis and must be considered. Initially, there is the status of the animal itself (age, genotype, immunity). Subsequently, there is consideration of the pathogenic agents involved (parasites, bacteria, viruses). Finally, there are environmental concerns including nutritional factors and breeding conditions such as hygiene and stress. Although many factors may provoke enteritis, the predominant and constant clinical sign observed is diarrhoea, which occurs in about 0.90 cases of enteritis (Licois et al., 1992). This may be related to the characteristics of the rabbit intestinal tract and its complex physiology.

The composition of the caecal contents as well as caecal function and caecal microbial activity is profoundly affected in cases of enteritis (Figs 11.1 and 11.2). The motility of the caecum is stimulated whereas that of the ileum and jejunum is inhibited in diarrhoea experimentally induced with coccidia (Fioramonti et al., 1981). Furthermore, Hodgson (1974) observed, in rabbits

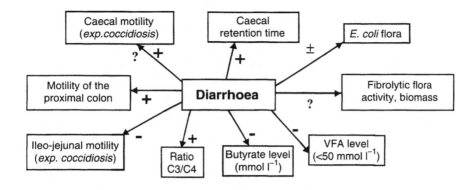

Fig. 11.1. Caecal changes occurring in cases of digestive disorders (diarrhoea) in the growing rabbit. ?, need for further studies; ±, inconsistent results.

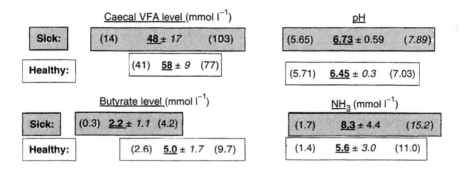

Fig. 11.2. *In vivo* caecal fermentation pattern of healthy and sick growing rabbits (means of 80 and 21 rabbits, respectively). Figures in parentheses are minimum and maximum values observed. The rabbits were 7–11 weeks old, and cannulated; sick rabbits were those with regular digestive troubles or abnormally low intake. (Data from Bellier, 1994.)

fed a low-fibre diet, an increased motility of the proximal colon that appeared contracted and thickened, and a greater retention of digesta in the total tract. This probably reflects a higher antiperistaltic activity of the proximal colon (see Chapter 1) induced by the high proportion of fine particles in a low-fibre diet. It is thus difficult to postulate that rabbit diarrhoea is characterized by hypomotility of the caeco-colic segment. In parallel, caecal fermentative activity is disturbed (Fig. 11.2); thus, for a 6-week-old rabbit, the concentration of caecal volatile fatty acids (VFA) falls to below 50 mmol l[-1]; butyrate is particularly affected (leading to a C3:C4 ratio in the range 1.5–8 instead of 0.5–0.8), and larger interindividual variations in the pattern of fermentation are observed. Higher pH (+0.5) and ammonia levels could also be observed. The composition of the caeco-colic flora may also be affected, but few results are available and they are not consistent, as they show both a decrease and an increase in *Escherichia coli* and/or *Clostridia*.

Fibre and starch supply

Many experiments have been designed to evaluate the effects of dietary fibre level on rabbit digestion because it is acknowledged that fibre deficiency greatly affects caecal metabolism and the health of the rabbit. In most cases, variation in dietary fibre content is negatively correlated with starch content. It could be concluded that an imbalance in the fibre/starch ratio acts as a predisposing factor in increasing the frequency of digestive disturbances. A deficiency in dietary fibre could be expressed in terms of quantity or quality of cell-wall constituents and, similarly, the effect of starch intake could differ according to the origin of the starch.

Consequences of a reduction in the intake of fibre

A lower dietary fibre intake, without major changes in the proportions of the cell-wall constituents (hemicelluloses or lignin), affects the composition of caecal contents. In healthy animals, the level of fibre in the caecum decreases, while the starch concentration remains low (around 0.015–0.04 g kg^{-1}) and there are no significant changes in the concentration of the end-products of fermentation (ammonia, VFA) and caecal pH (Fig. 11.3). However, the VFA molar proportion is affected by the fibre level. For instance, the proportion of butyrate generally increases significantly when the fibre/starch ratio decreases.

Several hypotheses have been suggested to explain how the dietary

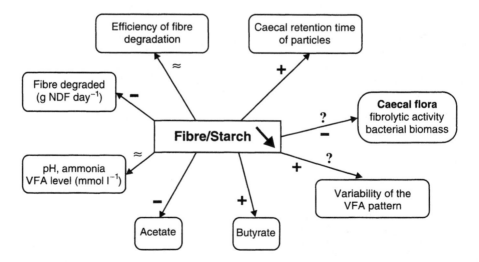

Fig. 11.3. Effect on several caecal digestive criteria of lowering the dietary fibre/starch ratio in the growing healthy rabbit. ?, need for further studies; ≈ effects insignificant.

supply of starch and fibre affects digestive physiology, but none has been completely validated by experimental results. Prohaszka (1980) put forward the antibacterial effect of caecal VFA originating from fibre fermentation, particularly in the case of *E. coli* infection. However, numerous studies have not observed a close relationship between the concentration of caecal VFA and the pH or between *E. coli* and caecal pH. In addition, Padilha *et al.* (1995) showed that, between 29 and 49 days of age, the caecal pH decreases whereas *E. coli* remains steady.

The favourable effect on health of rabbits of a high dietary level of low-digestibility fibre (lignocellulose or ADF) might correspond to a control in the rate of passage of digesta, particularly in the caeco-colic segment. Moreover, most of the results indicate that all the factors contributing to an increase in retention time (lowering the fibre level, reducing the particle size of the feed, feed restriction) contribute to destabilizing the caecal microbial activity and promote enteritis. It might be argued that a low caecal turnover of digesta would lead to an insufficient supply of substrates available for the fibrolytic flora (Fig. 11.4).

In addition, Morisse *et al.* (1985) suggested that a high-fibre and low-starch diet could impair caecal fermentation, because of a too low caecal input of fermentable sugars. Moreover, Cheeke and Patton (1980) postulated that an oversupply of readily available carbohydrates in the caecum could induce digestive problems. More specifically, caecal input of glucose (originating from starch degradation) seemed to favour pathogenic agents such as *Clostridium spiroforme* (Boriello and Carman, 1983).

Nevertheless, when the dietary fibre level decreases, the rabbit reduces its feed intake because of the associated increase in digestible energy

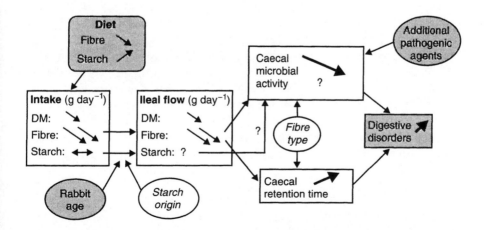

Fig. 11.4. Relationships between feeding the growing rabbit with low fibre and high starch diets and the incidence of digestive disorders.

concentration. Consequently the quantity of starch consumed increases slightly, whereas the fibre intake is reduced considerably. Under these conditions, the ileal flow of starch between a low- and a high-fibre diet would not be greatly affected, and would not be a primary factor in disturbances of caecal function. However, in the suckling rabbit it can be assumed that this feed intake regulation is not completely established and neither is pancreatic enzymatic activity (see Chapter 2). The combination of these two factors could lead to a high flow of starch into the caecum that could then promote digestive disturbances. For instance, Blas *et al.* (1994) reported that when the starch intake increased by 4.7 g day^{-1} and neutral detergent fibre (NDF) intake was reduced by 9.7 g day^{-1} in 6-week-old rabbits, the starch level could reach 0.13 of the DM of ileal contents and the mortality rate increased by 0.03 absolute.

In addition, it cannot be ignored that, if the ileal starch flow increased for a similar intake of fibre (for example if using low-digestible starch), it could then become an important factor in destabilization of the caecal flora. For instance, in rabbits fed a diet containing more than 100 g of crude potato starch kg^{-1}, Heckmann (1972) noticed a higher water content in the faeces, and suggested that this could be the first stage of diarrhoea. Moreover, Gidenne (1995) observed a higher incidence of diarrhoea for rabbits fed a diet containing a gluco-oligosaccharide that releases glucose after hydrolysis by caecal bacteria, while fructo- or galacto-oligosaccharides did not cause such effects (Peeters *et al.*, 1992; Aguilar *et al.*, 1996). Thus, dietary starch could play a substantial role, in interaction with fibre, in the control of caecal digestion in the growing rabbit. From a practical point of view, low starch diets (100–120 g kg^{-1}) seemed to reduce the risk of digestive disorders, particularly during the few weeks following weaning (Lebas and Maitre, 1989).

The substitution of starch for fibre was also studied in diets for rabbit does, using five isoenergetic diets (10.6 MJ digestible energy (DE) kg^{-1}) with increasing levels of NDF (from 278 to 371 g kg^{-1}) and fat (from 20 to 51 g kg^{-1}) at the expense of the level of starch, which decreased from 237 to 117 g kg^{-1} (de Blas *et al.*, 1995). Some impairment in performance of does was observed in rabbits fed the highest levels of fibre, which might be explained by higher fermentation losses in the caecum, together with an insufficient uptake of glucose from the gut to meet the requirements for pregnancy and milk lactose synthesis. Conversely, negative effects of high dietary starch concentrations were also mentioned and were related to an increase in diarrhoea mortality for does.

Effect of the type of cell wall constituents

Independently of its botanical origin, a decrease in the dietary acid detergent lignin (ADL) level (from 80 to 20 g kg^{-1}) leads to a higher digestibility and digesta retention time (Gidenne and Perez, 1994), but led to a significantly higher mortality rate in the growing rabbit (Perez *et al.*, 1994). Similar effects were observed when the 'cellulose' (ADF – ADL) level decreased (Gidenne and Perez, 1996; Perez *et al.*, 1996). A dietary 'cellulose' level of 120 g kg⁻¹ seemed sufficient to reduce mortality between 7 and 10 weeks of age, whereas a cellulose level of 160 g kg⁻¹ would be necessary from weaning until the rabbits were 7 weeks old.

An intake of more digestible fibres (hemicelluloses or pectins) such as those found in beet or citrus pulp induced a higher caecal content, higher VFA levels and a lower pH. Growing rabbits fed a high-starch, low-digestible-fibre diet showed a higher incidence of digestive disorders (diarrhoea) compared with those fed a low-starch, high-digestible-fibre diet (Jehl and Gidenne, 1996). Digestible fibre entering the caecum included pectins (mainly from beet pulp) and hemicelluloses (mainly from wheat bran), which are fermented, respectively, into galacturonic acids and pentoses. This induced a higher caecal microbial activity and a greater biomass production. The better health status of these rabbits could be the result of a barrier effect from the fibrolytic symbiotic caecal flora against pathogenic species. On the other hand, the level of glucose entering the caecum would be lower than for starchy diets, and this could also impair the development of glucose-dependent pathogenic bacteria.

Thus, a dietary recommendation for lignocellulose alone appears to be insufficient to prevent digestive disturbances in the rabbit. The starch level should also be considered, in addition to the ADF level, in the formulation of feed for growing rabbits and breeding does.

As a conclusion of the effects of fibre and starch on rabbit health, it must be emphasized that, if fibre is not a true nutrient for the nutrition of the rabbit, it is nevertheless a true nutrient for the rabbit as a whole, as there is a requirement for it in terms of adequate digestive physiology and rabbit health. As is the case for other nutrients, there is a minimum below which 'fibre deficiency' may be described. This minimum level can be expressed first as 120 g cellulose kg⁻¹ DM (ADF – ADL) for any rabbit older than 7 weeks, and 160 g for younger rabbits. In addition, a minimum of 45 g ADL kg⁻¹ can be considered as suitable to reduce the enteritis. Nevertheless these values must be modified according to the starch content of the diet, the optimum fibre/starch ratio being variable according to the age of the rabbits, and to the stage of production for the breeding does.

Protein

Evidence that a dietary deficiency of protein or amino acids affects the growth performance of the rabbit, together with optimum requirements, is reviewed in Chapters 3 and 12.

Conversely, protein supplied in excess did not affect growth itself, but would promote the incidence of diarrhoea. For instance, de Blas *et al.* (1981) observed an increased mortality during the fattening period with high protein diets. A level of 1.8–1.9 g crude protein MJ^{-1} DE seems optimum even if, with an increase up to 2.6 g, Kjaer and Jensen (1997) observed only a slight and non-significant increase of mortality.

Similarly, Catala and Bonafous (1979) demonstrated that a higher ileal flow of protein (obtained through a reduction of protein digestion by a ligature of the pancreatic duct) led to an increase in microbial proliferation in the hindgut. Furthermore, an excess of dietary protein could also favour the proliferation of *Clostridia* in the rabbit and could also increase slightly the prevalence of *E. coli* (Haffar *et al.*, 1988; Cortez *et al.*, 1992) which thus could lead to an increase in mortality associated with enteritis.

Accordingly, most feed manufacturers limit dietary protein levels in fattening diets because of the increased mortality rate observed when protein levels exceed by 10% the minimum levels recommended for maximum growth rate, a level itself related to DE concentration. Moreover, an excessive protein supply will probably become progressively more unlikely in Europe because of an increased dietary cost and principally because of the European policy in favour the reducing nitrogen excretion to the environment through the use of low protein diets (Maertens and Luzi, 1996).

Lipids

Additional fat incorporation in rabbit diets is limited to a maximum of 40–50 g kg^{-1} mainly because of feed pelleting constraints. Thus no interactions between the dietary fat level and the pathology of the growing rabbit have yet been noticed.

In contrast, the incorporation of fat in diets for breeding does could be of interest to increase their DE intake. However, contrasting results have been obtained indicating either a higher (Lebas and Fortun-Lamothe, 1996) or a trend for a lower (Fraga *et al.*, 1989; Fernandez-Carmona *et al.*, 1996) kit mortality. Furthermore, neither average weight of breeding does nor fertility or prolificacy was significantly affected by dietary fat incorporation.

Problems associated with dietary compounds present at toxic levels

Minerals and vitamins

Although recommendations for optimum and maximum levels of mineral and vitamins were discussed extensively in Chapter 9, it seems appropriate to consider maximum acceptable levels in diets. It is highly recommended during diet formulation that calculated nutrient levels, even if analyses are not available, are significantly below toxic levels.

The principal information available is summarized in Table 11.1. The

Table 11.1. Maximum levels of minerals or vitamins and levels known to induce signs of toxicity in the rabbit.

	Maximum level observed with no problems	Concentration with signs of toxicity	Period of life
Minerals			
Calcium	25 g kg⁻¹	40 g kg⁻¹	Growth
	19 g kg⁻¹	25 g kg⁻¹	Reproduction
Phosphorus	8 g kg⁻¹	—	Growth
	8 g kg⁻¹	10 g kg⁻¹	Reproduction
Magnesium	3.5 g kg⁻¹	4.2 g kg⁻¹	Growth
Sodium	6 g kg⁻¹	7 g kg⁻¹	Growth
Potassium	16 g kg⁻¹	15–20 g kg⁻¹	Growth
	16 g kg⁻¹	20 g kg⁻¹	Reproduction
Chlorine	4.2 g kg⁻¹	—	Growth
Copper	150–200 ppm	200–300 ppm	Growth
Fluorine	—	400 ppm	Growth
Iodine	10,000 ppm	—	Growth
	—	100 ppm	Reproduction
Iron	400 ppm	500 ppm	Growth
Manganese	—	50 ppm	Growth
Zinc	110 ppm	—	Growth
Vitamins			
Vitamin A	100,000 IU kg⁻¹	—	Growth
	40,000 IU kg⁻¹	75,000 IU kg⁻¹	Reproduction
Vitamin D	2,000 IU kg⁻¹	3,000 IU kg⁻¹	Reproduction
Vitamin E	200 mg kg⁻¹	—	Growth
Vitamin C	2 g kg⁻¹	—	Growth

values are those from Lebas *et al.* (1996) amended according to the most recent data obtained primarily from the 6th World Rabbit Congress: Bernardini *et al.* (1996) for vitamin E; Abd El-Rahim *et al.* (1996) for iron.

It must be pointed out that the maximum acceptable level is in general considerably higher than the recommended level, but with some noticeable exceptions such as potassium, phosphorus or vitamin D.

Mycotoxins

Mycotoxins are metabolites produced by certain fungi in the field on standing crops or harvest. The mould growth could also occur on stored grains or other raw materials because of non-hygienic storage conditions. The toxic substances may be contained within the spore or be secreted into the substrate on which the fungi are growing. Most of these substances have a high degree of animal toxicity. Feeding rabbits on mouldy diets (mixed toxin contamination) is responsible for many problems such as decreased feed intake, functional alteration of liver and genital tract, changes in blood constituents (Abdelhamid, 1990). Mycotoxicoses appear in chronic or acute forms. The acute form is caused by the rapid ingestion of large amounts of toxin over a short period.

Aflatoxins are naturally occurring toxins produced in grains and other feedstuffs both before and after harvest by toxigenic strains of the fungi *Aspergillus flavus* and *Aspergillus parasiticus*. Aflatoxin B_1 (AFB_1) is of primary concern since it is the most abundant and the most toxic. Acute or chronic aflatoxicosis may occur depending on dietary concentration of toxins. Rabbits are extremely sensitive to aflatoxins. The acute oral single-dose LD_{50} is about 0.3 mg kg^{-1} body weight (BW) (Newberne and Butler, 1969), among the lowest of any animal species. Moderate-to-severe mortality can be encountered with diets containing even low concentrations of toxin (<100 ppb) (Krishna *et al.*, 1991). Signs of toxicity included anorexia, weight loss, emaciation followed by icterus in terminal stages (Morisse *et al.*, 1981). Acute aflatoxin poisoning (AFB_1 daily doses > 0.04 mg kg^{-1} BW) causes prolonged blood-clotting time, extensive liver damage and death from liver failure (Clark *et al.*, 1980, 1982, 1986).

Zearalenone (F-2 toxin) is an oestrogenic substance frequently found in maize and other grains contaminated by *Fusarium graminearum* (Perez and Leuillet, 1986). Zearalenone causes hypertrophic development of the genital tract of the female rabbit (Pompa *et al.*, 1986; Abdelhamid *et al.*, 1992). This substance can also affect components of uterine tubal fluid known to be of critical importance during the early pre-implantation period (Osborn *et al.*, 1988). Levels of zearalenone in feed as low as 1–2 ppm can interfere with normal reproductive activity of rabbits when fed for only a few days (1–2 weeks). The high sensitivity of rabbits to this mycotoxin could be related to the slow hepatic transformation of zearalenone mainly into α-zearalenol, a more uterotrophic metabolite (Pompa *et al.*, 1986).

Another group of toxins produced by *Fusarium* species are the trichothecenes: T-2 toxin and vomitoxin.

T-2 toxin is produced by some strains of the fungus *Fusarium tricinctum*. It is relatively common in fibrous raw materials harvested or stored under poor conditions. In affected rabbits, T-2 causes marked feed refusal, lesions of the digestive tract and impairment of blood-clotting mechanisms (Gentry, 1982; Fekete *et al.*, 1989). Long-term (4–7 weeks) feeding of sublethal quantities of T-2 (0.19 ppm) altered ovarian activity of sexually mature female rabbits (Fekete and Huszenicza, 1993). Administration *per os* of 4 mg kg^{-1} BW of T-2 toxin causes death within 24 h (Vanyi *et al.*, 1989).

Vomitoxin (4-deoxynivalenol) may occur on cereal grains. Contamination of rabbit feeds with this toxin results in feed refusal. Adverse effects on fetal development were also encountered in does. Thus, Khera *et al.* (1986) observed that a level of 0.24 g vomitoxin kg^{-1} diet resulted in a fetal resorption incidence of 1.00.

The nephrotoxins (ochratoxin and citrinin) have also been implicated in rabbit mycotoxicoses.

Ochratoxin is produced by toxigenic strains of *Aspergillus ochraceus*. Galtier *et al.* (1977) examined the excretion of ochratoxin A in female rabbits after a single intravenous administration (1–4 mg kg^{-1} BW) and demonstrated the passage of the toxin into the milk: the level in milk reached 1 ppm for the highest dose of administration. The actual toxicity for rabbits is unknown but it should be pointed out that, in the above-mentioned experiment, the lactating does were able to tolerate a single dose of 4 mg kg^{-1} BW.

Citrinin is found in mouldy cereals contaminated by various fungal species of *Aspergillus* and *Penicillium*. Ingestion of this toxin induces acute erosive gastritis, fluid diarrhoea, with some rabbits dying less than 24 h after oral administration of a single 100–130 mg kg^{-1} BW oral dose (Hanika *et al.*, 1983). It also causes renal damage in the rabbit, with tubular dysfunction and necrosis similar to that found in other animal species (Hanika *et al.*, 1984).

Water quality and pathology

In most of the texts on animal nutrition, the sections devoted to water quality are limited to comments that water employed for animals' watering must be 'drinkable' and the recommended values given are those for human consumption without further discussion.

If these values are effectively obtained at the watering point available for the animals, there is effectively no health problem linked to water quality. Nevertheless bacterial and chemical composition of water does not always meet all of the recommended criteria.

Water polluted with bacteria can never be recommended for rabbits even if it is known that animals are more tolerant than humans. As very simple low-cost systems are available, the solution is disinfection. The classical criteria for bacterial quality of drinkable water are summarized in Table 11.2.

For minerals, removing the excess is technically possible in most cases, but the cost is very high, and the critical question relates to whether this is

Table 11.2. Recommended bacteriological status of drinkable water for human consumption.

Microorganisms	Maximum count
Salmonella sp.	0 in 5000 ml
Staphylococcus sp.	0 in 100 ml
Enteroviruses	0 in 10,000 ml
Faecal *Streptococcus* sp.	0 in 100 ml
Thermo-tolerant *E. coli*	0 in 100 ml
Clostridium sp.	1 in 20 ml

Table 11.3. Chemical composition of drinkable water for rabbits.

	Official recommendations for human consumption		Maximum experimented on rabbits without any trouble	
	Recommended maximum	Maximum tolerable	Value	Authors
Physical parameter (units)				
pH	7–8.5	6.5–9.2	3.5–9.0	Porter *et al.* (1988)
Chemical parameters (ppm)				
Total soluble salts	500	1500	3000	Abdel-Samee and El-Masry (1992)
Sodium	100	150	900	Ayyat *et al.* (1991)
Potassium	10	12	140	Ayyat *et al.* (1991)
Phosphorus	2	5		
Calcium	75	200	400	Porter *et al.* (1988)
Magnesium	30	150		
Iron	0.2	1.0		
Copper	0.1	1.5	60	Abo El-Ezz *et al.* (1996)
Manganese	0.05	0.5	12	Abdel-Samee and El-Masry (1992)
Zinc	5	15	55	Abdel-Samee and El-Masry (1992)
Aluminium	0.2	—	250	Rémois and Rouillière (1998)
Antimony	0.01	—		
Arsenic	0.05	0.20		
Cadmium	0.005	0.05		
Chromium	0.05	1.0		
Cobalt	—	1.0		
Fluoride	1.5	2.0		
Lead	0.05	0.10	0.40	Habeeb *et al.* (1998)
Mercury	0.001	0.01		
Nickel	0.05	1.00		
Selenium	0.01	—		
Silver	0.01	—		
Vanadium	—	0.10		
Chloride as Cl	250	600	1100	Habeeb *et al.* (1998)
Sulphate as SO_4	200	400	1340	Rémois and Rouillière (1998)
Nitrate as NO_3	45	50	600	Kammerer and Pinault (1989)
Nitrite as NO_2	0.05	0.10	11	Morisse *et al.* (1989)
Ammonium as NH_4	0.05	0.50		
H_2S	0.05	0.10		
Bicarbonate			400	Ayyat *et al.* (1991)
Nitrogen (N from NO_2 and NO_3 excluded)	2	—		
Cyanide as CN	0.05	—		

necessary for health in rabbits. Different experiments were conducted to establish the real tolerance of rabbits to mineral concentrations in drinking water, mainly in hot sub-tropical areas or in intensive animal production systems. The results are summarized in Table 11.3. Values are indicated for each mineral, but this does not mean that water with all criteria at the maximum would be accepted by rabbits.

It should be mentioned that, when known, the tolerance limits of rabbits vary considerably from the maximum 'officially' acceptable values for human consumption. One of the most significant is the tolerance of rabbits to high levels of nitrates or nitrites which is 10-fold the maximum accepted for human consumption. This is the subject of much debate in intensive animal production regions such as the Netherlands or Brittany in France. None of the maxima for rabbits are lower than those recommended for human consumption. This is why no specific chemical control is necessary if water provided for rabbits is from the same source as that for human consumption through a controlled public system. On the other hand, alteration of water quality by increasing the concentrations of some minerals can be illegal for human consumption, but is not necessarily injurious for rabbit health (Table 11.3).

References

Abdelhamid, A.M. (1990) Effect of feeding rabbits on naturally moulded and mycotoxin-contaminated diet. *Archives of Animal Nutrition* 40, 55–63.

Abdelhamid, A.M., Kelada, I.P., Ali, M.M. and El-Ayouty, S.A. (1992) Influence of zearalenone on some metabolic physiological and pathological aspects of female rabbits at two different ages. *Archives of Animal Nutrition* 42, 63–70.

Abd El-Rahim, M.I., El-Kerdawy Dawlat, A., El-Kerdawy, H.M. and Abdallah Fatma R. (1996) Bioavailability of iron in growing rabbits fed excess levels of dietary iron, under Egyptian conditions. In: Lebas, F. (ed.) *Proceedings of the 6th World Rabbit Congress*, Vol. 1. Association Française de Cuniculture, Lempdes, France, pp. 51–57.

Abdel-Samee, A.M. and El-Masry, K.A. (1992) Effect of drinking natural saline well water on some productive and reproductive performance of California and New-Zealand White rabbits maintained under north Sinai conditions. *Egyptian Journal of Rabbit Science* 2, 1–11.

Abo El-Ezz, Z.R., Salem, M.H., Hassan, G.A., El-Komy, A.G. and Abd El-Moula, E. (1996) Effect of different levels of copper sulphate supplementation on some physical traits of rabbits. In: Lebas, F. (ed.) *Proceedings of the 6th World Rabbit Congress*, Vol. 1. Association Française de Cuniculture, Lempdes, France, pp. 59–65.

Aguilar, J.C. Roca, T. Sanz, E. (1996) Fructo-oligo-saccharides in rabbit diet. Study of efficiency in suckling and fattening periods. In: Lebas, F. (ed.) *Proceedings of the 6th World Rabbit Congress*, Vol. 1. Association Française de Cuniculture, Lempdes, France, pp. 73–77.

210   F. Lebas et al.

Ayyat, M.S., Habeeb, A.A. and Bassuny, S.M. (1991) Effects of water salinity on growth performance, carcass traits and some physiological aspect of growing rabbits in summer season. *Egyptian Journal of Rabbit Science* 1, 21–34.

Bellier, R. (1994) Contrôle nutritionnel de l'activité fermentaire caecale chez le lapin. Thèse Doctorat, Institut National Polytechnique de Toulouse, France.

Bernardini, M., Dal Bosco, A., Castellini, C. and Miggiano, G. (1996) Dietary vitamin E supplement in rabbit: antioxidant capacity and meat quality. In: Lebas, F. (ed.) *Proceedings of the 6th World Rabbit Congress*, Vol. 3. Association Française de Cuniculture, Lempdes, France, pp. 137–140.

Blas, E., Cervera, C. and Fernandez Carmona, J. (1994) Effect of two diets with varied starch and fibre levels on the performances of 4–7 weeks old rabbits. *World Rabbit Science* 2, 117–121.

de Blas, J.C., Pérez, E., Fraga, M.J., Rodriguez, M. and Galvez J.F. (1981) Effect of diet on feed intake and growth of rabbits from weaning to slaughter at different ages and weights. *Journal of Animal Science* 52, 1225–1232.

de Blas, C., Taboada, E., Mateos, G.G., Nicodemus, N. and Méndez, J. (1995) Effect of substitution of starch for fiber and fat in isoenergetic diets on nutrient digestibility and reproductive performance of rabbits. *Journal of Animal Science* 73, 1131–1137.

Boriello, S.P. and Carman, R.J. (1983) Association of iota-like toxin and *Clostridium spiroforme* with both spontaneous and antibiotic-associated diarrhea and colitis in rabbits. *Journal of Clinical Microbiology* 17, 414–418.

Catala, J. and Bonnafous, R. (1979) Modifications de la microflore quantitative, de l'excrétion fécale et du transit intestinal chez le lapin, après ligature du canal pancréatique. *Annales de Zootechnie* 28, 128.

Cheeke, P.R. and Patton, N.M., (1980) Carbohydrate overload of the hindgut. A probable cause of enteritis. *Journal of Applied Rabbit Research* 3, 20–23.

Clark, J.D., Jain, A.V., Hatch, R.C. and Mahaffey, E.A. (1980) Experimentally induced chronic aflatoxicosis in rabbits. *American Journal of Veterinary Research* 41, 1841–1845.

Clark, J.D., Jain, A.V. and Hatch, R.C. (1982) Effects of various treatments on induced chronic aflatoxicosis in rabbits. *American Journal of Veterinary Research* 4, 106–110.

Clark, J.D., Greene, C.E., Calpin, J.P., Hatch, R.C. and Jain, A.V. (1986) Induced aflatoxicosis in rabbits: blood coagulation defects. *Toxicology and Applied Pharmacology* 86, 353–361.

Cortez, S., Brandeburger, H., Greuel, E. and Sundrum, A. (1992) Investigations of the relationships between feed and health status on the intestinal flora of rabbits. *Tierärztliche Umschau* 47, 544–549.

Davidson, J. and Spreadbury, D. (1975) Nutrition of the New Zealand White rabbit. *Proceedings of the Nutrition Society* 34, 75–83.

Fekete, S. and Huszenicza, G. (1993) Effect of T-2 toxin on ovarian activity and some metabolic variables of rabbits. *Laboratory Animal Science* 43, 646–649.

Fekete, S., Tamas, J., Vanyi, A., Glavits, R. and Bata, A. (1989) Effect of T-2 toxin on feed intake, digestion and pathology of rabbits. *Laboratory Animal Science* 39, 603–606.

Fernandez-Carmona, J., Cervera, C. and Blas, E. (1996) High fat for rabbits breeding does housed at 30°C. In: Lebas, F. (ed.) *Proceedings of the 6th World Rabbit*

Congress, Vol. 1. Association Française de Cuniculture, Lempdes, France, pp. 167–169.

Fioramonti, J., Sorraing, J.M., Licois, D. and Bueno, L. (1981) Intestinal motor and transit disturbances associated with experimental coccidiosis (*Eimeria magna*) in the rabbit. *Annales de Recherches Vétérinaires* 12, 413–420.

Fraga, M.J., Lorente, M., Carabaño, R. and de Blas, J.C. (1989) Effect of diet and remating interval on milk production and milk composition of the doe rabbit. *Animal Production* 48, 459–466.

Galtier, P., Baradat, C. and Alvinerie, M. (1977) Etude de l'élimination d'ochratoxine A par le lait chez la lapine. *Annales de la Nutrition et de l'Alimentation* 31, 911–918.

Gentry, P.A. (1982) The effect of administration of a single dose of T-2 toxin on blood coagulation in the rabbit. *Canadian Journal of Comparative Medicine* 46, 414–419.

Gidenne, T. (1995) Effect of fibre level reduction and gluco-oligosaccharides addition on the growth performance and caecal fermentation in the growing rabbit. *Animal Feed Science and Technology* 56, 253–263.

Gidenne, T. and Perez, J.M. (1994) Apports de lignines et alimentation du lapin en croissance. I. Conséquences sur la digestion et le transit. *Annales de Zootechnie* 43, 313–322.

Gidenne, T. and Perez J.M. (1996) Apports de cellulose dans l'alimentation du lapin en croissance. Conséquences sur la digestion et le transit. *Annales de Zootechnie* 45, 289–298.

Habeeb, A.A.M., Marai, F.M., El-Maghawry, A.M. and Gad, A.E. (1998) Physiological response of growing rabbit to different concentrations of salinity in drinking water under winter and hot summer conditions. *World Rabbit Science* (in press).

Haffar, A., Laval, A. and Guillou, J.P. (1988) Entérotoxémie à *Clostridium spiroforme* chez des lapins adultes. *Le Point Vétérinaire* 20, 99–102.

Hanika, C., Carlton, W.W. and Tuite, J. (1983) Citrinin mycotoxicosis in the rabbit. *Food and Chemical Toxicology* 21, 487–496.

Hanika, C., Carlton, W.W., Boon, G.D. and Tuite, J. (1984) Citrinin mycotoxicosis in the rabbit: clinicopathological alterations. *Food and Chemical Toxicology* 22, 999–1008.

Heckmann, F.W. (1972) Potato meal in complete feed of the fattening young rabbit. *Archiv für Geflügelkunde* 36, 182–185.

Hodgson, J. (1974) Diverticular disease. Possible correlation between low residue diet and raised intracolonic pressures in the rabbit model. *American Journal of Gastroenterology* 62, 116–123.

Jehl, N. and Gidenne, T. (1996) Replacement of starch by digestible fibre in the feed for the growing rabbit. 2) Consequences on microbial activity in the caecum and on incidence on digestive disorders. *Animal Feed Science and Technology* 61, 193–204.

Kammerer, M. and Pinault, L. (1998) Pollution de l'eau d'abreuvement par les nitrates. Tolérance générale et influence sur la croissance pondérale chez le lapin. In: *7èmes Journées de la Recherche Cunicole en France, Lyon.* ITAVI, Paris, pp. 191–194.

Khera, K.S., Whalen, C. and Angers, G. (1986) A teratology study on vomitoxin

(4-deoxynivalenol) in rabbits. *Food and Chemical Toxicology* 24, 421–424.

Kjaer, K.B. and Jensen, J.A. (1997) Perirenal fat, carcass conformation, gain and feed efficiency of growing rabbits as affected by dietary protein and energy content. *World Rabbit Science* 5, 93–97.

Koehl, P.F (1997) GTE Renalap 96: une lapine produit 118 kg de viande par an. *Cuniculture* 24, 247–252.

Krishna, L., Dawra, R.K., Vaid, J. and Gupta, V.K. (1991) An outbreak of aflatoxicosis in Angora rabbits. *Veterinary and Human Toxicology* 33, 159–161.

Lebas, F. (1973) Effet de la teneur en proteines de rations à base de soja ou de sésame sur la croissance du lapin. *Annales de Zootechnie* 22, 83–92.

Lebas, F. (1989) Besoins nutritionnels des lapins. Revue bibliographique et perspectives. *Cuni-Sciences* 5, 1–28.

Lebas, F. and Fortun-Lamothe, L. (1996) Effects of dietary energy level and origin (starch vs. oil) on performance of rabbit does and their litters: average situation after 4 weanlings. In: Lebas, F. (ed.) *Proceedings of the 6th World Rabbit Congress*, Vol. 1. Association Française Cuniculture, Lempdes, France, pp. 217–222.

Lebas, F. and Maitre, I. (1989) Alimentation de présevrage: étude d'un aliment riche en énergie et pauvre en proteines, résultats de 2 essais. *Cuniculture* 16, 135–140.

Lebas, F., Coudert, P., de Rochambeau, H. and Thébault, R.G. (1996) *Le Lapin: Elevage et Pathologie*. FAO, Rome, 227 pp.

Licois, D. Guillot, J.F. Mouline, C. Reynaud, A. (1992) Susceptibility of the rabbit to an enteropathogenic strain of *Escherichia coli* O-103: effect of animal's age. *Annales Recherches Vétérinaires* 23, 225–232.

Maertens, L. and Luzi, F. (1996) Effect of dietary protein dilution on the performance and N-excretion in growing rabbits. In: Lebas, F. (ed.) *Proceedings of the 6th World Rabbit Congress, Toulouse*, Vol. 1. Association Française de Cuniculture, Lempdes, France, pp. 237–242.

Morisse, J.P., Wyers, M. and Drouin, P. (1981) Aflatoxicose chronique chez le lapin. Essai de reproduction expérimentale. *Recueil de Médecine Vétérinaire de l'Ecole d'Alfort* 157, 363–368.

Morisse, J.P., Boilletot, E. and Maurice, R. (1985) Alimentation et modification du milieu intestinal chez le lapin (AGV, NH$_3$, pH, flore). *Recueil de Médecine Vétérinaire de l'Ecole d'Alport* 161, 443–449.

Morisse, J.P., Boilletot, E. and Maurice, R. (1989) Incidence des nitrites chez les lapins. *Cuniculture* 16, 197–199.

Newberne, P.M. and Butler, W.H. (1969) Acute and chronic effects of aflatoxin on the liver of domestic and laboratory animals: a review. *Cancer Research* 29, 236–250.

Osborn, R.G., Osweiler, G.D. and Foley, C.W. (1988) Effects of zearalenone on various components of rabbit uterine tubal fluid. *American Journal of Veterinary Research* 49, 1382–1386.

Padilha, M.T.S., Licois, D., Gidenne, T., Carré, B. and Fonty, G. (1995) Relationships between microflora and caecal fermentation in rabbits before and after weaning. *Reproduction, Nutrition and Development* 35, 375–386.

Peeters, J.E., Maertens, L. and Geeroms, R. (1992) Influence of galactooligosaccharides on zootechnical performances, caecal biochemistry and experimental collibacillosis O-103/8+ in weanling rabbits. *Journal of Applied Rabbit Research* 15, 1129–1136.

Perez, J.M. and Leuillet, M. (1986) Composition et valeur nutritive des céréales. *Perspectives Agricoles* 105, 56–61.

Perez, J.M., Gidenne, T., Lebas, F., Caudron, I., Arveux, P., Bourdillon, A., Duperray, J. and Messager, B. (1994) Apports de lignines et alimentation du lapin en croissance. II. Conséquences sur les performances de croissance et la mortalité. *Annales de Zootechnie* 43, 323–332.

Perez, J.M., Gidenne, T., Bouvarel, I., Caudron, I., Arveux, P., Bourdillon, A., Briens, C., Le Naour, J., Messager, B. and Mirabito, L. (1996) Apports de cellulose dans l'alimentation du lapin en croissance. II. Conséquences sur les performances et la mortalité. *Annales de Zootechnie* 45, 299–309.

Pompa, G., Montesissa, C., Di Lauro, F.M. and Fadini, L. (1986) The metabolism of zearalenone in subcellular fractions from rabbit and hen hepatocytes and its estrogenic activity in rabbits. *Toxicology* 42, 69–75.

Porter, L.P., Borgman, R.F. and Lightsey, S.F. (1988) Effect of water hardness upon lipid and mineral metabolism in rabbits. *Nutrition Research* 8, 31–45.

Prohaszka, L. (1980) Antibacterial effect of volatile fatty acids in enteric *E. coli* infections of rabbits. *Zentralblatt Veterinar Medecine* B 27, 631–639.

Rémois, G. and Rouillère, H. (1998) Effet du sulfate d'aluminium sur les performances des lapins d'engraissement. In: *7ème Journées de la Recherche Cunicole en France, Lyon.* ITAVI, Paris, pp. 195–197.

Vanyi, A., Glavits, R., Bata, A., Fekete, S. and Tamas, J (1989) The pathological effects, metabolism and excretion of T-2 toxin in rabbits. *Journal of Applied Rabbit Research* 12, 194–200.

12. Feed Manufacturing

J. Mendez,[1] E. Rial[1] and G. Santomá[2]

*[1]Cooperativas Orensanas Sociedad Cooperativa Ltda, Juan XXIII 33,
32003 Orense, Spain; [2]Agrovic, C/Mejia Lequerica 22–24, 08028
Barcelona, Spain*

Introduction

Compound feed processing consists of the treatment, combination and
mixture of different raw materials which will satisfy the nutrient
requirements of the animals except for water.

This process starts with the selection of those raw materials available on
the market which are appropriate for use in rabbit feeding. It is necessary to
control these raw materials, to know their nutritive value and to check their
freshness, as well as their organoleptic characteristics through a quality
control system. Next, the manufacturing process begins, with particle size
reduction provided by grinding (Heidenrweich, 1992), followed by mixing
and homogenization of the raw materials included in the diet which, in some
cases, results in the manufactured compound feed. However, in rabbits, as
will be discussed, feed as meal is not suitable, so meals are pressed into
pellets, as a means of supplying the feed to the animals. Finally, quality
control on the manufactured feed must be undertaken.

Grinding

The raw materials available for feed production vary greatly according to the
original texture and particle size; e.g. cereals are obtained as whole grains,
without any previous treatment, while lucerne and cereal straw are acquired
as pellets after being ground at the processing plants.

The main functions of grinding are:

- to reduce particle size in order to increase nutrient digestibility in the
  gastro-intestinal tract of the rabbit;
- to obtain an appropriate particle size which allows a good mixture of the
  raw materials and favours the pelleting process.

According to the mill design, there are two possible approaches to grinding raw materials: post-grinding and pre-grinding (Fig. 12.1). With post-grinding, the raw materials arrive at the grinder together, whilst in pre-grinding, each ingredient is ground individually. Both systems have advantages and disadvantages.

Pre-grinding

Advantages

- Particle size distribution for each ingredient can be modified by changing the grinder sieves.
- Maximum grinding capacity is easier because a homogeneous product is ground, so grinding yield is optimized.
- The mixing plant does not depend upon the grinding plant, so grinding and mixing can operate separately, while with the post-grinding system this is not possible.

Disadvantages

- More silo bins are required in the mill, because the same raw material requires at least two bins, one for reception and another for the ground raw material.
- Risk of separation because of the different particle size distribution of the individual raw materials.
- Materials which are difficult to grind (e.g. full-fat sunflower) cannot be ground individually.
- Materials with a high percentage of hulls, like oats or barley, are ground more efficiently together with other ingredients, such as maize or oilseed meals.
- The storage life of ground products is shorter than that for whole products.

In post-grinding mills the advantages of pre-grinding are now disadvantages and vice versa. Therefore, if it is very important to modify the particle size distribution of each ingredient, it is advisable to use a pre-grinding system; if not, it is better to use a post-grinding system.

There are several types of grinders, but currently the most commonly used one is the hammer mill grinder which, according to the shaft layout, can be vertical or horizontal. The main advantages of the vertical hammer mill compared with the horizontal are: (i) it requires less space, (ii) it consumes less energy per unit of ground weight, (iii) it gives a lower range of particle sizes, and (iv) it has a lower maintenance cost.

In order to use a hammer mill grinder effectively, it is necessary to include an automated system which controls the feeding of this grinder according to its energy consumption. The feeding device should also include one separator of heavy particles and metals. The utilization of a sieve before

(a)

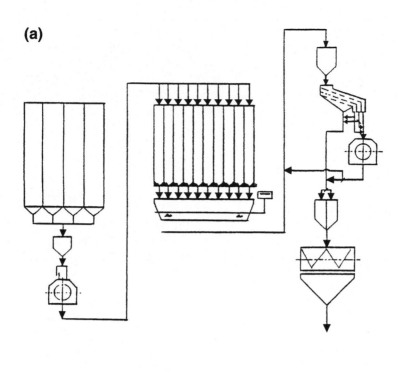

(b)

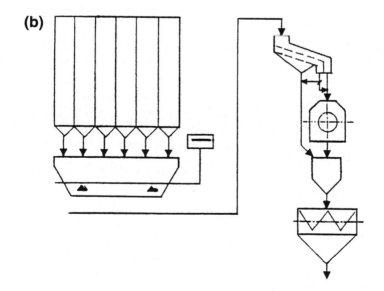

Fig. 12.1. Diagrams of pre-grinding (a) and post-grinding (b) feed mills.

the grinder also facilitates the grinding process and saves energy (because the particles which are already small bypass the grinder), it increases the productive life of the grinder (because materials such as minerals, which are very abrasive, can also bypass the grinder) and it can increase the efficiency of grinding by decreasing the plugging problems at the decompression hoses. A proper air intake will improve the grinding yield and will reduce the losses due to moisture reduction.

Particle size

Between certain limits, particle size reduction favours pelleting and consequently pellet quality, but the grinding yield decreases. From a physiological point of view, excessive grinding can increase the retention time of the digesta in the intestine (Lang, 1981), which leads to an increase in nutrient digestibility. In this way García *et al.* (1996) have found that neutral detergent fibre digestibility in rabbits is positively correlated with the percentage of particles with a size below 0.315 mm. However, an increase in retention time of digesta in the gut is apparently associated with digestive disturbances which can predispose to diarrhoea (Lebas and Laplace, 1975; Laplace and Lebas, 1977), because this increase in retention time occurs mainly at the caecum, which enlarges and undesirable fermentation patterns take place. In this way, irregularities of the ileocaecal valve motility with very fine grinding (1 mm sieve) in comparison with coarser grinding (4 mm sieve) have been reported (Pairet *et al.*, 1986).

In practice it is recommended to use sieves with a diameter between 2.5 and 4.0 mm (Mateos and Rial, 1989), because they permit a good balance between pellet quality and good intestinal motility. It is convenient, anyway, to run periodic controls on the particle size distribution to ensure that it is as required.

It has been suggested, without any strong experimental evidence, that two kinds of grinding might be useful: a fine one for low fibrous ingredients (e.g. cereals, soybean meal), and a coarse one for fibrous ingredients (e.g. lucerne, straw). It is thought that the former materials, when finely ground, have a higher digestibility, while the latter ones, when coarsely ground, could

Table 12.1. Particle size and grinding throughput vary according to some parameters.

		Throughput	Particle size
Hammer tip speed	High	Decrease	Finer
	Low	Increase	Coarser
Number of hammers	Small	Increase	Coarser
	Large	Decrease	Finer
Sieve hole diameter	Small	Decrease	Finer
	Large	Increase	Coarser

have a mechanical function on intestinal motility (Mateos and Rial, 1989).

Particle size and grinding throughput vary according to some parameters (see Table 12.1). Therefore, there is not a linear relationship between sieve hole diameter and particle size because, apart from the factors expressed in Table 12.1 (hammer tip speed and number of hammers), it also depends on the raw material. An identical technical condition of a hammer mill will give different particle size distributions depending on the raw material being ground.

In order to know accurately the particle size distribution, a group of different sieves is used, which classifies the ground material accordingly.

As an indication of a particle size distribution for rabbit feeds, the following table provides a general guideline:

Particle size (mm)	Proportion
> 1.5	0.15
1.0–1.5	0.20
0.5–1.0	0.40
< 0.5	0.25

Weighing

According to the kind of feed mill, weighing can take place before or after the grinding process. It should be carried out using precise scales with less than 0.5% error to prevent wide variations in the characteristics of the final product.

Within a feed mill, scales of different weighing ranges should be available according to the amounts to be weighed, as well as methods adapted to these amounts. For raw materials which are present in large amounts in the diet, such as lucerne, wheat bran, etc., the machinery and corresponding scale should allow weighing in a short period of time with a final period for weight adjustment. On the other hand, the weighing of other raw materials which enter the formulation at lower rates, such as limestone or dicalcium phosphate when they are stored as bulk in silo bins, should use more precise systems and scales in order to avoid oversupply.

Weighing equipment must include safety elements to halt the process, such that whenever a lack or an excess occurs during the weighing the mistake may be corrected. It is also important to have a printed register of all the weights determined.

Sometimes, raw materials which normally come into formulations at a high rate, for cost reasons must, enter at a low rate. In these circumstances there is a rapid weighing of amounts which are lower than 5 or 10 kg t^{-1}, so it is very easy to have large weight variations. To prevent this kind of error, the feed formulation software normally includes the option of fixing the minimum amount of a raw material that can be added in such a way that, if the amount is lower than this figure, the formulation is adjusted to exclude this ingredient.

Vitamins, minerals, amino acids, coccidiostats and other feed additives should be weighed in small scales. It is advisable to prepare a premix with these products before adding them to the main mixer. Under European Community law it is compulsory to add products to the main mixer at a concentration higher than 2 kg t^{-1}. If a product is to be included at less than 2 kg t^{-1}, it is necessary to make a premix in order to arrive at this concentration for addition.

Mixing

The main purpose of the mixing process is to homogenize the different raw materials included in the feed. The mixer is the central machine at a feed mill, because its function is to mix as uniformly as possible particles of different size and different density. However, this equipment normally creates a few problems and its importance can be overlooked.

The mixer is the machine which really produces the feed, and therefore it is very important to be confident of its quality and of its supplier. It is essential to realize that raw materials which represent up to 400–500 kg t^{-1} of the feed are mixed with microingredients which, in extreme cases, as with biotin, vitamin B_{12} or selenium, are included at levels of 10–100 mg t^{-1} and, in the case of selenium, toxic levels are close to the inclusion levels.

In order to check that the mixer is functioning properly and therefore that the mixture is uniform, homogeneity tests must be carried out. Such a test consists of taking 10–20 samples from the mixer itself, or at the outlet, at regular intervals of time. These samples are ground and divided in the laboratory into 10–20 g samples. These samples are analysed either for a specific chemical component of the feed (e.g. salt, manganese) or for a microtracer component that is specially added for this purpose. This latter component is a special indicator which is added in the feed at a concentration of around 10 g t^{-1} and analysis for it is conducted with each one of the samples.

A good quality mixture is one in which these analyses discover a coefficient of variation of less than 5% (Bülher, 1996), although this figure depends on the quality standard that the company establishes as its objective.

In the case of microingredients, it is necessary to use products with a minimum number of particles per gram or a maximum particle size in order to ensure homogeneous distribution in the feed. In this way, according to Beumer (1991), when an additive is added at 50 g t^{-1}, and 20 g samples are collected, the coefficient of variation falls from 5.2 to 1% and the number of particles in 20 g increases from 377 to 10,186 when the particle size changes from 150 µg to 50 µg.

According to Rial et al. (1993) the main factors which affect the quality of the mixture are:

1. Mixture time, which depends on the kind of mixer and on the ingredients to be mixed. The standard mixture time is around 4 min. A longer time does not mean a better mixture, and sometimes it is even detrimental to the quality of the mixture. The optimum mixture time can be determined through homogeneity tests.

2. Particle size distribution chart. Too small or too large a particle size negatively affects the quality of the mixture.

3. Density and particle shape. The heaviest particles tend to go to the bottom and the round particles have a better flowability.

4. The degree to which the mixer is full. Depending on the average density of the raw materials to be mixed, the weight of material in the mixer is variable. The optimum level is that which allows a horizontal profile of the upper level of the complete mixture. If there is too much or too little material in the mixture, the product accumulates at one extreme or the other which leads to a poor mixture quality.

5. Liquid addition promotes adherence; therefore it is important to establish a regular mixer cleaning programme.

Other considerations to be taken into account to ensure a high quality mixture include:

1. Before any liquid is injected into the mixer, at least 15 s should elapse from the beginning of the process, in order to ensure that the liquid is added to a more homogeneous environment.

2. To facilitate distribution in the mixture, liquids must be injected at several points (at least three).

3. When the fat level in the diet is over 20–30 kg t^{-1}, it is advisable to use mechanical devices to add the extra fat after pelleting, because high fat presence in the mixer impairs pellet quality.

The main mixer as well as all the previous mixers (if they exist at the feed mill) are places prone to cross-contamination if precautions are not taken. These cross-contaminations are particularly important in the case of some feed additives and some pharmacological products, not only because of possible toxicity for the rabbit, but in terms of human health, due to the residues that can accumulate in the rabbit carcass. To prevent these situations, there should be an independent manufacturing line for medicated feeds. However, in medium–small feed mills this is rarely possible, and the only way to minimize dangerous cross-contamination is to manufacture feeds for animals which are not susceptible to the medicines or additives which have been included in previous manufactures, and for animals whose products are not the final products to be consumed by humans (e.g. feeds for breeder animals).

Rabbits are especially sensitive to certain pharmacological medicines such as ampicillin or lincomycin, so special measures must be taken when organizing manufacturing sequences which include these antibiotics. Today, it is possible to prevent these situations through the use of software which includes all the incompatibilities to be considered, before manufacturing.

Molasses addition

Apart from the mixer, where small amounts of molasses can be added (20–30 kg t⁻¹), the addition of molasses is undertaken in a machine whose design is similar to that of a conventional conditioner, and it is placed before or after the mixer. Other liquids can also be added at this point but the final homogeneity is not as good as if they were added in the mixer.

The inclusion of molasses must be controlled automatically because this is a continuous process and the meal flow through the machine determines the amount of molasses to be added. The longer the mixture remains in the machine and the more regular the addition of the product, the better the quality of the mixture.

Because of its high level of sugars, molasses can be caramelized to a solid state. This process occurs when temperatures increase beyond 50–60°C, therefore these temperatures must not be exceeded when handling molasses.

Pelleting

The purpose of pelleting is to turn meals into compact pellets, normally of cylindrical shape. This process consists initially of meal conditioning by mixing it with steam in a conditioner and, subsequently, this conditioned product is pressed by rolls to pass through the holes of the pelleting die which gives the meal the pellet shape.

When pelleting, different parameters associated with the meal, with the pellet mill or with the process itself must be considered. All of them affect pelleting and therefore they affect pellet quality to an extent.

The first factor associated with the meal to be taken into account is the particle size at the pellet mill. Generally speaking, a large particle size (over 1.5 mm) hinders pelleting. On the other hand, as has been mentioned, a very small particle size can promote digestive problems. As a consequence, a compromise between both extremes must be assumed.

According to the physico-chemical characteristics of the raw materials which enter formulations, the pelleting capacity of the feed will vary. As a rule of thumb, ingredients with a high level of fat will promote an inferior pellet quality, because fat acts as a lubricant and decreases the pelleting resistance. Chemically speaking, the concept of fibre is ambiguous, and thus the effects on pellet quality will be variable depending upon the kind of fibre used. If it is a lignified fibre, it will tend to impair pellet quality, although if the fibre contains a high cellulose level, the trend is towards a better pellet quality because it is less rigid.

Starch gelatinizes at around 60°C, and this process favours pelleting. However, starch sources vary in their pelleting ability; among the cereals, wheat gives the best pellet quality and maize gives the worst, with barley intermediate. Because of their high fibre content, oats are abrasive for the die. Manioc can contain significant amounts of silica which is also very abrasive for the die and shortens its operating life. Although the fat content of soybean meal is normally low, variation can affect pellet quality. This factor is more important with full fat seeds such as soybean, sunflower and other oilseeds, because, as described, fat acts as a lubricant which impairs pellet quality. Among fibrous sources, straw promotes a low quality pellet, while lucerne and beet pulp favour high quality. At low levels of inclusion and at high temperatures, molasses improves pellet quality through sugar caramelization. Minerals are abrasive, especially limestone, therefore small particle size limestone must be used (see Table 12.2).

As far as the influence of raw materials on pellet quality is concerned, it must be remembered that feed mills today have a larger range of raw materials

Table 12.2. Pelleting properties of different raw materials (Bühler, 1996).

Raw material	Quality[a]	Capacity[a]	Abrasiveness[a]
Wheat	7	6	4
Maize	5	7	4
Barley	6	7	5
Oats	3	4	7
Full fat soybean	3	8	4
Soybean meal	6	6	5
Sunflower meal	6	5	5
Rapeseed meal	6	6	6
Palm kernel meal	6	4	6
Coconut meal	6	6	6
Peanut meal	8	6	5
Cottonseed meal	6	6	5
Carob meal	5	2	9
Des. dried grains and solubles	3	4	5
Brewers' by-products	7	6	5
Meat and bone meal	4	5	7
Citrus pulp	7	4	6
Beet pulp	8	3	6
Lucerne meal	7	3	7
Wheat bran	6	4	3
Minerals	2	4	10

[a]1 = low, 10 = high.

than some years ago. Therefore mills must update technologically in order to be able to make pellets with a wide range of ingredients. The pelleting process acounts for around 0.02–0.03 of the total feed cost, but the possibility of using a bigger or a narrower range of raw materials can imply a much higher cost.

Another factor which influences this process is meal density. Because pelleting is a compaction process, it can be assumed that the lower the density the greater the difficulties with pellet manufacture.

Conditioning previous to pelleting is very important to obtain a good quality pellet. Steam added to the meal will result in a partial gelatinization of starch, it will make the fibre less rigid and, generally speaking, it will prepare the meal for facilitation of pelleting.

Currently there are pellet mills on the market with double conditioners, which permit, on one hand, a closer contact between meal and steam for a longer period of time and, on the other hand, a better thermal treatment, both of which lead to better conditioning. The chamber where steam is injected is externally heated to avoid condensation and to permit a greater amount of steam to be added. In rabbit feeds this conditioning is very important because feeds with a very low amount of fines must be produced. Classical technology can produce a pellet with a very low proportion of fines, but with an excessive pellet hardness. Conversely, thanks to double conditioning, a less rigid pellet can be obtained which is more difficult to break but less hard, with a similar level of fines as with single conditioning. Generally, any good conditioning system will improve pellet quality.

Probably the most important factor associated with pelleting is the die, because of the large number of variables associated with the number and positioning of the holes, and overall die diameter and length. These last two parameters and their relationship will give the pellet mill compression capacity. For a given die length, the wider the hole, the lower the pressure and the lower the compaction. Equally, the longer the hole, the longer the time the meal will take to leave and the bigger the compaction. The energy consumed by the pellet mill is directly related to compaction capacity and therefore it is related to die length and hole diameter.

The rollers also greatly affect yield and compaction. It is generally advisable to use a set of rollers with the same die because they fit better. The space between the rollers and the die must be between very narrow limits, around 0.2 mm in order to reach a maximum pellet mill yield. If the gap is too wide the pressure on the meal decreases, the product remains longer in the hole, and therefore the yield decreases and the pellets can burn.

Another topic to consider is fines falling through the sieve that is generally positioned after the cooler. This material does not have the same nutrient and physical composition as the rest of the meal, generally it has a small particle size and its flow rate to the pellet mill is irregular, which makes the pelleting process more difficult. The amount of fines which returns to the pellet mill must be controlled because, when high, energy wastage is increased on repelleting the same material several times. The cause of this high level of fines must be found and, if possible, corrected.

When the hot pellet comes out from the pellet mill, before it cools and dries, it has a high humidity and is very fragile and deteriorates very easily. It is therefore very important to prevent as much as possible pellets having to drop significant distances and violent impacts in the tubes or in the subsequent equipment, otherwise fines will increase.

As shown, there are many factors which affect pellet quality, but most of them operate before pelleting, therefore once pelleting has started, the only possible means of changing pellet quality are:

1. If the meal allows it, the amount of steam can be increased, as long as the critical temperatures of the additives used are not surpassed.
2. To adjust the distance between the roller and the die.
3. To decrease speed of the pellet mill or production to increase the retention time in the die.

These three measures lead to an increase in pellet hardness. If there is a problem of excessive hardness, they must be applied in the opposite way.

When feeds with low pelleting ability are used, products which favour the pelleting process, known as *binders*, can be used. There is a large variety of products available in the market, but the most frequently used are ligno-sulphonates, vegetable gums, bentonites and sepiolites.

When poor pellets are obtained, an analysis of the place or places where the defect occurs must be undertaken, by examining the whole milling process, grinding, steam, pelleting, by taking samples at the outlet of pellet mill, at the cooler, at the sieve, at the unloading silo bin and even at the unloading of batch lorries because at this point, production of significant quantities of fines can take place.

Several defects are detectable by examining the shape of the pellet (Fig. 12.2):

A   Curved and fissured pellet: possibly poorly regulated cutters. The pellet length is three to four times its diameter and breaks by 'turning over'.

B   Fir-shaped pellet: this generally happens with high-fibre diets. It can be promoted by poor grinding (check the grinding sieves), by a lack of moisture or by too short a die length.

C   Pellet with a longitudinal fissure: poorly homogenized feed or high die speed. Check the meal entry at the pellet mill inlet. Increase liquid addition in the mixer.

D   Pellet with coarse particles: can be due to too coarse grinding or a broken sieve. Check the sieves frequently. Place a magnet and a sieve before the grinder.

E   Misshapen pellet with fissures: due to too coarse grinding. Grind finer. Aspiration throughout the sieve. Check the filters. Feed the grinder across the whole width of the sieve.

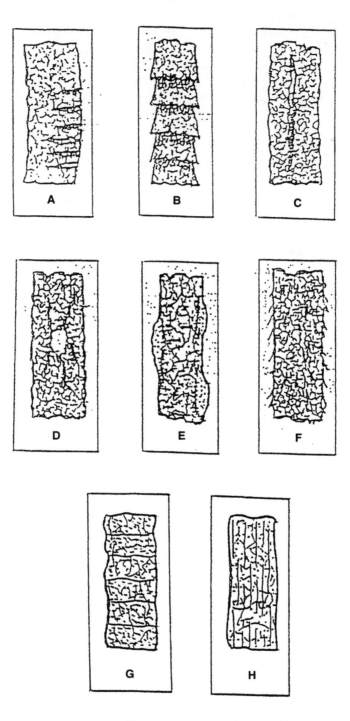

Fig. 12.2. Different types of defective pellets (McMahon and Payne, 1991).

F   Hairy pellet: an excess of steam, or temperature, or coarse particles. Reduce steam pressure and check meal particle size.
G   Pill-shaped pellet: too long a die length or a steam deficiency can be the cause. Add more fat in the mixer or reduce the die length.
H   Seamed pellet: die hole deformation. Change the die.

The pellet leaves the pellet mill at a high temperature and with an excess of water (steam) which must be reduced by passing through the cooler. There are three types of coolers: vertical, horizontal and countercurrent. The pellet enters the cooler at 60–90°C and with a moisture content of 140–180 g kg^{-1}. At the cooler outlet, the moisture must be less than 140 g kg^{-1} to facilitate preservation. The acceptable temperature will depend on the ambient temperature; as a rule, the difference between the pellet and the ambient temperature should not be higher than 5–7°C.

Other processing

Conditioning

This technique is based on long-term contact (around 20 min) between the meal, the water (steam) and the added liquids at a temperature of approximately 40°C. The objective is to obtain a greater liquid–meal interaction and to facilitate the pelleting process when a large amount of liquid is added. This addition is achieved in a conditioner placed before the ripener.

The conditioner consists of a vertical tank with one or several levels and a vertical axle to stir the product. In the feed mill it is placed immediately before the pellet mill. This equipment is very bulky and therefore requires space for its installation, furthermore, it is necessary to calculate a sufficient flow to prevent it becoming the limiting factor in overall pelleting yield.

Extrusion

This technique has been adapted from the food industry and nowadays is used in the feed industry, basically for the denaturation of the antinutritive factors (ANF) of legumes and also in pet food manufacture. This process implies a high energy consumption, so its cost is too high for the treatment of products with a limited gross margin such as rabbit feeds.

The extruder contains one or two axles which subject the meal to high pressures against the end of the machine where a fixed die, which shapes the feed, is placed.

The two basic nutritional improvements are:

1. Protein denaturation, without affecting amino acids. With this process, proteins partially lose their tertiary and quaternary structure, but amino acid availability is not modified if the process is run properly. The immediate consequence is a deactivation of the thermolabile ANF with proteinaceous composition. In addition some pathogenic microorganisms such as *Salmonella* sp. are eliminated.

2. Starch gelatinization, which makes this important compound more easily available, especially to young animals with an immature endogeneous enzyme production.

Because rabbit feeds have a low starch content and, normally, protein sources such as soybean meal have been thermally pretreated, extrusion does not appear to offer benefits for feeds for rabbits. On the other hand, the effects of the structural changes occurring during the extrusion process on the fermentation pattern in the rabbit hindgut have not been extensively studied.

Whenever a treatment implies a high temperature, the thermal stability of the vitamins and other additives added to the feed must be known. As shown in Table 12.3, extrusion is rather severe, so the addition of vitamins to the feed must be increased, especially of those which are more thermolabile. The expansion process, which will be described below, has an intermediate effect on vitamins, and pelleting has a low effect, except in relation to ascorbic acid, although this vitamin is not normally added to rabbit feeds. Vitamin K destruction will depend on the source employed, so either the most stable form is used (MNB) or the levels are increased to reach the desired levels in the final feed. Vitamin stability is also affected by other factors such as the presence of choline chloride in the vitamin–mineral premix. Vitamin K is

Table 12.3. Average vitamin stability according to the technical process (Coelho, 1996).

Vitamin	Pelleting 71–75°C	Extrusion 141–145°C	Expanding 111–115°C
A	87	61	85
D	93	78	84
E	96	88	95
K (MNB)	84	53	80
K (MSB)	66	28	43
B_1	91	70	86
B_2	94	80	88
B_6	93	84	91
B_{12}	98	91	96
Pantothenate	94	85	92
Folic acid	94	76	91
Biotin	94	76	91
Niacin	95	76	89
Ascorbic acid	60	25	50
Choline chloride	99	96	98

MNB, menadione nicotinamide bisulphite; MSB, menadione sodium bisulphite.

especially susceptible to choline chloride and, in this way, Coelho (1996) determined that vitamin K content decreased to 47% of the original activity after 2 months of storage when choline chloride was included in the premix, whilst the activity remained at 98% of the original value when choline chloride was excluded from the premix. This effect will vary according to the choline chloride level in the premix, as well as the amount of carrier in the overall premix. However, it is possible currently to add choline chloride as liquid in the mixer and, even for medium feed mills, this is a profitable investment.

Expansion

The expander is a thick-walled mixer tube with an axle supporting structures for mixing and kneading. The tube has internal bolts and steam injector valves. The pressure is mantained thanks to the final screw which modifies the annular gap at the end of the tube. The expander must be installed before the pellet mill with a bypass circuit to allow the double option of pelleting the product after the expander or to carry the product directly to the cooler. Energy consumption is lower than for the extrusion process, but higher than for the pelleting process alone, although this higher energy consumption is partially compensated for through a greater throughput of the pellet mill after expanding the meal. Expansion also subjects the meal to high temperature and pressure (Fig. 12.3) which greatly facilitates the subsequent pelleting, even when including raw materials with a low pelleting ability, and high amounts of liquids can be added (e.g. fats).

Expansion is not so severe a process as extrusion, so the benefits and disadvantages of expansion are also more limited than for extrusion. Protein denaturation is insignificant, so there is not a clear effect on the thermolabile ANF of legumes. In the same way, starch is only partially gelatinized. One of the advantages of expansion is the better bacteriological quality of the feed (Pipa and Frank, 1989).

Double pelleting

As the name implies, with double pelleting the meal goes through two pellet mills. The first one acts only as a conditioner, with a low compression, and the second one, which is the main pellet mill, receives the meal under optimal conditions to produce a high-quality pellet (Shultz, 1990).

This equipment also allows the inclusion in the formulation of raw materials with a low pelleting ability, as well as the inclusion of higher amounts of liquids. Double pelleting is recommended when a high-quality pellet is required, as happens with rabbits. The extra energy cost is around 15–20% in relation to normal pelleting.

As far as nutrient characteristics and feed presentation are concerned,

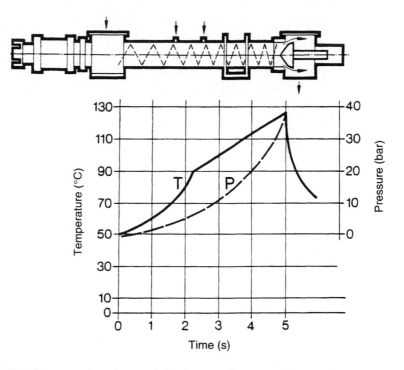

Fig. 12.3. Pressure and temperature during the expanding process (Pipa and Frank, 1989).

double pelleting has the same advantages and disadvantages as single pelleting, because the thermal treatment applied to the meal is basically the same in both cases.

Pellet quality

It is very important to supply feed for rabbits as pellets with a very small amount of fines. Rabbits are exceptionally sensitive to meals or fines in the feeder, because of their particular way of consuming feed; when fines are present they can enter the respiratory system, promoting respiratory problems. To resolve this situation, feeders with small holes at the bottom are used which allow the separation of fines, but unfortunately this implies a feed loss, because the fines fall into the faeces pit. This loss will increase the feed conversion ratio proportionally to the fines content of the feed.

This is the reason why a high durability pellet must be produced at the feed mill, by increasing the die compression capacity. However, if pellet hardness is too high rabbits will refuse the feed, especially rabbit pups.

There are different methods for measuring pellet durability and pellet hardness. Hardness is defined as the pellet resistance to pressure breakage.

This parameter is measured by means of a spring device (hardmeter) where the pellet is placed and, through a screw, the pressure is gradually increased until breakage occurs. The pressure resisted by the pellet is recorded. This parameter estimates pellet hardness but not durability. Thus a very hard pellet but with a low elasticity can be very fragile and therefore it can produce a high amount of fines.

To determine pellet durability there is a simple mechanism developed originally by Professor H.B. Pfost at Kansas State University and currently found in many feed mills with minor adjustments (Schultz, 1990). The equipment consists of a standard box which turns at a rate of 50 revolutions per minute over 10 min, with a specific amount of feed. After that time the pellets and the fines obtained are weighed. As a general rule, pellet durability for rabbit feed must not be below 0.97–0.98. There are also devices such as the Holmen pellet tester, which can take on-line measurements when the pellets are produced and, through this information, corrections can be made for one or several parameters. These automated devices are controversial because their reliability improves the more experienced the person undertaking the analysis is (Schultz, 1990).

Liquid additives

The objectives of liquid addition at a feed mill are many, i.e. provision of:

- an energy supply (animal and vegetable fats);
- a sugar supply (molasses);
- amino acids (methionine, lysine);
- vitamins (choline);
- enzymes;
- fungicides;
- flavours.

Interest in liquid addition is increasing, not only for economic reasons but also in terms of feed mill convenience, as the inclusion of these products can be automated, where otherwise they would be added by hand.

Because of its lubricant effect, it is convenient to distribute the fat supply at several places, when the inclusion levels are over 20–40 kg t^{-1}. Generally, fat is added initially in the mixer. A second addition point can be at the pellet mill outlet to take advantage of the fact that the pellet is still hot and its absorption capacity is high. At this level, 20 kg t^{-1} fat can be added. Another possible fat addition stage is after the pellet has cooled, but in this case a special mechanism is necessary, based on a specific drum where the fat is sprayed on to the pellet. At this stage the fat absorption capacity of the pellet is low because it is already cool; in order to increase this capacity the equipment can be warmed slightly. Pellet thickness is directly related to its absorption capacity: a 3 mm pellet shows a specific surface which is double that of a 6 mm pellet, and its absorption capacity is 50–100% higher (Walter, 1990).

Molasses can be added at the mixer, but it is more effective to add it in the specific conditioner for this purpose. Data for the liquid absorption characteristics of several raw materials are shown in Table 12.4.

Thermoresistant liquids, which are added in small amounts but which are very important qualitatively speaking, such as amino acids or choline, must be added in the mixer. Because of its negative effect on other vitamins, whenever possible it is appropriate to add choline chloride as a liquid at this point.

When liquid flavours are added, the ideal stage is after pelleting, in order to maintain their aromatic profiles. However, the flavour only covers the pellet surface, not the pellet interior, making an irregular taste across the pellet as a whole. The implications of this situation are not well known but, in practical terms, the most convenient approach would be to add a small amount of flavouring in the mixer and the rest after pelleting.

An important benefit for the poultry industry has been the introduction of

Table 12.4. Liquid absorption capacity of feedstuffs (Walter, 1990).

Raw material	Proportion of reflectance
Fish meal	0
Field beans	0.016
Rye	0.080
Raw rice bran	0.086
Barley	0.173
Hi pro soya meal	0.177
Whole rape	0.202
Wheat	0.220
Maize distillers	0.240
Rape/fat protein	0.584
Cotton (expellers)	0.610
Corn gluten	0.610
Shea nut	0.638
Sugar beet pulp pellets	0.659
Extracted rape	0.674
Meal from balanced feeds	0.700
Rape meal	0.706
Lucerne pellets	0.714
Wheatmeal pellets	0.763
Soya hulls	0.772
NIS pellets	0.805
Cocoa husk	0.822
Rice bran	0.822
Wheat meal	0.926

Source: United Molasses.

enzymes designed to improve the digestibility of diets including barley or wheat. These benefits are not so clear in rabbits, but it is possible that, in the near future, new enzyme activities will be developed for this species. Enzymes can be added as powder or as liquids. With powder, there is uncertainty about the stability of the product after thermal treatment (pelleting, expansion). To avoid this problem, apart from technological improvements in the thermal stability of these additives, it is possible to add them as liquids at the cooler outlet. Commercial equipment is already available for this.

The liquid dosage is normally run through volumetric devices, so some measures must be considered: (i) to determine periodically the precise product density because, when formulating, only weights are used, never volumes; (ii) to control the amounts used, by weighing those amounts added; (iii) to have available the proper equipment for measuring flow rate.

Feed presentation

One of the classical topics of rabbit nutrition is the importance of a good quality pellet, because of the aversion of rabbits to fines (Lebas, 1975), such that, if a very fine feed is provided, rabbits almost completely stop eating within 2–3 days. This is why discussions on methods of feed presentation to the rabbit are very uncommon; it is accepted worldwide that food must be presented as pellets, which are the only physical presentation form currently used commercially in industrial rabbit production.

When considering small-scale rabbit farms with no feed-buying possibility, it is more profitable either to use home-grown raw materials or to supply the raw materials without any grinding. Whenever particle size reduction is considered to facilitate feed manipulation, grinding must be very coarse, to prevent as much as possible the presence of fine particles.

An important practical point to consider is the ideal pellet size for the rabbit. This topic is discussed in other chapters but, as a general rule and according to several authors (Lebas, 1975; Maertens, 1994; Maertens and Vermeulen, 1995), it is advisable to use a pellet diameter between 3 and 5 mm. Pellet length should be 2–2.5 times pellet thickness, that is to say from 6 to 10 mm. If the length is greater, because of poor cutter adjustment, the animal can waste feed; when trying to eat it, after biting, the rest of the pellet falls down. Another practical rule is to use the same pellet diameter for all rabbit ages and physiological stages.

Quality controls

The main objective of a feed mill is to supply compounded feeds properly manufactured to the customers, whenever they require them and consistently including the nutrients necessary to satisfy animal requirements according to the type of production (Jones, 1996). Consequently, quality control must cover all the processes and services which facilitate the fulfilment of this objective, not only those based on chemical analysis of raw materials and final feeds. This concept represents a quality assurance more than quality control.

Quality assurance is expanding in all economic sectors, and the feed sector is not excluded. Currently there are several quality control systems which try to guarantee this quality to the consumer. The ISO 9001 and 9002 rules, and others, guarantee to the consumer that a processing methodology exists and that this methodology is externally audited.

There are also other sectorial systems, such as the Dutch rules on good manufacturing practice (GMP), which have the advantage that they are more orientated to the feed mill industry. It is a company decision to ask for this kind of certification but, in any case, it is fundamental to include these working principles in feed milling. GMP is very wide but the main aspects to point out are that the following pieces of documentation should be available:

1. Specifications for the raw materials received, premixes, additives and medicated premixes.
2. Specifications for the feeds to be supplied.
3. A descriptive diagram of the feed manufacturing process with identification of the critical points.
4. A precise description of all the controls and inspections at the critical points of the production process from raw material to the final product.
5. Description of the measurement methods to be used in the controls, indicating, if necessary, the regulations or bibliography on which they were based.

The development of these principles, as well as the regulations on residues, ambient pollution and other aspects of feed milling, will be very important for quality assurance, especially in the case of a feed such as that for rabbits, which presents so many difficulties in a feed mill.

Control of raw materials and feeds

As previously mentioned, the feed mill process consists of particle size reduction (grinding), raw materials blending (mixing) and feed processing (pelleting), but no process which drastically transforms the raw materials; thus only partial starch gelatinization and/or protein denaturation occurs. Therefore, the nutritive quality of feeds will depend on the nutritive quality of the raw materials and consequently their control will be decisive in feed manufacturing.

Raw material quality must be evaluated through two different aspects: by chemical values and by estimates of purity. All batches of raw materials which come to the feed mill, either by truck, railway or by ship, must be sampled. The first evaluation of the raw material before unloading is the *evaluation of purity*:

1. To check if the identification of the load is correct.

2. To detect the presence of foreign material, such as other raw materials, soil, metals, and the great diversity of substances which can contaminate the raw material during its harvest, its transport or its production process.

3. To detect the presence of insects.

4. To detect colours which do not correspond to the raw material, which may be due to defects of the process.

5. Off-odours or strange odours, which may be due to previous fermentations.

It is important that the operator at the raw material reception is well acquainted with this subject, because it is a basic tool for assurance of the final quality of the feed.

The *chemical analysis* of the raw materials will assist in their nutritional evaluation, either directly or indirectly to complete the quality control. The information obtained can be classified according to the origin, the supplier, etc., which will be very helpful for the purchasing department.

Classical analyses to be undertaken routinely in raw materials, as well as in final feeds, are for moisture, crude protein, fibre, ether extract and ash as well as urease activity in soybean meal. Occasionally other analyses can be undertaken either at the feed mill or in independent laboratories for amino acids, minerals, vitamins, fatty acid profiles, other type of carbohydrates, etc. It is also convenient to undertake periodic estimates of different ANF of the raw materials used, such as tannins, antitrypsin factors, alkaloids, glucosinolates, etc., depending on the feedstuff used.

- *Moisture*: water content higher than 140 g kg^{-1} stimulates microorganism growth, particularly fungus which can develop mycotoxins. Moisture also implies dilution of the nutritive value of the feedstuff, and therefore a moisture level over that specified in the contract implies economic losses.
- *Crude protein*: protein variation in a raw material as well as in the feed, implies amino acid composition changes which can be important in terms of expected performance results of the animals. In rabbits both a protein deficiency and also a protein excess are important because of the negative influence on the digestive process of this animal.
- *Crude fibre*: high fibre decreases the digestibility and therefore the nutritive value. Because of the specific fibre requirements of rabbits to regulate intestinal motility, this is an important parameter to control not only in the feedstuffs, but also in the finished feed. As discussed in other chapters, other kinds of fibre analyses (acid detergent fibre, neutral detergent fibre), or related fractions (lignin), may be useful, especially for raw materials with high content of any of these fractions (e.g. beet pulp high in hemicellulose, grape marc rich in lignin) where abnormal gastro-intestinal behaviour can result.
- *Ether extract*: this analysis is an indicator of the fat content. Because of the high energetic value of fats, it is important to know precisely, not only

the amount but also the quality of this fat. Raw materials which are very rich in fat, such as oilseeds, must be evaluated frequently because of variation according to their origin. At the feed mill, when volumetric systems for fat addition are used, it is necessary to have more evaluations to minimize errors.

- *Ash*: this analysis is an indication of the mineral content. When figures above the normal ones are obtained, this can indicate some degree of contamination, and a complementary analysis on material insoluble in hydrochloric acid would indicate the presence of silica (soil). In feeds, figures under or over the expected ones indicate incorrect additions of limestone, phosphate or salt. It is remarkable how important a reliable system for salt addition is, because this can be the main reason for feed rejection by rabbits: an excess as well as a deficit can cause difficulties.

The information obtained from the chemical analysis must be available as quickly as possible to be effective, in order to make the changes or the actions required, when the analytical figures deviate from the expected values, because after an excessive time lag the data are not so useful (Larsen, 1992). Furthermore, it is useful to have historical values, even if not so recent, in order to appreciate the trend according to the supplier, or the year of harvest, to avoid some adverse situations occurring that might be based on these trends. As far as rapid analysis is concerned, *near infrared reflectance* allows on-line information if connected to the production line and, if not, supplies information in a few minutes. The disadvantages of this technique are that it is not as reliable as classical laboratory analyses and there is the requirement for an accurate calibration of the apparatus for each parameter and for each raw material which requires traditional wet chemistry.

Another fast and cheap technique is *microscopy*, which allows a qualitative and sometimes semi-quantitative evaluation of feed ingredients. This technique is very useful for detection of contaminations or adulterations in raw materials (Bates, 1994).

Microbiological analysis (moulds, yeasts, *E. coli*, enterobacteria, *Salmonella*, etc.) is essential, because rabbits are very sensitive to the bacteriological quality of feed (Mateos and Rial, 1989). As for other kinds of feeds, it is necessary to ensure the absence of microorganisms which can cause pathological problems, and especially those which are zoonoses transmissible to the consumer.

Mycotoxins are metabolites produced by different fungi, which can cause different problems to the animals which ingest them. These substances can cause disturbances at very low rates (at the parts per billion, ppb, scale). There is a vast range of fungi, but those which can really cause problems are species from the genuses *Aspergillus*, *Fusarium* and *Penicillium* (Meronuck and Concibido, 1996). The most frequent mycotoxins produced by *Aspergillus* are aflatoxins, and, among them, B1 and B2 are common. The most frequent ones from *Fusarium* are the trichothecenes, especially T-2, as well as zearalenone and deoxynivalenol. There is not much information available on the effects of mycotoxins in rabbits; most of the data are from other species but, generally speaking, the first ones discovered and included in the international regulations were the aflatoxins, where a maximum limit has been established in order to allow a raw material to be marketed. Different countries include limits for other mycotoxins, but currently it is still difficult to establish safe limits both for rabbit health and residues. Sometimes a mycotoxin detection is just an indicator of mould growth in the raw material or in the feed analysed, but many other mycotoxin activities even more deleterious than those detected can be present. The only practical ways to employ a mycotoxin-contaminated raw material are to use it in feed for animals less susceptible to mycotoxins (ruminants), to dilute it in small amounts in the feed, to increase somewhat the vitamin content of the feed and, in some cases, some kind of absorbent can be useful (aluminosilicates).

Mycotoxin control can be undertaken by chromatography, but currently there are different commercial kits available to determine quantitatively over time the content of the most frequent and dangerous ones.

Feed labelling

There is considerable legislation relating to feeds and their labelling which, within the European Union, is practically the same across countries. The objective is to inform the consumer on the composition of feeds, on the additives used, the recommended periods of usage and the withdrawal period if necessary. Another basic objective of the legislation is to guarantee that the feed does not contain substances or microorganisms which are undesirable either for the animals or for the consumer.

There is specific legislation for medicated feeds, the main aim of which is control of the use of licensed products and their inclusion in the feed through veterinary prescription.

Processing control

Each feed mill will establish its own processing control adapted to its own requirements but, as a general rule, it is necessary to take into account the following aspects (Jones, 1996):

- A raw materials inventory, which must be done at least once a day. This will detect if any raw material has not been properly used.
- Silo-bin cleaning. The frequency will vary according to the raw material stored.
- Cleaning stage checking of the equipment.
- Grinding.

- Weighing system and volumetric measurements.
- Mixing.
- Pelleting and cooling.
- Conditioning temperatures.
- Truck cleaning and control.

According to Jones (1996), the following steps must be followed when a problem arises:

1. Is the analysis correct? Initially the laboratory result must be checked or the analysis repeated.
2. How was the sample taken? The sample can be taken incorrectly and it is advisable to repeat the analysis on another sample from the same batch.
3. Is there only one or several nutrients present from the established range? This could indicate the absence of a raw material from the formulated feed.
4. Was the process undertaken by the usual person?
5. Check the inventories to rule out discrepancies which could indicate mistaken identifications.
6. Check the measurement equipment.
7. Check the raw material and finished feed silo bins.
8. Re-evaluate the mixing times.
9. Check the amounts and concentrations of the raw materials used to detect a possible problem with delivery.
10. Check the formulation raw material matrix and the formulation itself to verify that the figures are updated.

It is obvious that, if the issues are serious, prompt action is necessary to avoid animals consuming a defective feed, because rabbits are probably one of the most sensitive species to dietary problems.

References

Bates, L.S. (1994) Microscopia de alimentos: secretos para un rapido control de calidad. *Avicultura Profesional* 11, 152–156.
Beumer, H. (1991) Quality assurance as a tool to reduce losses in animal feed production. *Advances in Feed Technology* 6, 6–23.
Bühler (1996) *Manual de Tecnologia de Fabricacion.* Schule für Futtermitteltechnik, Uzwill.
Coelho, M. (1996) Stability of vitamins affected by feed processing. *Feedstuffs* 68, 9–14.
García, J., Carabaño, R., Perez Alba, L. and de Blas, C. (1996) Effect of fibre source on neutral detergent fibre digestion and cecal transit in rabbit. In: Lebas, F. (ed.) *Proceedings of the 6th World Rabbit Congress.* INRA, Toulouse, pp. 175–180.
Heidenrweich, E. (1992) The feed meal in the year 2002. *Advances in Feed Technology* 8, 4–16.

Jones, F. (1996) Quality control in feed manufacturing. *Feedstuffs Reference Issue* 68, 135–138.

Lang, J. (1981) The nutrition of commercial rabbit. Part 1. Physiology, digestibility and nutrition requirements. *Nutrition Abstracts and Reviews, Series B* 51(4), 192–225.

Laplace, J.P. and Lebas, F. (1977) Digestive transit in the rabbit. 7. Effect of fineness of grind of ingredients before pelleting. *Annales de Zootechnie* 26, 413–420.

Lebas, F. (1975) *Le Lapin de Chair, les Besoins Nutritionales et son Alimentation Pratique*. Documents ITAVI, Paris.

Lebas, F. and Laplace, J.P. (1975) Digestive transit in the rabbit. 5. Variation in fecal excretion according to time of feeding and level of restriction of feed during five consecutive days. *Annales de Zootechnie* 24, 613–627.

Larsen, J.G. (1992) On line quality control. *Feed Magazine* 1/92, 4–15.

McMahon, M.J. and Payne, J.D. (1991) *The Pellet Handbook. A Guide for Production Staff in the Compound Feed Industry*. Holmen.

Maertens, L. (1994) Effect of pellet diameter on the growth performance of rabbits before and after weaning. In: *6èmes Journées de la Recherche Cunicole en France*. INRA-ITAVI, La Rochelle, pp. 325–332.

Maertens, L. and Vermeulen, A. (1995) Influence du diametre du granule sur le croissance des jeunes. *Cunicultures* 237–241.

Mateos, G.G. and Rial, E. (1989) Tecnologia de fabricacion de piensos compuestos. In: *Alimentacion del Conejo*. Mundi-Prensa, Madrid, pp. 101–132.

Meronuck, R. and Concibido, V. (1996) Mycotoxins in feed. *Feedstuffs Reference Issue* 68, 139–145.

Pairet, M., Bouyssou, T., Auvergne, A., Landau, M. and Ruckebusch, Y. (1986) Stimulation physico-chimique d'origine alimentarire et motricité digestive chez le lapin. *Reproduction, Nutrition and Development* 26, 85–95.

Pipa, F. and Frank, G. (1989) High-pressure conditioning with annular gap expander. A new way of feed processing. *Advances in Feed Technology* 2, 22–31.

Rial, E., Mendez, J. and Larraga, L. (1993) Nuevas tecnologias de fabricacion de piensos: doble granulacion, expander y adicion de liquidos. In: *Proceedings de F.E.D.N.A.*, Vol. 9. FEDNA, Barcelona, pp. 85–106.

Schultz, R. (1990) The progressive animal feed production and its fundamentals. Pelleting in practice. *Advances in Feed Technology* 3, 6–33.

Walter, M. (1990) The inclusion of liquids in compound feeds. *Advances in Feed Technology* 4, 36–48.

13.   Feed Formulation

C. de Blas and G.G. Mateos

*Departamento de Producción Animal, Universidad Politécnica de Madrid,
Ciudad Universitaria, 28040 Madrid, Spain*

Introduction

This chapter deals with the nutritive allowances in practical feed formulation
for intensive meat rabbit production. In recent years, performance of
intensively reared rabbits has greatly increased because of improvements in
genetics, management and pathology. Productivity levels, measured as
reproductive yield, milk production or growth rate in the fattening period, are
comparable with those obtained in other intensively farmed domestic species.
However, rabbits are herbivorous animals and require a high dietary fibre
content (about one-third of cell wall constituents on an as-fed basis) to
prevent digestive disorders.

The average composition of commercial feeds in Spain (Table 13.1)
reflects this situation, as they typically contain simultaneously a high
proportion both of fibrous and highly concentrated ingredients. Furthermore,
rabbit diets must be primarily designed to allow a sufficient nutrient intake to
meet their high nutritive requirements per unit of body weight. Therefore,
factors affecting feed consumption, such as nutrient imbalances, raw material
composition and pellet quality, are major concerns in this species.

Table 13.1. Ingredient composition of feeds for rabbits (g kg⁻¹).

Cereal grains[a]	150–250
Animal and vegetable fats	10–30
Molasses	10–30
Beet pulp	0–100
Cereal by-products	150–250
Lucerne hay	250–350
Fibrous by-products	50–100
Protein concentrates[b]	150–200

[a] Mostly barley.
[b] Mostly soybean and sunflower meal.

Prior to recommending practical feeding standards, the effect of varying dietary nutrient content on rabbit performance will be discussed. This information allows the formulation of diets on a performance–cost basis according to market prices. Effects on meat quality and pathology should also be considered, as reviewed in Chapters 10 and 11.

Substitution of fibre for starch

Rabbits are capable of achieving good performance on high-fibre diets as a result of their peculiar digestive physiology. As shown in Fig. 13.1, maximal growth rates are reached with diets containing around 180–210 acid detergent fibre (ADF) g kg^{-1}, which corresponds to approximately 9.7–10.3 MJ digestible energy (DE) kg^{-1}, when no fat is added. Above this fibre level, fattening rabbits are not able to maintain DE intake. High-fibre diets (350 g ADF kg^{-1} dry matter (DM)) decrease average daily gain and feed conversion rate by 30 and 50%, respectively, in comparison with the optimal values. Conversely, an excess of dietary starch and/or a deficit of fibre promotes

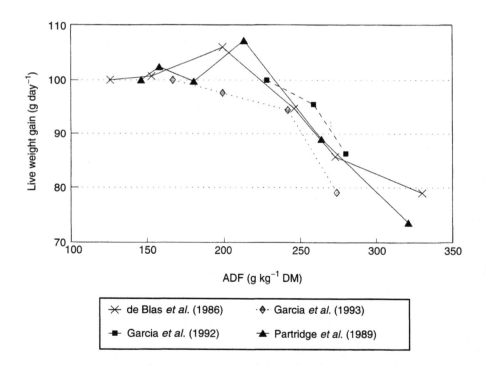

Fig. 13.1. Effect of dietary ADF content on average daily gain during the fattening period. 100 = control diet.

digestive disorders (Cheeke and Patton, 1980; de Blas *et al.*, 1986) and a slight decrease in DE intake (Fig. 13.1).

Three long-term studies (> 1 year) conducted with rabbit does have compared seven diets containing from 162 to 216 g ADF kg^{-1} and no fat added (Méndez *et al.*, 1986; Barreto and de Blas, 1993; Cervera *et al.*, 1993). The results have shown that, over this period, rabbit does are able to maintain DE intake by increasing consumption when dietary fibre content is increased. Type of feed had no influence on reproductive performance, but litter weight at weaning decreased (by about 11%) when dietary ADF content was above 180 g (equivalent to about 10 MJ DE kg^{-1}).

In other work, de Blas *et al.* (1995) have studied the effect of the substitution of starch for fibre in rabbit does using isoenergetic diets (10.6 MJ DE kg^{-1}). Five diets were formulated with increasing levels of neutral detergent fibre (NDF; from 278 to 371 g kg^{-1}) and ether extract (from 20 to 51 g kg^{-1}) at the expense of the level of starch which decreased from 237 to 117 g kg^{-1}. The type of diet had little effect on DM intake. However, the regression analyses indicated that dietary levels of NDF, ADF and starch of around 320, 170 and 180 g kg^{-1} respectively were optimal for maximal reproductive performance, growth of young rabbits and feed efficiency (see Fig. 13.2). The impairment observed in rabbits fed the highest levels of fibre might be explained by higher fermentation losses in the caecum, together with an insufficient uptake of glucose from the gut to meet the requirements for pregnancy and milk lactose synthesis. The negative effects of high starch concentrations in the diet were related to an increase in mortality through diarrhoea.

Type of fibre

Several studies have shown that cell wall composition and physical structure influence feed digestion in isofibrous diets (see Chapter 5).

Lucerne hay is the source of fibre most widely used in rabbit diets, accounting for around one-third of commercial feeds in Spain (see Table 13.1). Lucerne hay is highly palatable and provides both long and digestible fibre, which allows an adequate transit time of the digesta and a balanced caecal flora to be maintained.

Moderate inclusion in the diet (100–150 g kg^{-1}) of fibrous by-products has little effect on rabbit performance (Motta, 1990; García *et al.*, 1993). However, an excessive substitution of lucerne hay by highly lignified sources of fibre leads to a significant impairment both of average daily gain and feed efficiency (by about 10 and 20%, respectively, in diets with a 50:50 ratio of lucerne hay and grape marc; Parigi Bini and Chiericato, 1980; Motta, 1990). A high dietary lignin content depresses energy digestibility and caecal fermentative activity (García *et al.*, 1996). Therefore, a similar impairment of

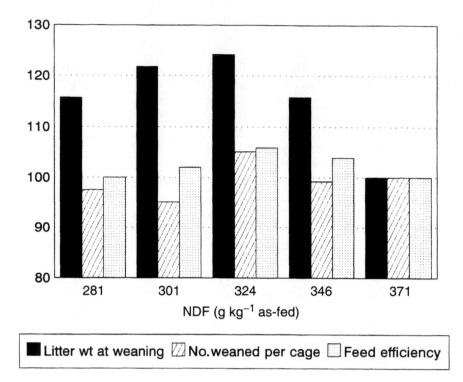

Fig. 13.2. Effect of dietary NDF content on performance of rabbit does and feed conversion rate (base 100 = diet containing 371 g NDF kg⁻¹) (de Blas *et al.*, 1995).

performance can be expected when wheat straw or sunflower hulls are included in the diet.

Conversely, high levels of substitution of lucerne hay by highly digestible or small-sized fibre sources (such as beet and citrus pulps, soybean hulls, paprika meal or rice hulls) promotes an increase in the weight of caecal contents, a longer retention time in the gut and a decrease of feed intake and performance (Fraga *et al.*, 1991; García *et al.*, 1993, 1996).

These results indicate a benefit of combining different sources of fibre when trying to substitute a high proportion of lucerne hay in the diet. However, further research is needed to establish feeding recommendations based upon this issue.

Fat supplementation

The effect of the addition of 30 g of different sources of fat (tallow, lard, deodorized oleins or sunflower oil) kg⁻¹ in isofibrous diets for fattening rabbits has been studied by several authors (Partridge *et al.,* 1986; Santomá *et al.,* 1987; Fernández and Fraga, 1992). In these studies dietary protein (DP) content was increased with fat addition to keep the DE:DP ratio as constant as possible. Results showed that fat inclusion had a positive effect on energy digestibility (5% on average) and feed efficiency (7%), but not on growth rate, as feed intake decreased by 6%. No interaction was found between type and level of supplemental fat. Therefore, the value of fat addition should be established on an energy–cost basis, taking also into account the effects of fat quality on carcass and pellet stability (see Chapters 10 and 12).

Several long-term (9–24 months) studies (Fraga *et al.,* 1987; Maertens and De Groote, 1988a; Barreto and de Blas, 1993; Cervera *et al.,* 1993) have studied the effect of fat addition in isofibrous diets (200 g ADF kg^{-1}) on performance of breeding does. Responses were higher for does than in growing rabbits. The inclusion of 35 g fat kg^{-1} in diets of does increased DE intake (by 14.5% on average), which promoted an increase in milk yield, and litter weight at weaning (by 8.5%). Neither average weight of breeding does nor fertility or prolificacy was significantly affected by type of diet, although a trend was found for fat supplementation to lower mortality in litters containing more than nine pups (Fraga *et al.,* 1987). These results indicate that the use of fat to increase energy concentration of feeds (over 11–11.5 MJ DE kg^{-1}) allows maximum milk production and litter growth in highly productive rabbits, when the remaining constituents of the diet (fibre, protein and starch) are kept in balance.

Optimal protein to energy ratio

Energy concentrations of rabbit diets may vary widely. Therefore, it is advisable to express total protein requirements as a ratio between DP and DE.

The effect of a variation in this ratio on the performance of fattening rabbits has been studied by de Blas *et al.* (1981) and Fraga *et al.* (1983) using 12 diets containing from 7.9 to 11.7 g DP MJ^{-1} DE. Maximal DE intake and average daily gain were obtained for diets having a ratio of 10 g MJ^{-1} (see Fig. 13.3). Accordingly, optimal DP content (g kg^{-1}) should be increased from 95 to 115 when dietary DE increases from 9.5 to 11.5 MJ kg^{-1}. Dietary digestible protein/energy ratios below and above this optimum led to an impairment of fattening performance, mortality and feed efficiency. Low values of this ratio (below 10 g MJ^{-1}) also promoted a curvilinear decrease in water and protein and an increase in body fat (see Fig. 13.3).

The effect of the protein/energy ratio in breeding does has been reviewed by Santomá *et al.* (1989) and Xiccato (1996). Optimal values are in the range 11.0–12.5 g DP MJ^{-1} DE, so that they are about 20% higher than that for fattening rabbits. The higher values correspond to females following an intensive breeding system. Dietary protein contents below the optimum led to

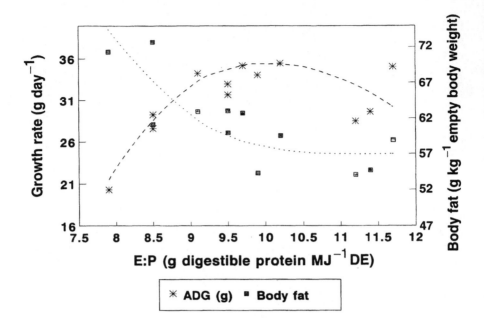

Fig. 13.3. Effect of the dietary protein/energy ratio on the average growth rate in the fattening period and content of fat in the empty body of rabbits at 2.25 kg (de Blas *et al.*, 1981; Fraga *et al.*, 1983).

a decrease in milk production, growth of young rabbits, fertility and weight of does. An excess of protein content related to energy decreases performance and increases diarrhoea incidence and environmental pollution.

Amino acid requirements

Until recently, no consideration was given to the quality of protein in rabbit feeds, because all the essential amino acid requirements were believed to be supplied through caecotrophy. However, as pointed out in Chapter 3, soft faeces represent only about 0.14 of the total protein intake in intensively reared rabbits. Accordingly, essential amino acid requirements, along with total protein, must be considered in practical feed formulation.

Several authors have studied the total amino acid requirements for rabbits on a dose–response basis (Tables 13.2 and 13.3). In some cases, dietary amino acid content had a quadratic effect on productivity for some of the traits studied (see Fig. 13.4). This type of response indicates the negative effects of an excess of amino acid and was especially important for threonine. For this amino acid, a level slightly greater than the optimal reduced performance, which indicates the importance of establishing a maximal concentration for this nutrient in the diet.

Table 13.2. Total amino acid requirements of growing-fattening rabbits (g kg⁻¹ as-fed).

Reference	DE (MJ kg⁻¹)	Growth[a] rate (g day⁻¹)	Optimal dietary concentrations			
			Lys	TSAA[b]	Thr	Trp
Adamson and Fisher (1971)	—	25.5	7.0	6.0	5.0	1.5
Colin (1975)	9.41	39.2	5.8	—	—	—
Colin (1975)	11.13	37.6	—	6.3	—	—
Davidson and Spreadbury (1975)	10.46[c]	36.5	9.0	5.5	6.0	2.0
Colin and Allain (1978)	10.88	35.0	6.2	—	—	—
Spreadbury (1978)	—	41.0	9.4	6.2	—	—
Berchiche and Lebas (1994)	11.17	40.2	—	6.2	—	—
Taboada *et al.* (1994)	10.70	40.7	7.6	—	—	—
Taboada *et al.* (1996)	10.75	40.4	—	5.4	—	—
De Blas *et al.* (1996)	10.13	43.2	—	—	6.0	—

[a]At the optimal amino acid concentration.
[b]Methionine must represent at least 0.35 of total sulphur amino acids (TSAA; Colin, 1978).
[c]Metabolizable energy.

Table 13.3. Total amino acid requirements of breeding does (g kg⁻¹ as-fed).

Reference	DE (MJ kg⁻¹ DM)	Optimal dietary concentrations		
		Lys	TSAA	Thr
Maertens and De Groote (1988b)	10.46	8.0	—	—
Taboada *et al.* (1994)	10.70	8.0[a]	—	—
Taboada *et al.* (1996)	10.75	—	6.3	—
De Blas *et al.* (1996)	10.13	—	—	6.4

[a]For maximal milk production. Reproductive performance did not improve above 6.8 g kg⁻¹.

As for other species, there are more available data for growing rabbits than for breeding does, as well as considerable variation between different studies. Part of this variation can be explained by differences in the methods used: purified vs. commercial diets, genetic potential of the animals and energy concentration of the diets.

Other causes of variability are related to the different availability of the sources of amino acids used (see Chapter 6). To take into account this effect, recent studies (Taboada *et al.*, 1994, 1996; de Blas *et al.*, 1996) have determined the lysine, sulphur and threonine requirements, expressed in digestible (apparent faecal) instead of crude units. Results are shown in Table 13.4. Optimal values for growth were consistent with those obtained by Moughan *et al.* (1988) based on the amino acid composition of the whole body of 53-day-old rabbits (Table 13.5), although the latter method does not consider the amino acid requirements for maintenance or the amino acid

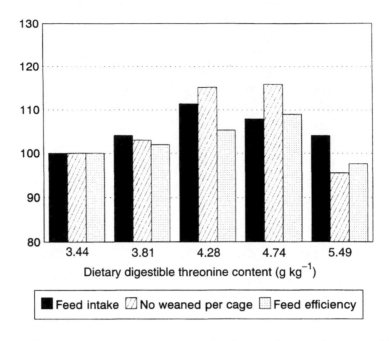

Fig. 13.4. Effect of dietary threonine content on feed intake, reproductive performance and feed efficiency of breeding does (base 100 = diet containing 3.44 g of digestible threonine kg⁻¹) (de Blas *et al.*, 1996).

Table 13.4. Digestible (faecal apparent) amino acid requirements of rabbits (g kg⁻¹ as-fed basis).

Amino acid	Optimal values		Reference
	Breeding does	Fattening rabbits	
Lysine	6.4[a]	6.0	Taboada *et al.* (1994)
Methionine + cystine	4.9	4.0	Taboada *et al.* (1996)
Threonine	4.4	4.0	de Blas *et al.* 1996)

[a]For maximal milk production. Reproductive performance did not improve above 5.2 g kg⁻¹.

supply by the caecotrophs. The use of digestible amino acids in practical feed formulation is still limited because of the lack of information on the amino acid digestibilities for the main ingredients used in rabbit diets.

Recommended nutrient concentration of diets

Nutrient requirements of intensively reared rabbits are presented in Tables 13.6 and 13.7. Values are given for the three types of diets more commonly

Table 13.5. Amino acid composition (mg g^{-1} N) of the whole body of 53-day-old New Zealand rabbits (Moughan *et al.*, 1988).

Amino acids	Absolute value	Relative to lysine
Lysine	383	100
Methionine	77.5	20.2
Cystine	158	41.3
Arginine	415	108
Histidine	193	50.4
Threonine	245	64
Leucine	429	112
Isoleucine	194	50.7
Valine	239	62.4
Phenylalanine	249	65
Tyrosine	192	50.1

used: breeding does, fattening rabbits and a mixed feed for all the animals. When rabbits are slaughtered at 2.5 kg, more than one fattening feed might be used. In this case, Maertens and Luzi (1996) proposed decreasing the dietary protein and amino acid content with age, to reduce N excretion without altering performance. Several recent studies have studied the possibility of formulating a special type of feed for starter rabbits (from 21 days of age). However, the use of such diets is limited because of practical problems of feed management. Breeding does and starting rabbits are commonly fed from the same feeder in commercial farms. Therefore, the possibility of using two different feeds is limited.

Energy concentrations in Table 13.6 have been determined from estimates based on the optimal proposed levels of carbohydrates and fat. Essential nutrient recommendations have then been referred to those concentrations. However, DE content of fattening feeds can vary from 9.7 to more than 11.5 MJ kg^{-1} with no effects on rabbit performance. Changes in DE concentration with respect to the values given in this table should be accompanied by proportional parallel corrections in the contents of essential nutrients.

Minimal levels of fibre and maximum levels of starch are more critical than maximum levels of fibre and minimum of starch, as they affect not only performance but also mortality.

Recommendations for type of fibre include an optimal concentration for lignin and a minimum level for long fibre particles. Both restrictions should be followed simultaneously, as some highly lignified by-products can have an insufficient content of long fibre.

Only the better established amino acid requirements are presented in Table 13.6. Dietary tryptophan content can be estimated at 0.18–0.20 of the optimal lysine concentration. For other essential amino acids, the ideal protein pattern (Table 13.5) can be followed.

Table 13.6. Nutrient requirements of intensively reared rabbits as concentration kg⁻¹ corrected to a dry matter content of 900 g kg⁻¹.

Nutrient	Unit	Breeding does	Fattening rabbits	Mixed feed
Digestible energy	MJ	11.1	10.5	10.5
Metabolizable energy	MJ	10.6	10.0	10.0
NDF[a]	g	31.5	33.5	33.0
		(30.0–34.0)[b]	(32.0–35.0)	(32.0–34.0)
ADF	g	16.5	17.5	17.0
		(15.0–18.0)	(16.0–18.5)	(16.0–18.0)
Crude fibre	g	13.5	14.5	14.0
		(12.5–14.5)	(13.5–15.0)	(13.5–14.5)
ADL	g	*5.0*[c]	*5.5*	*5.5*
Starch	g	18.0	16.0	16.0
		(15.0–21.0)	(14.5–17.5)	(15.0–17.0)
Ether extract	g	5.5	Free	Free
Crude protein	g	18.4	15.3	15.9
		(16.3–19.8)	(14.5–16.2)	(15.4–16.2)
Digestible protein	g	12.9	10.7	11.1
		(11.4–13.9)	(10.2–11.3)	(10.8–11.3)
Lysine[d]				
Total	g	8.4	7.5	8.0
Digestible	g	6.6	5.9	6.3
Sulphur[e]				
Total	g	6.5	5.4	6.0
Digestible	g	5.0	4.1	4.6
Threonine[f]				
Total	g	7.0	6.4	6.8
Digestible	g	4.8	4.4	4.7
Calcium	g	11.5	6.0	11.5
Phosphorus	g	6.0	4.0	6.0
Sodium	g	2.2	2.2	2.2
Chloride	g	2.8	2.8	2.8

[a]Proportion of long fibre particles (> 0.3 mm) should be higher than 0.25.
[b]Values in parentheses indicate range of minimal and maximal values recommended.
[c]Values in italics are provisional estimates.
[d]Total amino acid requirements have been calculated for a contribution of synthetic amino acids of 0.15.
[e]Methionine should provide a minimum of 35% of the total TSAA requirements.
[f]Maximal levels of 5.2 and 7.6 g kg⁻¹ of digestible and total threonine, respectively, are recommended for breeding does.

There is a lack of research on mineral and vitamin requirements. Standards proposed in Tables 13.6 and 13.7 are mostly based on practical levels used by the industry.

Table 13.7. Trace element and vitamin requirements of intensively reared rabbits as concentration kg^{-1} corrected to a dry matter content of 900 g kg^{-1}.

Nutrient	Unit	Breeding does	Fattening rabbits	Mixed feed
Cobalt	mg	0.3	0.3	0.3
Copper	mg	10	6	10
Iron	mg	50	30	50
Iodine	mg	1.1	0.4	1.1
Manganese	mg	15	8	15
Selenium	mg	0.05	0.05	0.05
Zinc	mg	60	35	60
Vitamin A	mIU	10	6	10
Vitamin D	mIU	0.9	0.9	0.9
Vitamin E	IU	50	15	50
Vitamin K$_3$	mg	2	1	2
Vitamin B$_1$	mg	1	0.8	1
Vitamin B$_2$	mg	5	3	5
Vitamin B$_6$	mg	1.5	0.5	1.5
Vitamin B$_{12}$	mg	12	9	12
Folic acid	mg	1.5	0.1	1.5
Niacin	mg	35	35	35
Pantothenic acid	mg	15	8	15
Biotin	mg	100	10	100
Choline	mg	200	100	200

References

Adamson, I. and Fisher, H. (1973) Amino acid requirements of the growing rabbit: an estimate of quantitative needs. *Journal of Nutrition* 103, 1306–1310.

Barreto, G. and de Blas, C. (1993) Effect of dietary fibre and fat content on the reproductive performance of rabbit does bred at two remating times during two seasons. *World Rabbit Science* 1, 77–81.

Berchiche, M. and Lebas, F. (1994) Supplémentation en méthionine d'un aliment à base de feverole: effets sur la croissance, le rendement à l'abattage et la composition de la carcasse chez du lapin. *World Rabbit Science* 2, 135–140.

de Blas, C., Pérez, E., Fraga, M.J., Rodríguez, M. and Gálvez, J. (1981) Effect of diet on feed intake and growth of rabbits from weaning to slaughter at different ages and weights. *Journal of Animal Science* 52, 1225–1232.

de Blas, C., Santomá, G., Carabaño, R. and Fraga, M.J. (1986) Fiber and starch levels in fattening rabbit diets. *Journal of Animal Science* 63, 1897–1904.

de Blas, C., Taboada, E., Mateos, G.G., Nicodemus, N. and Méndez, J. (1995) Effect of substitution of starch for fiber and fat in isoenergetic diets on nutrient digestibility and reproductive performance of rabbits. *Journal of Animal Science* 73, 1131–1137.

de Blas, C., Taboada, E., Nicodemus, N., Campos, R. and Méndez, J. (1996) The response of highly productive rabbits to dietary threonine content for reproduction and growth. In: Lebas, F. (ed.) *Proceedings of the 6th World Rabbit Congress, Toulouse*. Association Française de Cuniculture, Lempdes, pp. 139–144.

Cervera, C., Fernández, J., Viudes, P. and Blas, E. (1993) Effect of remating interval and diet on performance of female rabbits and their litters. *Animal Production* 56, 399–405.

Cheeke, P.R. and Patton, N.M. (1980) Carbohydrate overload of the hindgut. A probable cause of enteritis. *Journal of Applied Rabbit Research* 3, 20–23.

Colin, M. (1975) Effets sur la croissance du lapin de la supplémentation en L-lysine et en DL-méthionine de régimes végétaux simplifiés. *Annales de Zootechnie* 24, 465–474.

Colin, M. (1978) Effets d'une supplémentation en méthionine ou en cystine de régimes carencés en acides aminés soufrés sur les performances de croissance du lapin. *Annales de Zootechnie* 27, 9–16.

Colin, M. and Allain, D. (1978) Etude du besoin en lysine du lapin en croissance en relation avec la concentration énérgetique de l'aliment. *Annales de Zootechnie* 27, 17–31.

Davidson, J. and Spreadbury, D. (1975) Nutrition of the New Zealand White rabbit. *Proceedings of the Nutrition Society* 34, 75–83.

Fernández, C. and Fraga, M.J. (1992) The effect of source and inclusion level of fat on growth performance. In: Cheeke, P.R. (ed.) *Proceedings of the 5th World Rabbit Congress*. Oregon State University, Corvallis, Oregon, pp. 1071–1078.

Fraga, M.J., de Blas, C., Pérez, E., Rodríguez, J.M., Pérez, C. and Gálvez, J. (1983) Effects of diet on chemical composition of rabbits slaughtered at fixed body weights. *Journal of Animal Science* 56, 1097–1104.

Fraga, M.J., Lorente, M., Carabaño, R. and de Blas, C. (1987) Effect of diet and of remating interval on milk production and milk composition of the doe rabbit. *Animal Production* 48, 459–466.

Fraga, M.J., Pérez, P., Carabaño, R. and de Blas, C. (1991) Effect of type of fiber on the rate of passage and on the contribution of soft faeces to nutrient intake of finishing rabbits. *Journal of Animal Science* 69, 1566–1574.

García, G., Gálvez, J.F. and de Blas, C. (1992) Substitution of barley grain by sugar-beet pulp in diets for finishing rabbits. 2. Effect on growth performance. *Journal of Applied Rabbit Research* 15, 1017–1024.

García, G., Gálvez, J. and de Blas, C. (1993) Effect of substitution of sugarbeet pulp for barley in diets for finishing rabbits on growth performance and on energy and nitrogen efficiency. *Journal of Animal Science* 71, 1823–1830.

García, J., Carabaño, R., Pérez, L. and de Blas, C. (1996) Effect of type of fibre on neutral detergent fibre digestion and caecal traits in rabbits. In: Lebas, F. (ed.) *Proceedings of the 6th World Rabbit Congress, Toulouse*. Association Française de Cuniculture, Lempdes, pp. 175–180.

Maertens, L. and De Groote, G. (1988a) The influence of the dietary energy content on the performance of post-partum breeding does. In: *Proceedings of the 4th World Rabbit Congress, Budapest*. Sandar Holdas, Hercegalom, Hungary, pp. 42–52.

Maertens, L. and De Groote, G. (1988b) The effect of the dietary protein:energy ratio and the lysine content on the breeding results of does. *Archiv für Geflügelkunde* 52, 89–95.

Maertens, L. and Luzi, F. (1996) Effect of dietary protein dilution on the performance and N excretion of growing rabbits. In: Lebas, F. (ed.) *Proceedings of the 6th World Rabbit Congress, Toulouse*. Association Française de Cuniculture, Lempdes, pp. 237–242.

Méndez, J., de Blas, C. and Fraga, M.J. (1986) The effects of diet and remating interval after parturition on the reproductive performance of the commercial doe rabbit. *Journal of Animal Science* 86, 1624–1634.

Motta, W. (1990) Efectos de la sustitución parcial de heno de alfalfa por orujo de uva o pulpa de remolacha sobre la utilización de la dieta y los rendimientos productivos en conejos en crecimiento. PhD Thesis, Universidad Politécnica de Madrid, Spain.

Moughan, P.J., Schultze, W.H. and Smith, W.C. (1988) Amino acid requirements of the growing meat rabbit. 1. The amino acid composition of rabbit whole body tissue – a theoretical estimate of ideal amino acid balance. *Animal Production* 47, 297–301.

Parigi Bini, R. and Chiericato, G.M. (1980) Utilization of various fruit pomaces products by growing rabbits. In: Camps, J. (ed.) *Proceedings of the 2nd World Rabbit Congress, Barcelona*. WRSA, pp. 204–213.

Partridge, G.G., Findlay, M. and Fordyce, R.A. (1986) Fat supplementation of diets for growing rabbits. *Animal Feed Science and Technology* 16, 109–117.

Partridge, G.G., Garthwaite, P.H. and Findlay, M. (1989) Protein and energy retention by growing rabbits offered diets with increasing proportions of fibre. *Journal of Agricultural Science* 112, 171–178.

Santomá, G., de Blas, C., Carabaño, R. and Fraga, M.J. (1987) The effects of different fats and their inclusion level in diets for growing rabbits. *Animal Production* 45, 291–300.

Santomá, G., de Blas, C., Carabaño, R. and Fraga, M.J. (1989) Nutrition of rabbits. In: Haresign, W. and Cole, D.J.A. (eds) *Recent Advances in Animal Nutrition*. Butterworths, London, pp. 109–138.

Spreadbury, D. (1978) A study of the protein and amino acid requirements of the growing New Zealand White rabbit with emphasis on lysine and sulphur-containing amino acids. *British Journal of Nutrition* 39, 601–603.

Taboada, E., Méndez, J., Mateos, G.G. and de Blas, C. (1994) The response of highly productive rabbits to dietary lysine content. *Livestock Production Science* 40, 329–337.

Taboada, E., Méndez, J. and de Blas, C. (1996) The response of highly productive rabbits to dietary sulphur amino acid content for reproduction and growth. *Reproduction, Nutrition and Development* 36, 191–203.

Xiccato, G. (1996) Nutrition of lactating does. In: Lebas, F. (ed.) *Proceedings of the 6th World Rabbit Congress, Toulouse*. Association Française de Cuniculture, Lempdes, pp. 29–50.

14.   Feeding Systems for Intensive Production

L. Maertens[1] and M.J. Villamide[2]

[1]Agricultural Research Centre–Ghent, Rijksstation voor Kleinveeteelt, Burg. Van Gansberghelaan 92, 9820 Merelbeke, Belgium; [2]Departamento de Producción Animal, Universidad Politécnica de Madrid, 28040 Madrid, Spain

Diet presentation

In intensive rabbit production dried and ground raw materials are used to prepare balanced compound diets. These concentrated diets are generally pelleted because rabbits show a strong preference for pellets over the same diet in meal or mash form. The processing costs for pelleting rabbit diets are more than compensated for by a number of benefits. Significantly lower amounts are consumed on meal diets, resulting in lower daily weight gain (DWG), inferior feed conversion ratio (FCR) and also lower slaughter yield (Table 14.1). It has even been shown that the circadian cycle of feed intake is disturbed when meal diets are fed (Lebas and Laplace, 1977). When offered a

Table 14.1. Effect of diet presentation on the performances of growing rabbits (as a percentage of pellet).

Reference	Presentation form	DWG	DFI	FCR
Lebas (1973)	Pellet	= 100	= 100	= 100
	Meal	87	83	106
King (1974)	Pellet	= 100	= 100	= 100
	Meal	93	90	103
Machin et al. (1980)	Pellet	= 100	= 100	= 100
	Meal	98	80	123
	Mash (0.40 water)	75	84	89
Candau et al. (1986)	Pellet	= 100	= 100	= 100
	Meal	60	75	123
Sánchez et al. (1984)	Pellet	= 100	= 100	= 100
	Meal	64	52	279

DWG, daily weight gain; DFI, daily feed intake; FCR, feed conversion ratio.

choice between meal and pellet form, 0.97 of total feed intake was of the pelleted diet (Harris *et al.*, 1983).

Other benefits of pelleting are comparable with those for other animals: segregation or selection between the different raw materials is impossible, higher amounts of by-products can be fed and feed wastage is minimal. Pellets further reduce dust problems in the rabbitry and automatic or semi-automatic rabbit feeders work much more easily than with meal or mash.

Several studies have been undertaken with other forms of presentation. When a limited number of raw materials are mixed and supplied, rabbits select those raw materials according to their palatability. As a result of this unbalanced intake, performance deteriorates (Schlolaut, 1995). When the cereals and the protein sources were covered with molasses, rabbits were less able to sort between the different dietary components (Goby and Rochon, 1990). Compared with a commercial pelleted diet, weight gains of fatteners were similar on the meal diet. However, the FCR was unfavourable due to the high wastage of feed (Sánchez *et al.*, 1984; Goby and Rochon, 1990).

With lower cereal prices, efforts have been made to feed high amounts of whole grains together with a concentrated pellet. Rommers *et al.* (1996) compared pellets with a mixture of 0.85 pellets and 0.15 of wheat or barley for fatteners. The biological performance was not significantly different. However, the cereals accumulated in the feeders because the rabbits showed preference for the pellets. Because wastage of feed was not observed, mainly due to the construction of the feeders, it was concluded that a mixture of pellets and grains could reduce feeding costs. However, attention has to be paid to the need to avoid feed wastage and to feed a homogeneous mixture.

In addition, combinations of a concentrate (pellet or cereals) and fresh vegetable materials can be fed. However, a significantly reduced daily dry matter (DM) intake is obtained even when feedstuffs with a high palatability such as roots, cabbage or even grass silage were fed, resulting in 20–40% decreased DWG (Partridge *et al.*, 1985).

Recently, Maertens and Luzi (1995) compared pelleted with extruded presentation. The purpose was to try to modify starch hydrolysis of maize-based diets by extrusion and consequently improve ileal digestibility. Performances before or after weaning were not positively influenced by extrusion (Table 14.2). The decreased DWG could probably be related to the degradation of the protein quality due to the high temperature during extrusion. However, the intended higher starch digestibility was not obtained and consequently extrusion failed to reduce the mortality rate of the high starch containing diet. Similarly, Fernández-Carmona *et al.* (1993) obtained lower intake and growth rate and a great variability in FCR when rabbits from 18 to 42 days of age were fed an extruded diet as opposed to the pelleted diet. The authors explained these inferior results by the lower quality of the pellets of the extruded diet (lower durability and hardness).

In intensive rabbit meat production, a pelleted balanced diet is the basis

Table 14.2. Effect of diet and dietary extrusion on performance and starch digestibility (Maertens and Luzi, 1995).

Starch level (g kg⁻¹)	120		216		Significance	
Technological treatment	Pelleted	Extruded	Pelleted	Extruded	Diet	Treatment
Before weaning (21–28 days)						
DWG (g)	36.1	32.7	35.1	34.6	NS	NS
FI/cage (g day⁻¹)	574	518	546	551	NS	NS
After weaning (28–49 days)						
DWG (g)	40.1	39.1	40.5	38.0	NS	P<0.01
FI/cage (g day⁻¹)	91.7	90.9	93.1	88.9	NS	P<0.01
Feed/gain	2.63	2.33	2.30	2.34	NS	NS
Mortality	0.016	0.0008	0.056	0.057	P<0.01	NS
Starch digestibility						
28–32 days	0.980	0.987	0.923	0.903	P<0.001	NS
49–53 days	0.990	0.985	0.939	0.911	P<0.001	P<0.05

DWG, daily weight gain; FI, feed intake; NS, not significant.

for meeting nutrient requirements for maximizing biological performance. All 'alternative' methods, e.g. meal, mash, roughages, mixture of raw materials, etc., decrease the daily dry matter intake. Most of these methods are labour-consuming because diets have to be fed daily, are difficult to distribute automatically and are therefore not suitable for large-scale production.

Grinding of raw materials

The raw materials used in rabbit diets have a very different structure; some of them are available in pellets (e.g. dehydrated lucerne), others in fine meal (e.g. cassava meal) or as grain. In order to have a homogeneous mixture and to favour the pelleting process, grinding of the raw materials is necessary.

The technique of grinding of raw materials destined for rabbit diets has been the topic of much discussion. Finely ground feedstuffs (<1 mm) lead to an increased retention time in the gut due to the higher amount of fines (<0.3 mm) in the diet (Laplace and Lebas, 1977; Gidenne, 1992). García *et al.* (1996), comparing different fibre sources, found negative relationships between the proportion of fine particles (<0.3 mm) and pH of caecal contents ($r = -0.47$) and between neutral detergent fibre (NDF) digestibility and proportion of particles larger than 1.25 mm ($r = -0.63$). Although finely ground ingredients favour digestibility, reduced motility is not likely in rabbits because an increased incidence of diarrhoea, weight loss and mortality were noted, especially with highly digestible fibre sources (Auvergne *et al.*,

1977). However, when commercial screens (2–7 mm) were used in the mill, differences in particle size due to this were small and effects on both retention time and health disturbance were not detected (Lebas *et al.*, 1986).

Pellet quality

Durability and hardness are the major quality characteristics of rabbit pellets because rabbits do not eat the fines between the pellets. Several types of device for measuring pellet quality have been used by the industry. Generally these devices can be classified into those testing the resistance of pellets to crushing (hardness), or fragmentation when rubbed or shaken (durability). The pneumatic-powered hardness testers determine the power (in kg) for crushing pellets. Although this method is quick, sufficient pellets (>10) have to be tested to give a good repeatability. Testers using a motor drive instead of manual handling are preferable because they exclude effects due to the operator.

In 1962, a committee under the Presidency of Prof. Phost developed a standard method of pellet testing using a square tumbling can (Phost, 1963). The can is rotated at a speed of 50 rpm using a perpendicular axis centred on both sides. A quantity of 500 g of pellets, after screening out the fines, is used for the test. The sample is placed in the tumbling can and rotated for 10 min. After screening again (standard mesh size just smaller than the nominal pellet diameter), the remaining pellets or fines are weighed and the pellet durability index (g pellets after/g pellets before tumbling × 10) or the proportion of fines is calculated. For purposes of quality control in the mill, it is preferable to test pellets before they enter the cooler because the cooling process influences the test results. Under correct processing and handling conditions, less than 0.02 of 'fines' may be produced in quality pellets during transport, in silos or bags and in tubes and rabbit feeders.

Steam is generally used to facilitate the pelleting process and tends to favour feed intake and DWG compared with dry pelleting (Bombeke *et al.*, 1978). Binders like lignosulphonates, bentonites or sepiolites are frequently used at 10–20 kg t⁻¹ to increase pellet durability; their use is more important when high levels of fat are added (Angulo *et al.*, 1995). Molasses is known as a good pellet binder and therefore is introduced at a level of 40–60 kg t⁻¹.

Double pelleting increases the hardness of the pellets (from 5.5 to 10.3 kg) but equal DWG and feed intake were obtained although the rabbits showed preference for the pellets with the highest hardness (L. Maertens, unpublished). The hardness of pellets, however, when the resistance to crushing is between 7 and 13 kg, does not influence the biological performance (Morisse *et al.*, 1985).

After pelleting, pellets have to be cooled immediately and sieved before final packaging. Cooling increases the durability and avoids predisposing conditions for moulds.

Pellet size

Small pellet diameters (<2.5 mm) decrease feed intake and consequently DWG (Maertens, 1994). No benefits were observed through feeding small pellets to young rabbits before weaning. Ideal size seems to be in the range of 3.5–4.5 mm (Harris *et al.*, 1984; Maertens, 1994). At diameters above 5 mm, the risk of pellet wastage increases (Lebas, 1975). Changing from a large pellet diameter (4.8 mm) to a small pellet size (2.5 mm) at weaning has a negative impact on feed intake and weight gain (Maertens, 1994) (Fig. 14.1). The increased feeding time with thin pellets could be responsible for this phenomenon.

The length of pellets is normally between 0.8 and 1 cm. If longer, there is a higher risk of breaking during handling while losses of single pellets or parts of them are more frequent.

Storage and conservation of feeds

With increasing size of rabbit breeding units, feeds are mainly delivered in bulk. Packaging in bags is still used for small units or for special feeds (e.g. weaning diet). Storage time should be limited to 3–4 weeks, particularly with outdoor silos. Due to temperature variations, feed may sweat and become

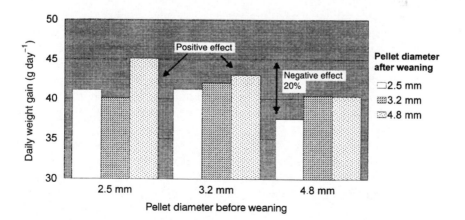

Fig. 14.1. Effect of pellet diameter fed before weaning on daily weight gain after weaning (Maertens, 1994).

mouldy. Taking into account that about 1 tonne of feed per week is consumed in a unit of 100 does and corresponding fatteners, a silo of 4000 kg capacity is necessary to use bulk feed, when only one diet is used for all rabbits. When different diets are supplied for lactation and growth the silo size for each 100 does is 1000 and 3000 kg for does and fatteners, respectively.

If stored in a dry location, rabbit feed with a DM content of at least 890 g kg^{-1} can be stored for several months, which implies lower diet changes for a growing cycle. Cages should have feeders able to contain at least the quantity consumed daily. When using automatic feeding, tubes supplying a number of cages rather than individual feeders are used. In such a system, feed is distributed several times per day. Although it is claimed that rabbits eat more when fresh feed is served, no experiments have demonstrated this. Cavani *et al.* (1991) found a comparable feed consumption and weight gain when food was offered many times a day but FCRs tended to be more favourable.

Number of diets

With increasing knowledge of the specific requirements of the different categories of rabbits, a series of diets can be proposed. However, based on practical considerations and because of the relatively small differences in nutrient concentrations between the different diets, the number of diets should be limited.

In practice, two or three silos (diets) are economically optimum for a medium-sized rabbitry, otherwise the quantities involved are too small to operate with bulk feed. Furthermore, automatic or semi-automatic feeding systems are increasingly being used in large units. They do not allow distribution of a different diet to each category. However, rabbit management is changing from individual handling to a batch system. In this way animals in the same reproductive phase or the same age are grouped together in one building or battery. Such a management system allows the use of phase feeding.

Does and fatteners consume more than 0.90 of the total feed (Fig. 14.2a). Young parent stock, males and does in pre-gestation cages do not have very divergent requirements. They can be fed a fattening diet, sometimes on a restricted basis.

After weaning, about 0.65 of the feed is consumed in the second half of the fattening period (Fig. 14.2b). A specific finishing diet (high energy level, reduced protein content) favours feed efficiency and reduces the nitrogen output to the environment (Maertens and Luzi, 1996).

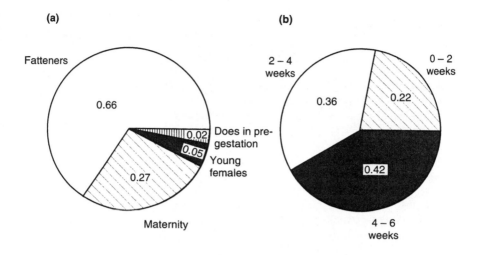

Fig. 14.2. Distribution of feed consumption in a rabbitry using a 42 day reproduction cycle, artificial insemination and weaning at 30 days. (a) Closed farm; (b) fattening unit.

Feed management of the different categories of rabbits

In commercial rabbitries, as a rule, rabbits are fed on an *ad libitum* basis. This is not only for practical considerations but because of the rapid reproductive rate and the adjustment of the voluntary feed intake in response to changes in dietary energy concentration. However, some categories are fed on a restricted basis to prevent excessive fattening. In Table 14.3, a feeding scheme for commercial rabbit meat production is presented.

For young does, the feeding regime depends to a large extent on the age of the first desired mating. Although common agreement exists in the literature that *ad libitum* feeding together with early mating (0.75–0.80 of adult weight) leads to favourable results in obtaining a first litter (Maertens and Okerman, 1987), in practice it is recommended to restrict young does and to postpone the first mating until the age of 17 weeks. This leads to a longer reproductive life and a lower replacement rate of does (M. Baumier, personal communication). A flushing of 5 days prior to mating or insemination leads to oestrus synchronization and high pregnancy rates (Van der Broeck and Lampo, 1979), and greater number of follicles (Gosalvez *et al.*, 1994).

Males increase their voluntary feed intake until the age of 5 months (Fig. 14.3). Afterwards, feed intake drops by about 30% or a natural feed restriction takes place. In comparison with restricted-fed litter mates (0.20–0.25 of *ad libitum*), libido or semen characteristics were not negatively influenced by *ad libitum* feeding (Luzi *et al.*, 1996). Feed restriction of males is therefore not recommended. However, males from a heavy line frequently show sore hocks in wire mesh cages; feed restriction reduces their adult weight by about 0.5 kg

Table 14.3. Feeding scheme for commercial rabbit meat production.

Category	Quantity	Diet
Young does		
Early mating (15–16 weeks)	*Ad libitum*	Fatteners
Late mating (17–18 weeks)	Restricted	Fatteners
	(35 g kg⁻¹ live weight, followed by a 4 day flushing before insemination)	
Does		
Late gestation	*Ad libitum*	Lactation
Lactating	*Ad libitum* : kits <3 weeks	Lactation
	kits >3 weeks	Weaning
In pre-gestation cages	*Ad libitum*	
except early pregnancy	Restricted	Fatteners
	35 g kg⁻¹ live weight	
Males		
Young (until 18 weeks)	*Ad libitum*	Fatteners
Old	Restricted	Fatteners
	35 g kg⁻¹ live weight	
Weaned rabbits		
3–6/7 weeks	*Ad libitum*	Weaning
6/7–10/11 weeks	*Ad libitum*	Fatteners/finishing

and consequently favourable effects on their longevity might be expected.

Lactating does and their young eat out of the same feeder. Specific starter diets (creep feeding) as for piglets are not commonly used in rabbits. During the first 3 weeks of lactation, a lactation diet adapted to the requirements of the

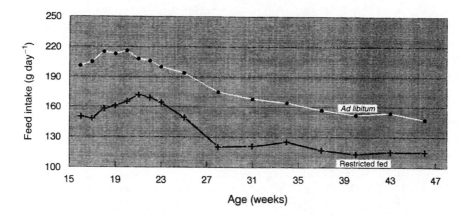

Fig. 14.3. Feed intake of males between 17 and 47 weeks *ad libitum* vs. restricted feeding (0.75 of *ad libitum*) (Luzi *et al.*, 1996).

doe is fed. Once the young start to eat solid feed, preference can be given to the young. A diet adapted to their requirements may be fed at the age of 3 weeks.

After weaning, does are further fed *ad libitum* if they are already in the second part of their pregnancy. Non-pregnant does or does in early pregnancy have to be restricted-fed to prevent excessive fattening, which would lead both to high perinatal mortality and to suppression of voluntary feed intake in early lactation (Partridge *et al.*, 1986).

After weaning, the young are fed *ad libitum* until slaughter weight. A moderate (0.10–0.15) feed restriction slightly improves feed efficiency and reduces the fat content of the carcass (Ledin, 1984; Perrier and Ouhayoun, 1996). However, it is very difficult to restrict fattening rabbits accurately because they are not individually caged. A phase feeding programme during the fattening period is designed to reduce mortality, to increase biological performance and to minimize mineral excretion in order to protect the environment (Maertens and Luzi, 1996). However, such a programme requires a scheduled production in large groups.

Feed and water intake

Average feed intake and feed efficiency of weaned rabbits until slaughter weight are given in Table 14.4. Intake data are relevant under moderate temperature conditions (15–22°C). A weaning age of 30 days is considered, because the most frequently used reproduction rhythm in commercial rabbitries is the 42-day cycle, with insemination or mating 11 days after parturition (Koehl and Mirabito, 1996). Growth, and consequently feed intake curves frequently show an irregular development during the fattening

Table 14.4. Average values of weight gain, feed consumption and technical FCR during the fattening period.[a]

Age (days)	Weight (g)	Weight gain (g day^{-1})	Feed intake		Feed/gain	
			g day^{-1}	g kg^{-1} LW	Per week	Cumulative
21–30	380–680	33	30 + milk	—	—	—
30–37	680–953	38	74	91	1.90	1.90
37–44	953–1247	42	102	93	2.43	2.17
44–51	1247–1583	49	132	94	2.69	2.39
51–58	1583–1905	46	147	85	3.20	2.60
58–65	1905–2199	42	165	81	3.93	2.86
65–72	2199–2479	40	176	76	4.40	3.10

[a]Moderate temperature conditions (15–22°C); diet: 10.10 MJ DE kg^{-1}.

period, therefore the data given in Table 14.4 are mean values of several batches of hybrid rabbits. Feed intake increases with age but not when expressed per kg live body weight. Highest feed intake per unit of weight is reached when maximal growth rate occurs.

Feed intake of does varies considerably during the reproductive cycle from 150 to 450 g day⁻¹. Feed intake patterns during the lactation period correspond largely with the milk yield of the doe. However during the first week, the feed intake/milk yield ratio exceeds 2, while it drops to 1.6 at peak lactation (Fig. 14.4).

Rabbits fed a dry feed have a water requirement which exceeds the DM intake. The ratio of feed to water consumption increases during the fattening period from 1.55 to 1.65 (Laffolay, 1985). This ratio is about 1.9 for non-lactating does or adults at rest. Lactating does have a water intake which is about twice the feed intake.

If water intake is restricted, feed intake drops quickly and will stop after 24 h of restriction. Limited drinking time leads to a reduced feed intake (Prud'hon *et al.*, 1975). Lebas and Delaveau (1975) gave rabbits only 0.5 h 24 h⁻¹ free access to drinking water. During the first 2 days rabbits consumed only about 0.50 of the voluntary feed intake. Although water and feed consumption increased thereafter, feed intake and weight gain were depressed and the feed/water ratio was 1. This low ratio leads to the risk of excretion problems in the kidneys.

The quality of water in terms of mineral content and microbiological profile should be revised for preventing health and productivity problems. There are variations in water quality and there could be external contamin-

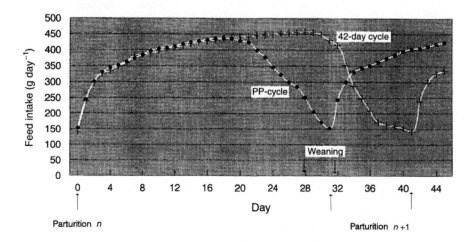

Fig. 14.4. Feed intake of does (hybrid, multiparous) during their reproduction cycle. Data from Maertens and De Groote (1988) and Maertens *et al.* (1988).

ation from dung collectors, crop areas with high organic or nitrogen fertilization, etc. Filtration through resin is recommended to lower high levels of nitrate. In the rabbitry water contamination could come from inefficient cleaning of tanks, pipes and drinkers. Thus, the first and most important treatment of water should be the cleanliness of water equipment. Moreover, chlorination of drinking water (10 mg chloride l^{-1}) is a simple and cheap way of decreasing the bacterial concentration. Table 14.5 shows the effect of cleanliness of water equipment, containing about six times lower bacteria count, and chloride treatment that eliminates the bacteria contamination. Chlorine also removes slime and algae build-up in water lines, which often blocks the drinkers, and eliminates harmful elements such as nitrites, iron, manganese and sulphur through precipitation or by converting them to less harmful components such as nitrites to nitrates (Berriot, 1992). Acidification of water with acetic acid is another common treatment for preventing enteritis, however water pH seems to have little effect on gut pH, because it has to cross the stomach with a pH of 2. With high doses of acetic acid (0.5– 1 ml l^{-1}) (pH about 5), pathogenic bacteria are almost completely eliminated and the rabbits drink freely.

Feed conversion ratio

FCR is the most extensively used parameter for estimation of feed efficiency in intensive systems. In practical situations, different calculations can be used according to the final objective. Thus, the technical FCR usually used in research work does not take into account the feed consumed by culled rabbits

Table 14.5. Effect of cleanliness of water equipment and treatment on level of contamination (García, 1984).

		Untreated water	Chlorine-treated water	
In tanks	Acceptable level	Dirty tank	Dirty tank	Clean tank
Total bacteria count (col ml^{-1})	200	7,600	1,300	0
Total coliform count (col 100 ml^{-1})	Nil	1,200	200	0
Total *E. coli* (col 100 ml^{-1})	Nil	13	1	0
Total *S. fecalis* (col 100 ml^{-1})	Nil	26	5	0
In drinkers (pan pot type)	Acceptable level	Dirty equipment	Dirty equipment	Clean equipment
Total bacteria count (col ml^{-1})	200	22,000	4,000	80
Total coliform count (col 100 ml^{-1})	10	1,400	100	2
Total *E. coli* (col 100 ml^{-1})	0	380	50	0
Total *S. fecalis* (col 100 ml^{-1})	10	1,300	400	3

(and/or rabbits with very low growth rate). However, in commercial rabbitries total feed consumed during growth is divided by weight of rabbits sold minus weight of rabbits weaned. This implies that high mortality rates produce worse FCR results. Thus, Maertens (1991) obtained an FCR of 3.01 for 2.4 kg of slaughter weight without mortality. When the mortality was 0.10 or 0.20 the FCR deteriorated to 3.51 and 4.15, respectively. Similarly, a global FCR can be defined as total feed consumed (feed bought) in the rabbitry (maternity and growth) divided by weight of rabbits sold. In this case, the reproductive cycle of does and other management techniques (culling rate, renewal) will have a very significant effect (Table 14.6).

The most important factors that influence the technical FCR for rabbits fed *ad libitum* are those linked to: (i) rabbits: slaughter weight and strain; and (ii) feed: energy concentration, feed additives, etc.

The first point, slaughter weight, can be shown by the increment of feed/gain ratio with age in Table 14.4. The FCR deteriorates with increasing slaughter weight, but the contribution of the maternal feeding per kg of rabbit produced decreases, so the differences between technical and global FCR decrease with slaughter weight and with doe productivity (Table 14.6). Although the slaughter weight is determined by the market demands, some variations can be introduced by producers according to the ratio of feed price/rabbit price or feed price/carcass price.

The effect of strain has also been shown by Maertens (1992), comparing two lines selected over 2 years for reproductive traits (dam line) and for growth rate and feed efficiency (sire line) (Table 14.7). The FCR was 11% lower for the sire than the dam line (2.86 vs. 3.18) despite its higher finishing weight (14%). However, a negative effect was shown in the dressing percentage of the sire line, so the feed conversion measured as feed intake/commercial carcass is more similar between lines. This indicates the need for including slaughter traits and meat quality parameters in the selection programmes (Blasco *et al.*, 1996).

The feed characteristic that most influences the FCR is energy concentration. Thus, Maertens and De Groote (1987) found a linear relationship between FCR and DE content of diets (FCR = $6.6 - 0.33$ DE MJ kg^{-1}), for

Table 14.6. Global FCR for different slaughter weights and number of rabbits produced per doe and per year.

Number of rabbits produced per doe and year			40	45	50
Slaughter weight (kg)	Weight gain (g day⁻¹)	Feed/gain	Global feed conversion ratio		
2.00	40	3.0	3.38	3.27	3.16
2.25	39	3.2	3.54	3.44	3.34
2.50	37	3.7	3.79	3.70	3.61

Table 14.7. Effect of strain (after 2 years of selection for reproductive traits, DWG and FCR for dam and sire lines, respectively) on growth parameters (Maertens, 1992).

	Dam line	Sire line
Weaning weight (g, 28 days)	642[a]	582[b]
Daily weight gain (g)	37.6[A]	46.6[B]
Feed conversion ratio	3.18[A]	2.86[B]
Finishing weight (g, 70 days)	2221[A]	2539[B]
Slaughter weight (g)	2358[A]	2812[B]
Dressing proportion	0.625[A]	0.606[B]

Rows with different letters differ significantly (A, B, $P < 0.01$; a, b, $P < 0.05$).

commercial rabbits from 4 to 11 weeks of age. The precise limits of this ratio are not clear. Although a regulation of intake, to achieve constant daily energy intake, is possible for DE concentrations higher than 9.3 MJ kg^{-1}, the daily energy intake for experimental diets with added fat tends to be higher.

Some feed additives are added to diets because of their beneficial effects on growth parameters mainly on FCR and/or mortality. However, there are conflicting results on this subject in the literature. The addition of fructo-oligosaccharides and probiotics produces variable results, but usually the most important effect is the reduction of mortality and sometimes an improvement in FCR, these effects being more pronounced under less favourable housing conditions. Growth promoters, such as flavophospholipol, have a favourable effect on feed efficiency (about 5%; Maertens *et al.*, 1992), although no effect was found recently by Jerome *et al.* (1996) for growth promoters used alone or in combination with oxytetracycline.

Other factors such as season, housing and environmental conditions affect the FCR because of their effect on the requirements for thermo-regulation. Table 14.8 shows the effect of season on growth parameters in the east of Spain. The best feed efficiency is obtained during the summer, despite the lowest growth rate. On the other hand, the highest growth rate and FCR is obtained during winter (Ramon *et al.*, 1996). Additionally, the stocking density of animals, the type of cages, the feeder space per rabbit and its design to prevent feeding wastage affect the FCR. Thus, a recent study (Taboada *et*

Table 14.8. Effect of season on feed efficiency and postweaning growth (Ramon *et al.*, 1996).

Season	Spring	Summer	Winter	SE
Weight at 63 days (g)	2298[b]	2134[a]	2356[c]	17.0
Daily weight gain (g day^{-1})	45.64[b]	41.62[a]	47.72[c]	0.34
Feed conversion ratio	2.93[b]	2.70[a]	3.08[c]	0.03

Rows with different letters differ significantly ($P < 0.05$).

al., 1996) comparing the same diets supplied to rabbits caged individually or in groups of eight (stocking density four times higher) obtained a better FCR (2.8 vs. 3.0) for rabbits housed individually.

References

Angulo, E., Brufau, J. and Esteve-García, E. (1995) Effect of sepiolite on pellet durability in feeds differing in fat and fibre content. *Animal Feed Science and Technology* 53, 233–241.
Auvergne, A., Bouyssou, T., Pairet, M., Bouillier-Oudot, M., Ruckebusch, Y. and Candau, M. (1977) Nature de l'aliment, finesse de mouture et données anato-fonctionelles du tube digestif proximal du lapin. *Reproduction, Nutrition and Development* 27, 755–768.
Berriot, C. (1992) Problems with drinking water in livestock production. *Misset World Poultry* 8, 25–27.
Blasco, A., Piles, M., Rodriguez, E. and Pla, M. (1996) The effect of selection for growth rate on the live weight growth curve in rabbits. In: Lebas, F. (ed.) *Proceedings of the 6th World Rabbit Congress, Toulouse*, Vol. 2. Association Française de Cuniculture, Lempdes, pp. 245–248.
Bombeke, A., Okerman, F. and Moermans, R. (1978) L'influence de la granulation à sec et à la vapeur de rations à teneurs différents en énergie sur les résultats de production des lapins de chair. *Revue de l'Agriculture* 31, 945–955.
Candau, M., Auvergne, A., Comes, F. and Bouillier-Oudot (1986) Influence de la forme de présentation et de la finesse de mouture de l'aliment sur les performances zootechniques et la fonction caecale chez le lapin en croissance. *Annales de Zootechnie* 35, 373–386.
Cavani, C., Bianconi, L. and Urrai, G. (1991) Distributione automatizzata e frazionata degli alimenti nel coniglio in accrescimento: 1. Influenza della modalità di distribuzione e del livello alimentare. In: *Proceedings IX Congresso National de ASPA*, Vol. 2. ASPA, Rome, pp. 857–864.
Fernández-Carmona, J., Cervera, C. and Blas, E. (1983) Utilización de piensos extrusionados en el destete de gazapos. In: ASESCU (ed.) *Proceedings XVIII Symposium de Cunicultura, Granollers*, pp. 55–57.
García, H.R. (1984) El agua de bebida. In: ASESCU (ed.) *Proceedings IX Symposium de Cunicultura, Figueras*, pp. 147–153.
García, J., Carabaño, R., Perez Alba., L. and de Blas, C. (1996) Effect of fibre source on neutral detergent fibre digestion and caecal traits in rabbits. In: Lebas, F. (ed.) *Proceedings of the 6th World Rabbit Congress, Toulouse*, Vol. 1. Association Française de Cuniculture, Lempdes, pp. 175–179.
Gidenne, T. (1992) Effect of fibre level, particle size and adaptation period on digestibility and rate of passage as measured at the ileum and in the faeces in the adult rabbit. *British Journal of Nutrition* 67, 133–146.
Goby, J.P. and Rochon, J.J. (1990) Utilisation d'un aliment fermier chez le lapin à l'engraissement: digestibilité et impact du tri alimentaire. In: *Proceedings 5ème Journées de la Recherche Cunicole, Paris*, Vol. 2. Communication no. 62.
Gosalvez, L.F., Alvariño, J.M.R., Estavillo, S. and Tor, M. (1994) Adelanto del incio

de la vida reproductiva de la coneja mediante estimulo alimentico. In: ASESCU (ed.) *Proceedings XIX Symposium de Cunicultura, Silleda*, pp. 149–154.

Harris, D.J., Cheeke, P.R. and Patton, N.M. (1983) Feed preference and growth performance of rabbits fed pelleted versus unpelleted diets. *Journal of Applied Rabbit Research* 6, 15–17.

Harris, D.J., Cheeke, P.R. and Patton, N.M. (1984) Effect of pellet size on the growth performance and feed preference of weaning rabbits. *Journal of Applied Rabbit Research* 7, 106–109.

Jerome, N., Mousset, J.L., Lebas, F. and Robart, P. (1996) Effect of diet supplementation with oxytetracycline combined or not with different feed-additives on fattening performance in the rabbit. In: Lebas, F. (ed.) *Proceedings of the 6th World Rabbit Congress, Toulouse*, Vol. 1. Association Française de Cuniculture, Lempdes, pp. 205–210.

King, J.O.L. (1974) The effects of pelleting rations with and without an antibiotic on the growth rate of rabbits. *Veterinary Record* 94, 586–588.

Koehl, P.F. and Mirabito, L. (1996) Working times in rabbit production system with batch. In: Lebas, F. (ed.) *Proceedings of the 6th World Rabbit Congress, Toulouse*, Vol. 3. Association Française de Cuniculture, Lempdes, pp. 369–372.

Laffolay, B. (1985) Croissance journalière du lapin. *Cuniculture* 12, 331–336.

Laplace, J.P. and Lebas, F. (1977) Le transit digestif chez le lapin. VII. Influence de la finesse de broyage des constituants d'un aliment granulé. *Annales de Zootechnie* 26, 575–584.

Lebas, F. (1973) Possibilités d'alimentation du lapin en croissance avec des régimes présentés sous forme de farine. *Annales de Zootechnie* 23, 249–251.

Lebas, F. (1975) *Le Lapin de Chair, ses Besoins Nutritionnels et son Alimentation Pratique*. ITAVI, Paris, 49 pp.

Lebas, F. and Delaveau, A. (1975) Influence de la restriction du temps d'accès à la boisson sur la consommation alimentaire et la croissance du lapin. *Annales de Zootechnie* 24, 311–313.

Lebas, F. and Laplace, J.F. (1977) Le transit digestif chez le lapin. VI. Influence de la granulation des aliments. *Annales de Zootechnie* 26, 83–91.

Lebas, F., Maitre, I., Seroux, M. and Franck, T. (1986) Influence du broyage des matières premières avant agglomération de deux aliments pour lapins, différant par leur taux de constituants membranaires: digestibilité et performances de croissance. In: *Proceedings 4ème Journées de la Recherche Cunicole, Paris*, Vol.1. Communication no. 9.

Ledin, L. (1984) Effect of restricted feeding and realimentation on compensatory growth, carcass composition and organs growth in rabbit. *Annales de Zootechnie* 33, 33–50.

Luzi, F., Maertens, L., Mijten, P. and Pizzi, F. (1996) Effect of feeding level and dietary protein content on libido and semen characteristics of bucks. In: Lebas, F. (ed.) *Proceedings of the 6th World Rabbit Congress, Toulouse*, Vol. 2. Association Française de Cuniculture, Lempdes, pp. 87–92.

Machin, D.H., Butcher, C., Owen, E., Gryant, M. and Owen, J.E. (1980) The effects of dietary metabolizable energy concentration and physical form of the diet on the performances of growing rabbits. In: *Proceedings 2nd World Rabbit Congress, Barcelona*, Vol. 2, pp. 65–75.

Maertens, L. (1991) Factors affecting the feed conversion ratio of commercial rabbits. Paper presented at the Conference of the Belgium Branch of the WRSA, Ghent, 10 November.

Maertens, L. (1992) Selection scheme, performance level and comparative test of two lines of meat rabbits. *Journal of Applied Rabbit Research* 15, 206–212.

Maertens, L. (1994) Influence du diamètre du granulé sur les performances des lapereaux avant et après sevrage. In: *Proceedings 6ème Journées de la Recherche Cunicole, La Rochelle*, Vol. 2. pp. 325–332.

Maertens, L. and De Groote, G. (1987) Quelques caractéristiques spécifiques de l'alimentation des lapins. *Revue de l'Agriculture* 40, 1185–1203.

Maertens, L. and De Groote, G. (1988) The influence of the dietary energy content on the performances of post-partum breeding does. In: *Proceedings of the 4th World Rabbit Congress, Budapest*, Vol. 3. Sandor Holdas, Hercegalom, pp. 42–53.

Maertens, L. and Luzi, F. (1995) The effect of extrusion in diets with different starch levels on the performance and digestibility of young rabbits. 9. Arbeitstagung über Haltung und Krankheiten der Kaninchen, Pelztiere und Heimtiere. Celle. *Deutsche Vet. Medizinische Gesellschaft e. V., Giessen*, pp. 131–138.

Maertens, L. and Luzi, F. (1996) Effect of dietary dilution on the performance and N-excretion of growing rabbits. In: Lebas, F. (ed.) *Proceedings of the 6th World Rabbit Congress, Toulouse*, Vol. 1. Association Française de Cuniculture, Lempdes, pp. 237–241.

Maertens, L. and Luzi, F. (1997) Reduction of N-excretion of growing rabbits using phase feeding. 10. Arbeitstagung über Haltung und Krankheiten der Kaninchen, Pelztiere und Heimtiere. Celle. *Deutsche Vet. Medizinische Gesellschafte e. V., Giessen*, pp. 136–142.

Maertens, L. and Okerman, F. (1987) L'influence de la méthode d'élevage sur les performances des jeunes lapines. *Revue de l'Agriculture* 40, 1171–1183.

Maertens, L., Vermeulen, A. and De Groote, G. (1988) Effect of post-partum breeding and litter management on the performances of hybrid does. In: *Proceedings of the 4th World Rabbit Congress, Budapest*, Vol. 1. Sandar Holdas, Hercegalom, pp. 141–150.

Maertens, L., Moermans, R. and De Groote, G. (1992) Flavomycin for early weaned rabbits: response of dose on zootechnical performances – effect on nutrient digestibility. *Journal of Applied Rabbit Research* 15, 1270–1277.

Morisse, J.P., Maurice, R. and Boilletot, E. (1985) La dureté du granulé chez le lapin. *Cuniculture* 12, 267–269.

Partridge, G., Allan, S.J. and Findlay, M. (1985) Studies on the nutritive value of roots, cabbage and grass silage for growing commercial rabbits. *Animal Feed Science and Technology* 13, 299–311.

Partridge, G., Daniels, Y. and Fordyce, R. (1986) The effects of energy intake during pregnancy in doe rabbits on pup birth weight, milk output and maternal body composition change in the ensuing lactation. *Journal of Agricultural Science Cambridge* 107, 679–708.

Perrier, G. and Ouhayoun, J. (1996) Growth and carcass traits of the rabbit. A comparative study of three modes of feed rationing during fattening. In: Lebas, F. (ed.) *Proceedings of the 6th World Rabbit Congress, Toulouse*, Vol. 3. Association Française de Cuniculture, Lempdes, pp. 225–232.

Phost, H.B. (1963) Testing the durability of pelleted feed. *Feedstuffs* 35 (13), 66–68.

Prud'hon, M., Cherubin, M., Carles, Y. and Goussopoulos, J. (1975) Effets de différents niveaux de restriction hydrique sur l'ingestion d'aliments solides par le lapin. *Annales de Zootechnie* 24, 299–310.

Ramon, J., Gomez, E.A., Perucho, O., Rafel, O. and Baselga, M. (1996) Feed efficiency and postweaning growth of several Spanish selected lines. In: Lebas, F. (ed.) *Proceedings of the 6th World Rabbit Congress, Toulouse,* Vol. 2. Association Française de Cuniculture, Lempdes, pp. 351–353.

Rommers, J.M., Meijerhof, G, Van Someren and Kranenburg, M. (1996) Effect van het bijvoeren van granen aan vleeskonijnen. *NOK Kontaktblad* 14, 88–94.

Sánchez, W.K., Cheeke, P.R. and Patton, N.M. (1984) The use of chopped alfalfa rations with varying levels of molasses for weanling rabbits. *Journal of Applied Rabbit Research* 7, 13–16.

Schlolaut, W. (1995) *Das große Buch vom Kaninchen.* DGL-Verlag, Frankfurt am Main, pp. 219–226.

Taboada, E., Mendez, J. and De Blas, J.C. (1996) The response of highly productive rabbits to dietary sulphur amino acid content for reproduction and growth. *Reproduction, Nutrition and Development* 36, 191–203.

Van der Broeck, L. and Lampo, P. (1979) Influence de l'âge au premier accouplement sur la fertilité de jeunes lapines et leurs performances en première portée. *Annales de Zootechnie* 28, 443–452.

15. Climatic Environment

C. Cervera and J. Fernández Carmona

Departamento de Ciencia Animal, Universidad Politécnica de Valencia, Camino de Vera, Apdo 22012, 46071 Valencia, Spain

Introduction

Biometeorology is the study of the relationship between the environment and living organisms. This chapter will consider one specific area by focusing on nutrition.

The variables which define an environment are temperature, humidity, light, altitude, smell and noise. Little is known about the last three, apart from some data relating levels of CO_2 or NH_3 to some productive parameters of animals, which are essential to the control of the frequency of air exchange in intensive farms. The reciprocal problem of the contaminating effect of the animals themselves on the atmosphere and soil also appears to be very interesting, and numerous studies have been undertaken on species other than rabbits which show the relationship between nutrition and the production of NH_3, CH_4, P and Na.

Lighting schedules are programmed to control reproduction and, together with some feed restriction, they could marginally improve feed efficiency in growing rabbits. This is a controversial idea because the effect depends on the age of the rabbit, length of feeding and time of feeding. Feeding usually occurs in the late afternoon and at night, so feed must be available at these times (Maertens, 1992).

Ambient temperature and humidity are the variables which most affect nutrition. Both directly influence the energy equilibrium of the animal, changing the flow of heat between the animal and the environment. In particular situations for species other than rabbits, correlations have been found between different functions of temperature and humidity which permit the most reliable to be selected; for example, the temperature humidity index (THI), wet bulb globe temperature, radiant temperature, equivalent temperature and effective temperature. Some other indices define the environment in terms of the animal response, including operative temperature, isoambient lines, thermoneutral zone (TNZ) and comfort zone. Air temperature alone may be acceptable in laboratory conditions but it becomes an inadequate measure of the degree of thermal stress on animals exposed to the natural environment (Young, 1977). Unfortunately insufficient information is available to express all the climatic variables in one unit, such as the effective ambient temperature or others, and mean air temperature is employed instead as the usual index.

The lack of research work on rabbits probably limits the discussion of climatic variables solely to the effect of temperature, daily minimum temperature (Fuquay, 1981) or daily average temperature. This can lead to substantial errors when other parameters have a significant effect, for example wind, solar radiation or ambient humidity. Taking these factors into account, the dry temperature is the most important and representative index, and humidity (wet bulb temperature) can also be used when its incidence has been measured. Future research will perhaps define a THI similar to that used for other non-sweating species.

The environment is sometimes defined by the season of the year, and even by the particular system of production, such as building interior vs. open air. This type of loose definition adds even more imprecision to the conditions, and makes it impossible to replicate the observations or the experimental work, although it may have practical value for some specific zones or purposes.

Another difficulty in measuring temperature in terms of minimum, maximum or mean results from diurnal and seasonal fluctuations. Many experiments have been designed with animals in climate chambers at a constant temperature, making comparisons with animals exposed to a changing environment of limited relevance. In a naturally fluctuating environment, animals appear to overcome partially the adverse effects by taking advantage of the more favourable part of the cycle. It follows that a constant temperature is less well tolerated than a cyclic one, and the NRC (1981) suggested that the effect should be similar to that of a constant temperature equivalent to the mean of the cycle.

As a consequence of the small and fragmentary amount of work undertaken on rabbits to date a review of experimental data raises more questions than it solves. This chapter has tried to include the most valuable published work describing a general approach to the complex nutrition–environment relationship.

Interest in the subject is not recent, but has increased lately because it is a crucial element in the nutrition of domestic animals in cold climates and even more in subtropical and tropical environments.

Tables of nutrient requirements have been developed for rabbits from work carried out in intensive production systems in temperate countries under moderate conditions relatively free of thermal stress, where animals can realize their full genetic potential. Application of these data can lead to considerable errors in feeding animals exposed to acute cold or heat stress, because the physiological responses involve changes in voluntary water intake, feed intake, maintenance energy requirements and level of production. It seems that some corrections need to be introduced, depending upon the environment to which animals are acclimatized or temporarily exposed.

Thermal stress directly affects reproduction, health and nutrition, and all of these interact with each other. The overall result for animals exposed to thermal stress is always a reduction in productivity, which varies according to the severity of the stress and the acclimatization of the animal.

Thermoneutral zone

Homeothermic animals maintain a constant core temperature, compensating for heat loss through morphological, physiological and behavioural mechanisms. 'The range of ambient temperature as an expression of thermal environment within which metabolic rates are at minimum and temperature regulation is achieved by non-evaporative physical processes alone' is the definition of thermal neutrality accepted by the International Union of Physiological Sciences (Bligh and Johnson, 1973). Including a phrase such as 'in normal productive animals' should make the concept of TNZ more acceptable to farmers. It implies that a non-stressful situation means not only normal body temperature and no shivering, sweating or panting but also that feed intake does not change. Net energy for productive purposes is the difference between metabolizable energy intake and heat production which in turn involves maintenance needs and the heat increment (HI) of food. Therefore level of nutrition and production of heat increase together, lowering both critical temperatures with maximum energy retention for a given energy intake expected to be in the range between them. HI contributes to maintenance of body temperature in cold conditions but compromises the heat balance in hot environments.

In the TNZ range the animal invokes mainly postural and vasomotor mechanisms to conserve or dissipate body heat. Rabbits take on a ball posture below 10°C to decrease their surface area for conduction or radiation losses. The spread posture at 30°C allows more sensible heat to be dissipated.

The energy-demanding behaviours of rabbits include maternal feeding and digging burrows. Many animals dig burrows underground to protect themselves in both tropical countries and the Arctic. Different types of burrows are connected with survival of rabbits in arid areas of New South Wales, as described by Hall and Myers (1978). Some systems in New Mexico (Gentry, 1983) and Tunisia have been designed to exchange conventional cages or burrows for concrete burrows. In these deep pits temperature remains up to 9°C lower than outside (Finzi *et al.*, 1989).

Vasoconstriction of peripheral blood vessels helps the rabbit to keep warm in cold conditions, so that metabolism need not increase to offset heat loss. In particular the amount of heat conserved in the ears (with an area of

0.026 m^2 in adults) is considerable. McEwen and Heath (1973) reported that heat loss was 0.2 W per °C linear over the range of 30–0°C ambient temperature, about half of the theoretical loss if the ears were maintained at core body temperature. Responses to cold stress rely on more carbohydrates being utilized and fat tissue being easily mobilized. Long-term adaptation to cold involves increased insulation.

In a hot environment rabbits have to dissipate metabolic heat when the thermal gradient between them and the ambient temperature is small or even negative. Firstly they use the least expensive heat loss procedure of skin vasodilation (Kruk and Davydov, 1977). When this becomes insufficient, a decrease in intake of food, which decreases HI, and evaporation of water from the respiratory tract are the mechanisms used to lose excess heat.

The lower and upper critical temperatures should be assessed from metabolic heat production figures. The great variability between published data may be due more to the experimental procedure than to any question related to breed (Table 15.1). The type of chamber, method of measurement, restlessness, adaptation of animals to the temperature and individual differences change the limits of the TNZ.

The TNZ given by several studies between the 1940s and 1960s varies from 20 to 30°C. More recent research has found heat production to decrease from 5 to 35°C (Gonzalez et al., 1971), 10 to 30 (Kluger et al., 1973), 0 to 30 (McEwen and Heath, 1973), 18 to 28 (Scheele et al., 1985) and 20 to 30 (Jin et al., 1990).

Upper critical temperature (UCT) is estimated by all of the cited studies to be lower than 30°C, when some physiological reactions such as feed intake,

Table 15.1. Heat production (kJ kg^{-1} body weight$^{0.75}$).

Source	Temperature °C								
	0	5	10	15	20	25	30	35	40
Johnson et al. (1958)[a]			660		630	580	575	570	675
Johnson et al. (1958)[b]					660		620	570	640
Gonzalez et al. (1971)[c]		613	520	428	398	364	352	421	
McEwen and Heath (1973)[d]	352		312		300		273		
Nichelmann et al. (1974a)[e]			570	520	490	425	400	400	485
Scheele et al. (1985)[f]				396			331		
Sanz et al. (1989)[g]					550	520	440	400	
Jin et al. (1990)[h]					488		378		

[a]4 kg fed females reared at 9°C; [b]4 kg fed females reared at 28°C (both sets of values approximate. Calculated from the original figure); [c]fasting animals (assuming 3 kg liveweight); [d]3.9 kg fasting animals; [e]fasting 42 day animals approximately 1 kg liveweight (points interpolated from the original response); [f]fed growing rabbits; [g]10-day-old sucking rabbits (interpolating original values); [h]fed growing rabbits approximately 1.6 kg (mean of two diets).

respiration rate, panting and body temperature were examined, but lack of information for intermediate values from 20 to 30°C makes it difficult to reduce the uncertainty. Most studies did find appreciable differences in heat production between 10 and 20°C, and therefore the lower critical temperature (LCT) should be in between these values. From a survey of the literature McEwen and Heath (1973) calculated a value of approximately 15°C for the LCT.

Heat production was also measured by Johnson *et al.* (1958) in fed mature New Zealand rabbits exposed to temperature levels from 10 to 40°C. Heat production decreased steadily up to 35°C and appeared to increase afterwards. From the data of heat production, feed intake and thyroid activity the authors suggested a UCT of around 24°C.

These results have been confirmed by a series of experiments conducted by Nichelmann *et al.* (1973a, 1974a). They showed that heat production had the shape of a parabola, with the point of inflection at 25°C for adults, 30°C for 42-day-old rabbits and 35°C for 10-day-old pups. The TNZ lay between 15 and 25°C for animals aged 80 days. This range coincides approximately with some estimates carried out on commercial farms. The shape of the response was not so clear in the work of Gonzalez *et al.* (1971) who reported a relatively constant heat production between 15 and 30°C.

It was recognized 60 years ago that a high ambient temperature causes

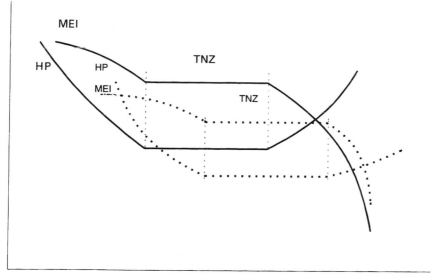

Fig. 15.1. Schematic representation of heat production and energy intake. NTZ, thermal neutral zone; MEI, metabolizable energy intake; HP, heat production; — non-acclimatized animals; . . . heat-acclimatized animals.

increased body temperature (Lee, 1939). The stress at 30°C and above seems to be very high, because rectal temperature starts to increase at 27°C, rises abruptly at 32°C, together with a sudden decrease in feed consumption (Johnson et al., 1958), and reaches 42.5°C at 35°C (Nichelmann et al., 1973a). Thereafter, cooling of the animals is based upon evaporation, the rabbit being particularly ineffective at this process compared with other species. Relatively low ratios of surface to skin vaporization, and respiratory tract evaporation to heat production indicate that the rabbit has little ability to dissipate heat. Panting alone, even at 400 breaths min^{-1} at 35°C, did not prevent an elevation of rectal temperature of 1.4°C (Gonzalez et al., 1971). In stressed non-acclimatized animals at 38°C, Johnson et al. (1958) reported that evaporation accounted for 0.30 (0.40 was recorded at 40°C by Nichelmann et al., 1973b) of the total heat produced by adult animals housed at 35°C, of which 0.60 is lost through panting; no increase was observed from 20 to 30°C in young animals acclimatized for 7 days (Nichelmann et al., 1974b). Total evaporative heat loss has been estimated to be 0.6, 1.26 and 2.0 W kg^{-1} at each ambient temperature of 15, 30 and 35°C. Using the latent heat of evaporation as 0.7 W $g^{-1} h^{-1}$, a 4 kg doe should evaporate around 7 g water h^{-1} at 30°C. The increased energy requirement during panting, with the additional rise in body temperature, seems to be non-linear as opposed to the apparent linear increase in cold conditions. In severe heat stress, feed intake is so depressed that the maintenance requirements cannot be properly calculated.

Ears are a means of dissipating heat. The heat exchange coefficient is 9.1 W m^{-2} °C^{-1}, about four times the coefficient for the whole animal (Kluger et al., 1973). In a wind of 60 m s^{-1}, fully dilated ears can lose twice as much heat as ears in non-forced convection. However, most rabbits die after a few days exposure to 40°C.

Stressed animals have lower productivity in the short term. The problem can be solved in temperate or even cold climates where highly productive breeds can be satisfactorily reared in heated or insulated buildings. However, the production of meat and milk, and reproduction, suffer in this increasing order as a consequence of hot climates. Multiple factors always cause the response of rabbits to a hot climate to be reduction in food intake, litters per year and offspring per litter. Normal production figures, as estimated by several experts, could be four or five litters, 20–25 animals weaned yearly and less than 20 g liveweight gain day^{-1} for growing animals (Colin, 1991, 1995).

A lower basal or resting heat production reported in animals continuously exposed to environments at around 30°C would be of value in adaptation to hot environments. Alliston and Rich (1973) suggested that, following an 18-day acclimation period, there was a positive relationship between thermoregulatory efficiency and pregnancy. The results of Johnson et al. (1958) suggest that the influence of rising environmental temperature was less in New Zealand rabbits reared at 28°C than in those reared at 9°C, as rectal minus air temperature, pulse rate and respiration rate were lower.

It is important to predict the future resistance to thermal stress of animals which are presumably to suffer it, and the selection of breeders must be based on a type of index linked to the production of heat, because this decreases in previously acclimatized animals. Probably the most useful are based on rectal temperature, such as the Iberia test (Rhoad, 1944), used on beef cattle and also of proven validity in rabbits (Alliston and Rich, 1973; Finzi *et al.*, 1988). Other indices could be some biochemical factors (Amici *et al.*, 1995) and certain nutritional parameters, such as nitrogen retention, body fluids or fat (Kamal and Johnson, 1971).

There has been very little work to date designed to anticipate and understand the degree of readaptation or recovery of animals which have suffered thermal stress. Such readaptive behaviour can void or change the conclusions obtained from short-term observations. Age, physiological condition, and duration and intensity of stress should affect recovery to the original level of production, which in female breeding rabbits was possible after 10 months at 30°C (Fernández Carmona *et al.*, 1994c).

Heat production of newborn rabbits housed in metabolism cages and subjected to several ambient temperatures has been determined both for 10-day-old animals suckling normally (Nichelmann *et al.*, 1974a), and as early as 2 h after birth in fasting animals (Cardasis and Sinclair, 1972). These studies demonstrated a cold-induced increment of heat production. Kits were able to raise their basal level at 35°C, when temperature dropped to 30, 20 or 10°C. Peak metabolism was achieved at around 15°C at 10 days.

Neonatal animals cannot maintain body temperature if unfed or if the ambient temperature drops abruptly (Hill, 1961), although they tend to group together and curl up. The metabolic responses to both cold and fasting are the result of complex interactions (Schenk *et al.*, 1975) which bring about a higher mortality rate compared with normal suckling rabbits.

Survival time and rate of weight loss of fasted newborn rabbits are strongly influenced by temperature, which is known to stimulate thermogenic activity of brown fat. The probability that large or small littermates survive could be related to the different lipid stores (Dawkins and Hull, 1964). Cardasis and Sinclair (1972) placed newborn rabbits at 35, 30 and 25°C, reporting that smaller animals survived for fewer hours. Young animals whose birth weight exceeded 50 g were able to maintain metabolic rate and body temperature for longer.

Heat production of newborn rabbits increases during the first days of life. A 10-day-old rabbit coordinates movement and has already improved insulation, gaining hair and tissue fat. The use of nest boxes with bedding, where young rabbits huddle, thereby reducing body-surface exposure, means that the optimal temperature range becomes wider with increasing age: 20–30°C at 20 days of life (Rafai and Papp, 1984). Practical observations conclude that a 20°C temperature indoors leads to high viability in farms (Delaveau, 1982), but maternal behaviour significantly affects the proper

used as an energy source. The practical approach in cold conditions is to increase dietary energy levels.

The consequences of hot environments on feed intake which means less protein being ingested and reduced growth, have generally resulted in recommending higher levels of protein in warm climates or seasons. Addition of some amino acids, particularly lysine, has alleviated the effect of heat on pigs and poultry. It may be observed that equal daily intakes at different temperature have resulted in comparable liveweight gains.

In fact, levels of protein in the diet should be corrected according to body gains or milk protein yield rather than being changed only as a result of the expected intake. The predicted performance of pigs and poultry is used to tabulate the adjustments for environmental stress, but for rabbits insufficient information is available.

Estimations of ME requirements based on heat production figures have been used for pigs and poultry exposed to different temperatures, but it would appear that no similar studies have been undertaken with rabbits.

High-energy diets have been reported to overcome the lower energy intake in hot environments when fed to cattle, pigs and poultry. Diets with minimal HI should be beneficial; Jin *et al.* (1990) found that evaporative losses in 1.9 kg rabbits at 30°C were higher for a diet containing 231 g NDF than for one based on 165 g NDF kg^{-1}. However, feed intake in that experiment was also higher, making it difficult to establish the relationship between fibre and stress.

Attempts to overcome the expected poor intake would require the use of high fat–low fibre diets. While the range of fibre is relatively narrow in feeding rabbits, the limit of fat incorporated into a diet seems to be controlled only by the physical structure of the pellet. Very few experiments have related the composition of rabbit diets with production outside the TNZ. However some results suggest that the response of rabbit females at high temperature was especially poor when they were fed on a low-energy diet (Simplicio *et al.*, 1991), and it was significantly improved when a high-energy diet was supplied to does (Simplicio *et al.*, 1991; Fernández Carmona *et al.*, 1996).

At this point it may be emphasized that two different ways of enhancing performance through improvement of a diet can be considered (Fig. 15.2). The first may be defined as a general improvement of performance, but it is questionable whether this is really what is wanted. Had the previous diet not been deficient, it might have promoted a similar response, although there is usually scope to improve a diet. In fact most diets are formulated using a least-cost programme, assuming optimal, non-maximal performance. The second approach can respond to actual attempts to obtain a better diet for severe climatic conditions. It should be noticed that a different response caused solely by some difference in temperature implies a significant diet–temperature interaction.

Increased voluntary intake in cold conditions tends to overcome any

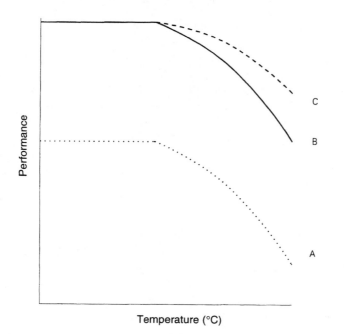
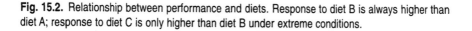

Fig. 15.2. Relationship between performance and diets. Response to diet B is always higher than diet A; response to diet C is only higher than diet B under extreme conditions.

marginal deficiency in nutrients although not energy. Compared with a given diet in TNZ, a smaller proportion of dietary protein is needed for the above reasons, and more protein is utilized as an additional energy source. In view of their heat increment, proteins may temporarily exert a favourable effect in a cold environment. For many years a high-fat diet has been proved to be the most effective means of maintaining body temperature, coincident with the needs arising from the two-stage adaptation of rabbits to cold: cold stress with depletion of fat reserves, and cold adaptation linked to lipid deposition.

The increment of ME requirement corresponds exactly to the extra heat produced. From data on heat production, it can be estimated as 0.15–0.20 of the requirements for a growing animals at 15–20°C (Table 15.1).

HI of roughage has been useful for feed-restricted or low-productive cattle. Ortiz *et al.* (1989) using indirect calorimetry trials on growing rabbits, have reported that the DE value overestimated by 5% the net energy value expected when ADF level increased from 180 to 240 g kg^{-1}. Fibre could then be considered for rabbits as a moderately effective means of enhancing heat production, and conversely of limiting the total net energy intake. No reference work is available for cold-stressed rabbits in any of the subjects mentioned here.

Obviously there is broad scope for research on the interaction between protein and energy allowances and the environment in rabbits fed *ad libitum* in hot or cold conditions.

Virtually no attempts have been undertaken to test inputs other than energy, and nutrients other than protein. There are some references on the addition of probiotics in hot climates, with little emphasis on the problem of temperature. Adding disodium or dipotassium carbonate has proved to be effective at high temperatures (Bonsembiante *et al.*, 1989; Fayez *et al.*, 1994), probably preventing the action of hyperventilation on the acid–base balance.

Performance of rabbits in tropical countries is likely to be maximized by providing free access to drinking water at all times (Thwaites *et al.*, 1990). Water-restricted rabbits, besides saving water output in faeces and urine, show a significant reduction in dry matter intake. Kits under 21 days of age seem to learn how to drink water at high temperatures (McNitt and Moody, 1992). A water/food intake ratio of about 2 was recorded for adult rabbits fed *ad libitum* by Prud'hon (1976) at 20°C, similar to that published by Jin *et al.* (1990) in growing females and Kasa *et al.* (1989) in fattening rabbits. There is a rise in the ratio of water intake to DM intake up to 2.4 between 20 and 30°C. Drinking cool water has sometimes been recommended in hot situations but it should be noted that, besides the practical difficulty, the cooling effect is very small compared with the loss of heat through water vaporizing from the respiratory tract.

Breeding does and litters

The effect of heat stress has been measured in experimental conditions on mating, fertility, embryo survival and litter size at birth (Howarth *et al.*, 1965; Alliston and Rich, 1973; Shafie *et al.*, 1979; Abo-Elezz, 1982; Kamar *et al.*, 1982; Tramell *et al.*, 1988). This poor productive performance of does has also been found in commercial farms: Masoero and Auxilia (1977) reported at 25°C a decline in mating and fertility. Summertime has brought about poor results in mating, numbers of live offspring, mortality and size of weaning litter (Sittmann *et al.*, 1964; Pagano-Toscano *et al.*, 1990). The worst overall productive results are mostly obtained in hot conditions (summer in southern European countries). In a rabbit mortality survey, Rosell (1996) found respiratory disorders to be the main cause of death in females, and certainly in winter time the mortality of lactating rabbits can increase and respiratory pathology may become severe (Battaglini *et al.*, 1986; Costantini and Castellini, 1990; Mori and Bagliacca, 1990). Reproductive traits in hot countries are far removed from European Standards (Cardelli, 1993), though some advances have recently been reported (Baselga and Marai, 1994).

High ambient temperature appears to act on reproduction directly and through the depression of voluntary feed intake (Fig. 15.3). Experimental research, where feed intake during heat stress was recorded, has confirmed

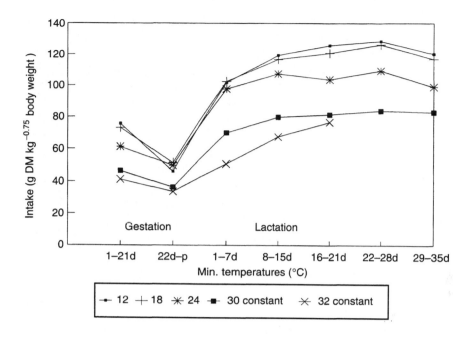

Fig. 15.3. Feed intake of does. 32°C constant from Wittorff *et al.* (1988); remainder from Fernández Carmona *et al.* (1994b). Regression line from original data for 22d-p and lactation: intake g DM kg$^{-0.75}$ body weight = 81.72 + 10.45 * W − 0.048 * T^2 (SE = 15.8, R^2 = 0.71; W 1–6 weeks starting from week 22-partum; T = temperature in °C).

low performance in summertime (Mendez *et al.*, 1986), or in a transitory temperature rise (Maertens and De Groote, 1990), and in a controlled environment (Rafai and Papp, 1984; Papp and Rafai, 1988; Wittorff *et al.*, 1988; Simplicio *et al.*, 1991; Fernández Carmona *et al.*, 1995).

The work of Rafai and Papp (1984) examined daily milk output and food intake for temperatures between 5 and 35°C. At 25°C a noticeable reduction of feed intake started, and subsequently milk, litter weight and doe weight were affected. Later work (Papp and Rafai, 1988) showed that does kept at 35°C die within 72 hours.

Angora rabbits suffer considerable stress when sheared at temperatures below 20°C (Schlolaut, 1987), and very low wool yield and quality have been reported at temperatures of 30°C (Stephan, 1980) or above (Kong *et al.*, 1987). Cheng *et al.* (1991) recorded in summer, with a 23.1°C monthly average temperature, 5.6% less intake and 10.6% less wool yield than in spring. Shorter intervals between shearings can be employed in high temperature conditions. Optimal temperatures for hair growth have been estimated at about 25°C after shearing, and not above 15°C before it (Schlolaut, 1987).

Cold seasons affect the mortality of suckling rabbits (Ferraz *et al.*, 1991) but Lukefahr *et al.* (1983) did not observe this relationship. A survey of the literature (Partridge, 1988) gave a mortality rate in pre-weaning rabbits from 0.15 to 0.30. Around 0.80 of the pups died as a result of chilling and starvation during the first days of life. The data demonstrated that survival increased up to 0.957 when nestboxes with a low-voltage heated floor were provided.

The possibility that modification of diets formulated for normal environments can alleviate thermal stress by avoiding nutrient deficiencies or increasing voluntary intake has hardly been explored. One logical course of action would seem to be to increase DE of diets, with more cereal or by adding fat as ingredients. Although does compensate in normal conditions for different diet density through corresponding changes in feed intake (Maertens and De Groote, 1988; Fraga *et al.*, 1989), long-term experiments carried out by Mendez *et al.* (1986), Simplicio *et al.* (1991) and Cervera *et al.* (1993) have shown that some added fat elicited a better response from does, perhaps related to high milk fat output (Christ *et al.*, 1996; Pascual *et al.*, 1996).

Barreto and de Blas (1993) did not detect any improvement during summer in Madrid with diets varying from 8.9 to 11.9 MJ DE g^{-1} DM, the last one containing 35 g lard kg^{-1}. In an experiment connected to this one the low-energy diets gave poorer response at 30°C constant temperature, while no statistical difference was found between the high-energy diets (Table 15.2). However, a diet containing 100 g fat kg^{-1} at 30°C promoted better weight gains in litters than a normal control diet (Cervera *et al.*, 1997). It has yet to be established whether this is a specific effect of fat or of energy intake.

The relationship between heat increment of food and nutrition of does can only be found in the study of Fernández Carmona *et al.* (1995), where 57 New Zealand crossbred does were fed on two isoenergetic (DE) diets, with 121 and 193 g crude fibre kg^{-1} DM. Feed intake results showed the second diet to be less efficient, suggesting that does compensated for higher HI in terms of milk production.

Table 15.2. Effect of diet and housing on daily feed intake of does and litter weight at 28 days.

	Diet DE MJ kg^{-1} DM			
	12.9[a]	11.3	10.4	9.0
30°C[b]				
Intake (g DM)	185	181	162	163
Litter weight (kg)	2.65	2.32	1.64	1.38
TB[c]				
Intake (g DM)	283	277	304	320
Litter weight (kg)	3.80	3.42	3.24	3.16

[a]Includes 35 g tallow kg^{-1}; [b]30°C constant (Simplicio *et al.*, 1991); [c]TB, traditional building (Cervera *et al.*, 1993).

Sanz *et al.* (1989) reported that milk intake was similar in rabbits aged 1, 10 or 20 days, reared at several temperatures between 20 and 36°C. Suckling rabbits start to ingest solid food at about the age of 18 days, and a measurable amount at 21 days. Fernández Carmona *et al.* (1991) reported that rabbits kept at 30°C ingested less DM in a 35-day lactation than those at normal temperatures. The composition of pellets seems to be unimportant at these ages, at least from the results of Fernández Carmona *et al.* (1991) who reported no differences in ingestion between diets both at 30°C and at normal temperatures, in agreement with Blas *et al.* (1990) who evaluated a diet with 100 g skimmed milk kg^{-1}.

Growing rabbits

The effect of temperature on the growth of weaned rabbits in the fattening period has often been measured, confirming the fact that voluntary feed intake varies according to whether conditions are cold or warm. DM intake in terms of metabolic weight is shown in Fig. 15.4, where a regression equation has been calculated from the original data, discarding diets with a high DE content. From published works by Zicarelli *et al.* (1979), Stephan (1980), Mori and Bagliacca (1985), Bordi (1986), Casamassima *et al.* (1988) and Samoggia *et al.* (1988) a temperature range from 13 to 20°C can be assumed to be suitable, though some authors have widened this range. Not all of them agree, but it is generally accepted that a reduction in intake occurs at 25°C, and perhaps above 22°C (Casamassima *et al.*, 1988; Fernández Carmona *et al.*, 1994a) and certainly impaired growth is assured around 30°C (Table 15.3).

This outcome is less predictable when seasons of the year are considered, because of the usually large degree of variation in temperature and humidity; regardless of this, there are a substantial number of controlled experiments that have recorded lower intake and growth rate in summer than during the rest of the year (Masoero and Auxilia, 1977; Simplicio *et al.*, 1988). It is hard to establish consistent differences between the results for winter, autumn and spring. Samoggia *et al.* (1988) found higher figures for intake and feed efficiency in winter, but other studies are far from consistent.

Open-air systems tend to be slightly worse in terms of feed efficiency than indoor systems (Blocher *et al.*, 1990) but some opposing results have also been published (Crimella *et al.*, 1996). It seems that a large rabbit farm should be provided with some means of regulating the internal environment, because the stress factors are strongly reinforced by a significant presence of pathogenic microbes and chemical pollutants in the air. Conversely, microbe colonies are non-existent in an open-air system, but climatic stress factors cannot be closely controlled.

The interaction between feeding and temperature remains almost unknown. A reduction of 25% in feed intake, comparable with the percentage

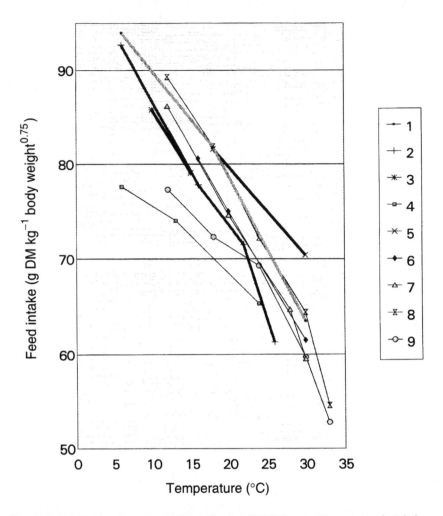

Fig. 15.4. Feed intake of growing rabbits. 1, Stephan (1980) (constant temperatures); 2, Lebas and Ouhayoun (1987) (mean temperatures); 3, Casamassima *et al.* (1988) (minimum temperatures); 4, Nizza *et al.* (1995), 12.5 MJ DE kg⁻¹ (mean temperatures); 5, Borgida and Duperray (1992) (mean temperatures); 6, Fernández Carmona *et al.* (1994a) (minimum temperatures); 7, Boiti *et al.* (1992), Chiericatto *et al.* (1995) (nearly constant temperatures); 8, Cervera *et al.* (1997), 11 MJ DE kg⁻¹ (minimum temperatures); 9, Cervera *et al.* (1997), 12.4 MJ DE kg⁻¹ (minimum temperatures).

observed in hot climates, should be balanced by about the same increment of dietary nutrients. Both Simplicio *et al.* (1988), increasing DE some 10%, and Borgida and Duperray (1992), increasing protein and lysine, found no improvement in average daily gain in rabbits kept at 30°C. Neither increasing protein from 130 to 200 g kg⁻¹ nor an increase from 57 to 62 total digestible nutrients improved the average daily gain when these diets were fed to

Table 15.3. Prediction of growth for 1.5 kg rabbits at two different temperatures.

	18°C	30°C
Feed intake[a] (g DM kg$^{-0.75}$ body weight)	80	60
Energy intake[b] (kJ DE)	1188	891
Maintenance requirements (kJ DE)	745[c]	633[d]
Growth allowances (kJ DE)	443	258
Growth[e] (g day^{-1})	37	21

[a]Approximate; from Fig. 15.4; [b]1.5$^{0.75}$ = 1.35; diet 11 MJ DE kg^{-1} DM; [c]552 kJ DE kg^{-1} body weight$^{0.75}$ (de Blas *et al.*, 1985); [d]15% lower requirements at 30°C; [e]approximately 12 kg DE g^{-1} live weight gain.

slow-growing rabbits in tropical conditions (Deshmukh and Pathak, 1991). Rabbits fattened at mean temperatures of 6, 16, 22 and 26°C, and fed on a diet with 210 g crude protein kg^{-1} (Lebas and Ouhayoun, 1987) gained about 5 g more at each temperature than those with 157 g kg^{-1}. When comparing low-fat and high-fat diets, while gains at moderate temperatures of 12 and 18°C were similar, at 24, 30 and 33°C, the use of high-fat diets slightly improved growth performance as a consequence of the small differences in DM intake (Cervera *et al.*, 1997), and at a temperature of 30°C it also led to around 50% more dissectable fat (M. Plá, personnal communication)

Carcass yield, carcass fat and the efficiency of energy for fattening alter the significance of liveweight gain as the sole predictor of growth performance. Slower growth leads to lighter and leaner carcasses, so that any diet should produce less carcass fat at high ambient temperature.

References

Abdel-Samee, A.M., El Gendy, K.M. and Ibrahim, H. (1994) Rabbit growth and reproductive performance as influenced by feeding desert forage (*Acacia saligna* and *Atriplex nummularia*) at North Sinai. *Egyptian Journal of Rabbit Science* 4, 25–36.

Abo-Elezz, Z.R. (1982) The effect of direct solar radiation during pregnancy on the postnatal growth of rabbits. *Beitrage Tropischen Lanwirtschaft Veterinar-medizin* 20, 69–74.

Alliston, C.W. and Rich, T.D. (1973) Influence of acclimation upon rectal temperatures of rabbits subjected to controlled environmental conditions. *Laboratory Animal Science* 23, 62–67.

Amici, A., Finzi, A., Mastroiacomo, P., Nardini, M. and Tomassi, G. (1995) Functional and metabolic changes in rabbits undergoing continuous heat stress for 24 days. *Animal Science* 61, 399–405.

Barreto, G. and de Blas, J.C. (1993) Effect of dietary fibre and fat content on the reproductive performance of rabbit does bred at two remating times during two seasons. *World Rabbit Science* 1, 77–81.

290   *C. Cervera and J. Fernández Carmona*

Baselga, M. and Marai, I.F.M. (eds) (1994) *Cahiers Options Mediterréennes.* Vol. 8: *Rabbit Production in Hot Climates.* Centre International de Hautes Etudes Agronomiques Méditerranéennes, Paris, 550 pp.

Battaglini, M., Penella, F. and Pauselli, M. (1986) Influenza del mese di parto sulla produttivitá del coniglio. *Rivista di Coniglicoltura* 8, 35–39.

de Blas, C., Fraga, M.J. and Rodriguez, J.M. (1985) Units for feed evaluation and requirements for commercially grown rabbits. *Journal of Animal Science* 60, 1021–1028.

Blas, E., Moya, A., Cervera, C. and Fernández-Carmona, J. (1990) Utilización de un pienso con leche en gazapos lactantes. *Alimentación y Mejora Animal* 30, 155–157.

Bligh, J. and Johnson, K.G. (1973) Glossary of terms for thermal physiology. *Journal of Applied Physiology* 35, 941–961.

Blocher, F., Kohl, P.F. and Strehler, J.F. (1990) Un engraissement en plein air: résultats techniques et economiques. *Rivista di Coniglicoltura* 17, 144–149.

Boiti, C., Chiericato, G.M., Filotto, U. and Canali, C. (1992) The effect of high environmental temperature on plasma testosterone, cortisol, T3 and T4 levels in the growing rabbits. *Journal of Applied Rabbit Research* 15, 447–455.

Bonsembiante, M., Chiericato, G.M. and Bailoni, L. (1989) Risultati sperimentali sull'impiego del bicarbonato di sodio in diete per conigli da carne allevati in condizioni di stress termico. *Rivista di Coniglicoltura* 9, 63–70.

Bordi, A. (1986) Aspetti fisioclimatici dell'allevamento del coniglio. *Rivista di Coniglicoltura* 23, 36–41.

Borgida, L.P. and Duperray, J. (1992) Summer complementary feding of rabbits. *Journal of Applied Rabbit Research* 15, 1063–1070.

Cardasis, C.A. and Sinclair, J.C. (1972) The effects of ambient temperature on the fasted newborn rabbit. 1. Survival time, weight loss, body temperature and oxygen consumption. *Biology of the Neonate* 21, 330–346.

Cardelli, M. (1993) Rabbit production in developing countries. *Rivista di Coniglicoltura* 30, 34–39.

Carew, S.N., Aoyoade, J.A. and Zungwe, E.N. (1989) Composition of plants fed to rabbits in Benue State of Nigeria. *Journal of Applied Rabbit Research* 12, 169–170.

Casamassima, D., Manera, C. and Mugnozza, G.S. (1988) Influenza del microclima sulla produtivita del coniglio. *Rivista di Coniglicoltura* 25, 31–35.

Cervera, C., Fernández Carmona, J., Viudes, P. and Blas, E. (1993) Effect of remating interval and diet on the performance of female rabbits and their litters. *Animal Production* 56, 399–405.

Cervera, C., Blas, E. and Fernández Carmona, J. (1997) Growth of rabbits under different environmental temperatures using high fat diets. *World Rabbit Science* 5, 71–75.

Cheng, C., Shimin, L., Li, Z. and Zhuang, X. (1991) The effect of nutrition, temperature and physiology on wool production of the Angora rabbit. *Journal of Applied Rabbit Research* 14, 89–92.

Chiericatto, G.M., Boiti, C., Canali, C., Rizzi, C. and Ravarotto, L. (1995) Effects of heat stress and age on growth performance and endocrine status of male rabbits. *World Rabbit Science* 3, 125–131.

Christ, B., Lange, K. and Jeroch, H. (1996) Effect of dietary fat on fat content and fatty acid composition of does milk. In: Lebas, F. (ed.) *Proceedings of the 6th World Rabbit Congress, Toulouse,* Vol. 3. Association Française de Cuniculture, Lempdes, pp. 135–138.

Colin, M. (1991) La cuniculture des pays méditerranéens. *Cuni-Sciences* 7, 73–96.

Colin, M. (1995) La cuniculture Sud-Américaine. Le Bresil. *World Rabbit Science* 3, 85–90.

Costantini, F. and Castellini, C. (1990) Effetti dell'habitat e del management. *Rivista di Coniglicoltura* 27, 31–36.

Crimella, C., Biffi, B. and Luzi, F. (1996) Allevamento plein-air: attualitá e prospettive. *Rivista di Coniglicoltura* 33, 15–23.

Dawkins, M.J.R. and Hull, D. (1964) Brown adipose tissue and the response of new born rabbits to cold. *Journal of Physiology, London* 172, 216–238.

Delaveau, A. (1982) Mortalité des lapereaux entre la naissance et le sevrage. 2. Resultats experimentaux. *Dossiers de l'Elevage* 5, 69–84.

Deshmukh, S.V. and Pathak, N.N. (1991) Effect of different dietary protein and energy levels on growth performance and nutrient utilization in New Zealand White rabbits. *Journal of Applied Rabbit Research* 14, 18–24.

Fayez, I., Marai, M., el-Masry, K.L. and Nasr, A.S. (1994) Heat stress and its amelioration with nutritional, buffering, hormonal and physical techniques for New Zealand White rabbits maintained under hot summer conditions of Egypt. In: Baselga, M. and Marai, I.F.M. (eds) *Cahiers Options Mediterréennes.* Vol. 8: *Rabbit Production in Hot Climates.* Centre International de Hautes Etudes Agronomiques Méditerranéennes, Paris, pp. 475–487.

Fernández Carmona, J., Cervera, C. and Sabater, C. (1991) Efecto del pienso y de una temperatura ambiente alta sobre la ingestion de pienso de gazapos lactantes y recien destetados. In: *Proceedings of the XVI Symposium Nacional de Cunicultura.* Asociación Española de Cunicultura, Castellón de la Plana, pp. 79–81.

Fernández Carmona, J., Cervera, C. and Blas, E. (1994a) Efecto de la inclusión de jabón cálcico en el pienso y de la temperatura ambiental sobre el crecimiento de conejos. *Investigación Agraria: Producción y Sanidad Animales* 9, 5–11.

Fernández Carmona, J., Cervera, C. and Blas, E. (1994b) Feed intake of does and their litters in different environmental temperature. In: Baselga, M. and Marai, I.F.M. (eds) *Cahiers Options Mediterréennes.* Vol. 8: *Rabbit Production in Hot Climates.* Centre International de Hautes Etudes Agronomiques Méditerranéennes, Paris, pp. 145–149.

Fernández Carmona, J., Cervera, C. and Blas, E. (1994c) Readapted does from high to normal ambient temperature. In: Baselga, M. and Marai, I.F.M. (eds) *Cahiers Options Mediterréennes.* Vol. 8: *Rabbit Production in Hot Climates.* Centre International de Hautes Etudes Agronomiques Méditerranéennes, Paris, pp. 469–470.

Fernández Carmona, J., Cervera, C., Sabater, C. and Blas, E. (1995) Effect of diet composition on the production of rabbit breeding does housed in a traditional building and at 30°C. *Animal Feed Science and Technology* 52, 289–297.

Fernández Carmona, J., Cervera, C. and Blas, E. (1996) High fat diets for rabbit breeding does housed at 30°C. In: Lebas, F. (ed.) *Proceedings of the 6th World Rabbit Congress, Toulouse,* Vol. 1. Association Française de Cuniculture, Lempdes, pp. 167–170.

Ferraz, J.B.S., Johnson, R.K. and Eler, J.P. (1991) Breed and environmental effects on reproductive traits of Californian and New Zealand rabbits. *Journal of Applied Rabbit Research* 14, 177–179.

Finzi, A., Kuzminsky, G. and Morera, P. (1988) Evaluation of thermotolerance parameters for selecting thermotolerant rabbit strains. In: *Proceedings of the 4th World Rabbit Congress*, Vol. ll. Sandor Holdas, Hercegalom, pp. 388–394.

Finzi, A., Tani, A. and Scappini, A. (1989) Tunisian non-conventional rabbit breeding systems. *Journal of Applied Rabbit Research* 12, 181–184.

Fraga, M.J., Lorente, M., Carabaño, R.M. and de Blas, J.C. (1989) Effect of diet and remating interval on milk production and milk composition of the doe rabbit. *Animal Production* 48, 459–466.

Fuquay, J.W. (1981) Heat stress as it affects animal production. *Journal of Animal Science* 52, 164–174.

García, J., Pérez-Alba, L., Alvarez, C., Rocha, R., Ramos, M. and de Blas, J.C. (1995) Prediction of the nutritive value of lucerne hay in diets for growing rabbits. *Animal Feed Science and Technology* 54, 33–44.

Gentry, J.W. (1983) Raising rabbits in domes. *Journal of Applied Rabbit Research* 6, 89.

Gonzalez, R.R., Kluger, M.J. and Hardy, J.D. (1971) Partitional calorimetry of the New Zealand White rabbit at temperature 5–35°C. *Journal of Applied Physiology* 31, 728–734.

Gray, R. and McCracken, K.J. (1974) Utilization of energy and protein by pigs adapted to different temperature levels. In: *Proceedings of the 6th Symposium on Energy Metabolism*. EAAP publ. No. 14, Stuttgart University, p. 161.

Hall, L.S. and Myers, K. (1978) Variations in the microclimate in rabbit warrens in semi-arid New South Wales. *Australian Journal of Ecology* 3, 187–194.

Harris, D.L., Cheeke, P.R., Patton, N.M. and Brewbaker, J.L. (1981) A note on the digestibility of leucaena leaf in rabbits. *Journal of Applied Rabbit Research* 4, 99.

Harris, D.L., Cheeke, P.R. and Patton, N.M. (1983) Food preference and growth performance of rabbits fed pelleted versus unpelleted diets. *Journal of Applied Rabbit Research* 6, 15–17.

Hill, J.R. (1961) Reaction of the newborn animal to environmental temperature. *British Medical Bulletin* 17, 164–167.

Howarth, B., Alliston, C.W. and Ulbera, L.C. (1965) Importance of uterine environment on rabbit sperm prior to fertilization. *Journal of Animal Science* 24, 1027–1032.

Jin, L.M., Thomson, E. and Farrell, D.J. (1990) Effects of temperature and diet on the water and energy metabolism of growing rabbits. *Journal of Agricultural Science, Cambridge* 115, 135–140.

Johnson, H.D., Cheng, C.S. and Ragsdale, A.C. (1958) *Environmental Physiology and Shelter Engineering*. Research Bulletin 648, Agricultural Experiment Station, University of Missouri, 26 pp.

Kamal, T.H. and Johnson, H.D. (1971) Total body solids as a measure of a short stress in cattle. *Journal of Animal Science* 32, 306–311.

Kamar, G.A.R., Shafie, M.M., Hassanein, A.M. and Borady, A.M. (1982) Effects of modification in environmental conditions on the reproductive activity of Giza rabbit does. In: *Proceedings of the 6th International Conference on Animal and Poultry Production*, Vol. 1. Zagazig, pp. 316–328.

Kasa, W., Thwaites, C.J., Jianke, X. and Farell, D.L. (1989) Rice bran in the diet of rabbits grown at 22° and 30°C. *Journal of Applied Rabbit Research* 12, 75–77.

Kluger, M.J., González, R.R. and Stolwijk, J.A.J. (1973) Temperature regulation in the exercising rabbit. *American Journal of Physiology* 224, 130–135.

Kong, P.L., Ping, F.Z., He, S.Q., Yuan, M.T., Huang, T.X. and Li, G.M. (1987) Effect of temperature on wool yield and quality in German Angora rabbits. *Chinese Journal of Rabbit Farming* 3, 22–25.

Kruk, B. and Davydov, F. (1977) Effect of ambient temperature on thermal sensitivity of POAH area in the rabbit. *Journal of Thermal Biology* 2, 75–78.

Lebas, F. and Ouhayoun, J. (1987) Incidence du niveau protéique de l'aliment, du milieu et de la saison sur la croissance et les qualités boucheres du lapin. *Annales de Zootechnie* 36, 421–432.

Lee, R.C. (1939) The rectal temperature of the normal rabbit. *American Journal of Physiology* 125, 521–529.

Lukefahr, S.D., Hohenboken, W.D., Cheeke, P.R. and Patton, N.M. (1983) Doe reproduction and preweaning litter performance of straight-bred and crossbred rabbits. *Journal of Animal Science* 57, 1090–1099.

McEwen, G.N. and Heath, J.E. (1973) Resting metabolism and thermoregulation in the unrestrained rabbit. *Journal of Applied Physiology* 35, 884–886.

McNitt, J.I. and Moody, G.L. (1992) A method for weaning rabbits pups at 14 days. *Journal of Applied Rabbit Research* 15, 661–665.

Maertens, L. (1992) Rabbit nutrition and feeding: a review of some recent developments. *Journal of Applied Rabbit Research* 15, 889–913.

Maertens, L. and De Groote, G. (1988) The influence of the dietary energy content on the performance of post-partum breeding does. In: *Proceedings of the 4th World Rabbit Congress*, Vol. 3. Sandor Holdas, Hercegalom, pp. 42–52.

Maertens, L. and De Groote, G. (1990) Comparison of feed intake and milk yield of does under normal and high ambient temperature. *Journal of Applied Rabbit Research* 13, 159–162.

Masoero, G. and Auxilia, M.T. (1977) Evoluzione della produttivita del coniglio nel curso di un anno. In: *Annali dell'Istituto Sperimentale per la Zootecnia*, Vol. 10. Istituto Sperimentale per la Zootecnia, Torino, pp. 93–111.

Mendez, J., de Blas, J.C. and Fraga, M.J. (1986) The effects of diet and remating interval after parturition on the reproductive performance of the commercial doe rabbit. *Journal of Animal Science* 62, 1624–1634.

Mori, B. and Bagliacca, M. (1985) Allevamento del coniglio: microclima e ritmo riproduttivo. *Rivista di Coniglicoltura* 22, 45–50.

Mori, B. and Bagliacca, M. (1990) Effetto dell'ambiente sulle produzioni cunicole. *Rivista di Coniglicoltura* 27, 17–21.

Nichelmann, M., Rohling, H. and Rott, M. (1973a) Der Einfluss der Umgebungs-temperatur auf die Hohe des Energieumsatzes erwachsener Kaninchen. *Archiv für Experimentelle Veterinarmedizin* 27, 499–505.

Nichelmann, M., Rohling, H. and Rott, M. (1973b). Der Einfluss der Umgebungs-temperatur auf die Warmeabgabe beim Kaninchen. *Archiv für Experimentelle Veterinarmedizin* 27, 507–512.

Nichelmann, M., Rott, M. and Rohling, H. (1974a) Beziehungen zwischen Energieumsatz und Umgebungstemperatur beim Kaninchen. *Monatshefte für Veterinarmedizin* 29, 257–261.

Nichelmann, M., Rott, M. and Rohling, H. (1974b) Untersuchungen zur evaporativen Warmeabgabe beim Kaninchen. *Monatshefte für Veterinarmedizin* 29, 261–266.

Nizza, A., Moniello, G. and Lella Di, T. (1995) Prestazioni produttive e metabolismo energetico di conigli in accrescimento in funzione della stagione e della fonte proteica alimentare. *Zootecnica e Nutrizione Animale* 21, 173–183.

NRC (1981) *Effect of Environment on Nutrient Requirements of Domestic Animals.* National Academy Press, Washington, DC, 152 pp.

Ortiz, V., de Blas, J.C. and Sanz, E. (1989) Effect of dietary fiber and fat content on energy balance in fattening rabbits. *Journal of Applied Rabbit Research* 12, 159–162.

Pagano-Toscano, G., Zoccarato, I., Benatti, G. and Lazzaroni, C. (1990) Fattori ambientali e prestazioni delle coniglie. *Rivista di Coniglicoltura* 2, 23–29.

Papp, Z. and Rafai, P. (1988) Role of the microclimate in intensive rabbit production. IV. Effects of environmental temperature stressors in pregnant does, embryonic development and viability of young rabbits. *Magyar Allatorvosok Lapja* 43, 529–534.

Partridge, G.G. (1988) Research of nutrition, reproduction and husbandry of commercial meat rabbits at the Rowett Institute, 1971–1985. *Journal of Applied Rabbit Research* 11, 136–141.

Pascual, J.J., Cervera, C., Blas, E. and Fernández Carmona, J. (1996) Milk yield and composition in rabbit diets using high fat diets. In: Lebas, F. (ed.) *Proceedings of the 6th World Rabbit Congress, Toulouse,* Vol. 1. Association Française de Cuniculture, Lempdes, pp. 259–262.

Platukhin, AI. and Konokhov, S.A. (1972) Winter kindling in one rabbit sheds. *Krolikovdstvo-i-Zverovodstvo* 5, 15–16.

Pote, L.M., Cheeke, P.R. and Patton, N.M. (1980) Use of greens as a supplement to a pelleted diet for growing rabbits. *Journal of Applied Rabbit Research* 3, 15–19.

Prud'hon, M. (1976) Comportement alimentaire du lapin soumis aux temperatures de 10, 20 et 30°C. In: Lebas, F. (ed.) *Proceedings of the 1er Congrès International Cunicole.* Communication no 14, Association Française de Cuniculture, Dijon.

Rafai, P. and Papp, Z. (1984) Temperature requirement of does for optimal performance. *Archiv für Experimentelle Veterinarmedizin, Leipzig* 38, 450–457.

Raharjo, Y.C., Cheeke, P.R. and Patton, N.M. (1988) Evaluation of tropical forages and rice by-products as rabbit feeds. *Journal of Applied Rabbit Research* 9, 201–211.

Raharjo, Y.C., Cheeke, P.R. and Patton, N.M. (1990) Effect of cecotrofy on the nutrient digestibility of lucerne and black locust leaves. *Journal of Applied Rabbit Research* 13, 56–61.

Rhoad, A.O. (1944) The Iberia heat tolerance test for cattle. *Tropical Agriculture (Trinidad)* 21, 162–164.

Rosell, J.M. (1996) Rabbit mortality survey. Necropsy findings in the field during the period 1989–1995. In: Lebas, F. (ed.) *Proceedings of the 6th World Rabbit Congress, Toulouse,* Vol. 3. Association Française de Cuniculture, Lempdes, pp. 107–111.

Samoggia, G., Bosi, P. and Scalabrini, C. (1988) Ambiente zootecnico e performances produttive del coniglio da carne. *Rivista di Coniglicoltura* 25, 37–40.

Sanchez, W.K., Cheeke, P.R. and Patton, N.M. (1984) The use of chopped alfalfa rations with varying levels of molasses for weanling rabbits. *Journal of Applied Rabbit Research* 7, 13–16.

Sanz, R., Fonollá, J. and Aguilera, J. (1973) Estudios de digestibilidad en conejos sometidos a alta temperatura. Utilización de antitérmicos. *Revista de Nutrición Animal* 11, 167–172.

Sanz, E., Ortiz, V., de Blas, J.C. and Fraga, M.J. (1989) A note on the critical temperature of sucking rabbits. *Animal Production* 49, 333–334.

Scheele, C.W., van der Broek, A. and Hendricks, F.A. (1985) Maintenance energy requirements and energy utilization of growing rabbits at different environmental temperatures. In: Moe, P.W., Tyrrell, H.F. and Reynolds, P.J. (eds) *Energy Metabolism of Farm Animals*. Rowman and Littlefield, NJ, pp. 202–204.

Schenk, H., Heim, T., Varga, F. and Goetze, E. (1975) Effects of cold and fasting on metabolism of serum free fatty acids in newborn rabbits. *Acta Biologica et Medica Germanica* 34, 613–623.

Schlolaut, M. (1987) Angora rabbit housing and management. *Journal of Applied Rabbit Research* 10, 164–168.

Shafie, M.M., Abdel-Malek, E.G., El-Issawi, H.F. and Kamar, G.A.R. (1979) Effect of environmental temperature on physiological body rections of rabbits under sub-tropical conditions. *Egyptian Journal of Animal Production* 10, 133–149.

Simplicio, J.B., Cervera, C. and Blas E. (1988) Effect of two different diets and temperatures on the growth of meat rabbit. In: *Proceedings of the 4th World Rabbit Congress*, Vol. 3. Sandor Holdas, Hercegalom, pp. 74–77.

Simplicio, J.B., Fernández-Carmona, J., Cervera, C. and Blas, E. (1991) Efecto del pienso sobre la produccion de la coneja a una temperatura ambiente alta. *Investigación Agraria: Producción y Sanidad Animales* 6, 67–74.

Sittman, D.B., Rollins, W.C., Sittman, K. and Casady, R.R. (1964) Seasonal variation on reproductive traits of the New-Zealand rabbits. *Journal of Reproduction and Fertility* 8, 29–37.

Stephan, E. (1980) The influence of environmental temperatures on meat rabits of different breeds. In: Camps, J. (ed.) *Proceedings of the 2nd World Rabbit Congress*, Vol. 1. Asociación Española de Cunicultura, Barcelona, pp. 399–409.

Thwaites, C.J., Baillie, N.B. and Kasa, W. (1990) Effects of dehydration on the heat tolerance of male and female New Zealand White rabbits. *Journal of Agricultural Science, Cambridge* 115, 437–440.

Tramell, T.L., Stallcup, O.T. and Harris, G.C. (1988) Effect of high temperature on certain blood hormones and metabolites on reproduction in rabbit does. *Journal of Applied Rabbit Research* 12, 101–102.

Wittorff, E.K., Heird, C.E., Rakes, J.M. and Johnson, Z.B. (1988) Growth and reproduction on nutrient restricted rabbits in a heat stressed environment. *Journal of Applied Rabbit Research* 11, 87–92.

Young, B.A. (1977) Effect of cold environments on nutrient requirements of ruminants. In: Fonnesbeck, F.V., Harris, L.E. and Kearl, L.C. (eds) *Proceedings of the 1st International Symposium. Feed Composition, Animal Nutrient Requirements and Computerization of Diets*. Utah Agricultural Experiment Station, Logan, Utah, pp. 491–496.

Zicarelli, L., Nizzi, A. and Perrucci, G. (1979) Influenza della temperatura ambientale su alcuni parametri produttivi del coniglio da carne. *Rivista Acta Medica Veterinaria* 25, 79–84.

16. Nutritional Recommendations and Feeding Management of Angora Rabbits

F. Lebas[1], R.G. Thébault[2] and D. Allain[3]

[1]Station de Recherches Cunicoles, INRA Centre de Toulouse, BP 27, 31326 Castanet-Tolosan, France; [2]INRA Centre Poitou-Charentes, U.E. Génétique Animale Phanères, Domaine du Magneraud, BP 52, 17700 Surgères, France; [3]INRA Centre de Toulouse, Station d'Amélioration Génétique des Animaux, BP 27, 31326 Castanet-Tolosan, France

Introduction

The Angora rabbit produces 1.0–1.4 kg year^{-1} of pure fine animal fibre without grease or plant material contamination, named 'Angora wool'. This represents some 0.30 of its live weight, the highest keratin production/live weight ratio found in any fibre-producing animal. In sheep, goat or camelids, this figure is generally less than 0.10.

The capacity of the Angora rabbit to convert food to keratin requires that particular attention be given to its nutrition. There are two important nutritional objectives:

1. To provide all the nutrients the rabbit needs to realize its genetic potential for wool production.
2. To avoid any disorder that may reduce the life-time performance of the animal.

Individual productive longevity (3–4 years on average) is an important economic parameter in the Angora production system.

There is a considerable paucity of information on Angora rabbit nutrition compared with published work on the production of meat from rabbits or wool from sheep. This applies to the genetics and physiology of Angora wool growth as well to other areas of study such as pathology. In practice, producers have observed that the nutrient requirements of Angora rabbits bred for wool production are similar to those of breeding does kept for meat production and consequently have used this knowledge as a basis for diet formulation. Nevertheless some specific modifications are necessary.

For a long time Angora rabbits were fed in the same way as rabbits kept for meat production on a mixture of cereals (oats, barley or wheat), lucerne hay plus some fresh forages such as cabbage or fodder beet. Since the 1960s, complete diets based on pelleted concentrates have been used extensively in rabbit meat systems; Angora rabbit farmers, however, continued with the traditional feeding method through the 1970s, while angora wool yields remained below 850 g year^{-1}. To improve wool yields, a mixture of 0.75 traditional feed and 0.25 supplementary feed pellets was subsequently used in some practical systems (Rougeot and Thébault, 1984). Other producers began using pelleted concentrates alone for Angora breeding does. By the beginning of the 1980s, as the genetic potential for wool production exceed 1 kg animal^{-1} year^{-1}, the use of specific pelleted diets formulated for Angora production became general practice as feed quality and safety (absence of induced disorders) were also improved. Schlolaut (1985) quantified the production advantages of concentrate feeds. Taking the angora wool yield obtained with these as 1.00, mixed feeding (raw products + cereals) and hay-based feeding reduced the yearly wool production to 0.85 and 0.72, respectively.

Nutritional requirements

Nutrient requirements are considered here for Angora wool-producing females, since males are not frequently employed because of their lower wool production (5–10% less). Animals producing Angora wool are assumed to be adults with no production other than wool. For breeding does, or growing animals, the recommendations are those proposed for meat production from rabbits (see Chapters 10 and 14).

Consequences of daily variations in wool production

The amount of hair covering the body plays an important role in thermal insulation and heat loss. In France, Angora rabbits are defleeced every 3 months and are consequently completely or relatively naked without thermal protection for 2–3 weeks. Vermorel *et al.* (1988) demonstrated a large increase in heat production just after the harvest (Tables 16.1 and 16.2). To reduce heat loss, some form of protection is often provided, either in the form of woollen jackets ('jersey rabbits') or by leaving a strip of fleece on the back ('strip rabbits'). Such techniques are less common with the German strain as the animals are shorn, i.e. a few millimetres of stubble is always left above the skin, which limits heat loss. In addition German Angora rabbits have a higher proportion of down in the fleece, which improves thermal insulation. Nevertheless, whatever the Angora strain, the period of 2–3 weeks after harvesting is when energy requirements for thermal regulation are at their highest.

A further source of variation for nutrient requirements is the hair growth rate, i.e. the rate of keratin synthesis. The highest growth rate is observed during the 4th week after harvesting (31.7 g week^{-1}). A reduction in the

Table 16.1. Skin temperature, total and net radiative heat flow of Angora rabbits before and after defleecing, with or without a strip of hair on the back or a jersey jacket (means of six different spots measured during the 2 days following harvest, and standard deviations) (from Vermorel *et al.*, 1988).

	Before harvest	After complete defleecing	'Strip rabbits'		'Jersey rabbits'	
			On the hair strip	Naked areas[a]	With woollen jacket	Without woollen jacket
No. of animals	6	6	9	9	6	6
Skin temperature at 10°C	38.8±0.4	36.5±0.6	—	37.9±0.6	31.9±0.5[b]	36.4±0.7
Total radiative heat flow at 15°C (W m²)	422±5	513±12	416±7	519±7	479±6	515±7
Net radiative heat flow at 15°C (W m²)	23±6	176±7	35±6	187±12	24±12	177±7

[a]Means of values obtained on the thigh, thorax and abdomen.
[b]Temperature on the jersey jacket.

Table 16.2. Heat production (kJ kg^{-1} W$^{0.75}$ h^{-1}) of 'strip rabbits' or 'jersey rabbits' before defleecing (at 10°C) and after defleecing, with the strip of hair or the jersey jacket (for 2 days) or after harvesting the strip of hair or taking off the jersey jacket (2 days) (means and standard deviations) (from Vermorel *et al.*, 1988).

		'Strip rabbits'				'Jersey rabbits'		
Environmental temp.		Before defleecing	Strip of hair			Before defleecing	Jersey jacket	
	n		With	Without	*n*		With	Without
15°C	2	18.1±2.0[a]	20.8±0.2[a]	29.9±1.3[b]	–			
10°C	4	16.7±0.3[a]	23.2±2.4[b]	32.6±1.5[c]	5	17.6±1.0[a]	28.9±1.7[b]	32.8±2.0[c]
5°C	3	16.3±2.6[a]	25.5±3.3[b]	35.6±2.7[c]	–			

[a, b, c]For the same type of animals and the same environmental temperature, values with different superscripts are significantly different ($P < 0.05$ or $P < 0.01$).

weekly wool output is observed afterwards (Fig. 16.1). Between the 4th and the 14th week, the wool output is halved.

According to these results, the first month following fleece harvesting appears to be the one during which nutrient and energy requirements are maximum for energy, protein and sulphur amino acids (SAA, the main components of keratin). The weekly requirements vary during the 3 months between two consecutive harvests (Table 16.3) and were summarized by Rougeot and Thébault (1984).

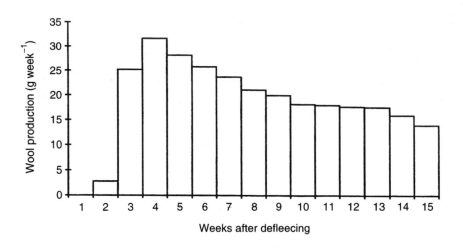

Fig. 16.1. Variations in wool production (g week⁻¹) between two harvests. From Rougeot and Thébault (1984).

Table 16.3. Weekly variations in nutrient and energy requirements for French Angora rabbits (recalculated from Rougeot and Thébault, 1984).

Month between two wool harvestings	1st month	2nd month	3rd month
Crude protein	190 g	175 g	160 g
Fat	37 g	34 g	31 g
Crude fibre	205 g	190 g	170 g
Sulphur amino acids	10 g	9 g	8 g
Digestible energy	12.6 MJ	11.6 MJ	11.5 MJ

Nutrient recommendations

As previously mentioned, Angora rabbits are now fed with balanced pelleted feeds. The desirable composition of such feeds has been the object of some specific experiments. The recommendations proposed by different authors are a combination of the results of a critical analysis of the available 'Angora' data and of the recommendations proposed for meat rabbits. Recommendations of German and Chinese authors, when available, are presented in Table 16.4, in comparison with the recommendations proposed by the authors of the current chapter.

Table 16.4. Nutrient recommendation for adult Angora rabbits.[a]

Nutrients	Unit kg⁻¹	Germany[b]	China[c]	Current work
Digestible energy	MJ	9.6–10.9	10.0–11.7	10.5
	kcal	2300–2600	2400–2800	2500
Lipids	g	20	30	30
Crude fibre	g	140–160	120–170	140
Crude protein	g	150–170	150–160	160
Digestible protein	g	—	110	122
Lysine	g	5	7	7
Methionine + cystine	g	7	7	8
Arginine	g	6	7	6
Minerals				
Calcium	g	10	—	8
Phosphorus	g	3–5	—	4
Sodium	g	2.5	—	3
Potassium	g	7	—	13 maximum
Chloride	g	4	—	4
Sulphur	mg	—	—	400
Magnesium	mg	300	—	300
Iron	mg	50	—	50
Copper	mg	10	—	50
Zinc	mg	50	—	50
Manganese	mg	10	—	10
Vitamins				
A	IU	6000	—	10,000
D₃	IU	500	—	800
E	mg	20	—	40
K	mg	1	—	1

[a]As-fed basis with 890 g of dry matter kg⁻¹; [b]Schlolaut (1985); [c]Liu *et al.* (1992).

Energy

Recommendations for dietary digestible energy (DE) content are in the same range for German (Schlolaut, 1985), Chinese (Liu *et al.*, 1992) and French authors (Rougeot and Thébault, 1984; Charlet-Lery *et al.*, 1985). Nevertheless, German and Chinese data are not very precise since the variation between the proposed minimum and maximum represents 13–17% of the minimum. The Hungarian recommendations are also in the same range: 10.7 MJ kg⁻¹ (Tossenberger and Henics, 1988; Henics *et al.*, 1989). The recommendation is accordingly 10.5 MJ kg⁻¹ of feed on an as-fed basis.

According to Charlet-Lery *et al.* (1985) metabolizable energy (ME) represents 0.95 of DE content; in addition, the same authors indicated that energy utilization by Angora rabbits, as DE or ME, is independent of season or of the time since the previous wool harvest.

Protein

Among German, Chinese and French data, there is agreement on dietary protein requirements which are of the order of 160 g kg^{-1}. On the other hand, the Hungarian recommendations are higher at 196 g kg^{-1} (Tossenberger and Henics, 1988; Henics *et al.*, 1989). However, this latter recommendation is based on an experiment where protein and SAA were studied simultaneously and where the lowest protein level tested was 175 g kg^{-1} with a relatively low SAA content. Thus, this recommendation for a very high concentration in protein seems unrealistic and was not retained.

Crude fibre

No specific study has been published on dietary fibre content as a possible source of variation in Angora wool production. The recommendations of various authors are easily calculated from the analysis of practical diets employed. The available recommendations are currently expressed only in terms of level of crude fibre. However, one of the roles of dietary fibre is to remove hair swallowed by the rabbit from the digestive tract. To achieve this objective, a significant proportion of dietary fibre must be non-digestible; a minimum level of lignin seems reasonable to reduce fibre digestibility and, to reduce this latter objective, a value of 40 g ADL (acid detergent lignin according to the Van Soest methodology) kg^{-1} may be proposed.

Amino acids

Lysine. German recommendations for dietary lysine in Angora rabbit production are 5 g kg^{-1} diet, significantly lower than the 7 g kg^{-1} suggested by Chinese data; however, neither of these figures are based on direct experiments. Lysine is not an important component of keratin but it plays a significant role in body protein turnover and assists the animal in restoring its live weight following body weight loss observed after defleecing. Therefore a level of 7 g of lysine kg^{-1} feed for Angora rabbits is recommended.

Methionine + cystine. Several studies have been undertaken in Germany (Schlolaut and Lange, 1983), in Hungary (Henics *et al.*, 1990) and in France (Lebas and Thébault, 1990) on the requirements for SAA. From this latter work (Table 16.5), it has been concluded that, for a level of wool production higher than 1000 g year^{-1}, SAA intake is an important limiting factor. Practical recommendations for SAA are 8 g kg^{-1} diet on an as-fed basis. A more recent study (F. Lebas and R.G. Thébault, unpublished data) indicates that efficient SAA supplementation can be achieved either with D,L-methionine or with L-cystine. Under some conditions a slight advantage can be attributed to cystine supplementation. Nevertheless for economic reasons SAA supple-

Table 16.5. Mean relative performances of Angora rabbits receiving SAA supplementation at different dietary levels (adjusted by covariance analysis to an initial live weight of 4.128 kg). Fleece weight 1.00 = 264.2 g harvest^{-1}, feed intake 1.00 = 15.57 kg between two harvests. Other nutrients as those recommended in Table 16.4 (from Lebas and Thébault, 1990).

Dietary SAA level (g kg^{-1})	5.6	6.4	7.2 (control)	8.0	Residual coefficient of variation	Statistical probability
Number of harvests	69	62	76	60	—	—
Fleece weight	0.948[a]	1.008[b]	1.000[b]	1.056[c]	0.135	0.002
Feed intake	0.991	1.000	1.000	1.006	0.043	NS
Feed efficiency[d]	0.951[a]	1.005[bc]	1.000[b]	1.049[c]	0.127	0.004
Live weight	0.990	1.020	1.000	1.006	0.07	NS

[a,b,c]Values with different superscripts are significantly different ($P < 0.05$ or $P < 0.01$).
[d]Calculated as g wool produced per g feed intake.

mentation, if necessary, is recommended in the form of D,L-methionine.

The Hungarian SAA recommendation is 9 g kg^{-1} diet (Henics *et al.*, 1990); but this figure was obtained after comparison of only two SAA dietary levels: 5.6 vs. 9.0 g kg^{-1}. Because in the French study (Lebas and Thébault, 1990) the highest level studied (8.8 g SAA kg^{-1}) failed to induce any improvement in wool production above that achieved with a level of 8.0 g kg^{-1}, the Hungarian recommendation of 9 g kg^{-1} leads to a significant SAA wastage, and is not considered practical.

Other amino acids. No specific evaluation was undertaken for the other amino acids. The current dietary recommendation for arginine (7 g kg^{-1}) is based only on the actual content observed in adequate Angora diets. In the absence of further information, the recommendations for growing rabbits are suitable.

Minerals and vitamins

As for most of the amino acids, the current recommendations for minerals (Table 16.4) are derived from the observed composition of Angora diets and from knowledge of the mineral requirements of growing and adult meat rabbits.

The German recommendation for vitamin A (6000 IU kg^{-1}; Schlolaut, 1985) is lower than that proposed in France by Rougeot and Thébault (1984), i.e. 10,000 IU kg^{-1}. Hungarian experimental results (Table 16.6) indicate clearly that 5000 IU kg^{-1}, which is very close to the German recommendation, is not adequate for Angora wool production (Kovácsné-Virányi, 1990). By comparison with meat rabbit reproduction it can be assumed that the maximum level employed in the Hungarian experiments is too large and the proposed recommendation is 10,000 IU vitamin A kg^{-1}, which is the same as for most meat rabbits. A complementary experiment included in the same

Table 16.6. Relative effect of dietary supplementation with vitamin A or β-carotene, on the quantity of hair produced by a surface of 14 cm² of skin shaved once a week during 8 consecutive weeks (from Kovácsné-Virányi, 1990). Value for the control 1.000 = 1.17 g.

	Control (5000 IU vit. A kg⁻¹)	Vitamin A + 15,000 IU kg⁻¹	β-Carotene + 45 μg kg⁻¹
No. of rabbits	5	7	7
Hair production	1.000ᵃ	1.132ᵇ	1.055ᵃᵇ

a ≠ b (P = 0.05)

Hungarian publication demonstrated that β-carotene can sometimes completely replace the supply of vitamin A, but the two experiments were not precise enough to support any calculation of the transformation of β-carotene into vitamin A.

It is important to note that dietary vitamin D levels should not exceed 800 IU kg⁻¹. Adult females which are not reproducing, lactating or growing are susceptible to heart valve and kidney calcification with D hypervitaminosis (Thébault and Allain, 1995).

Feeding management

As mentioned in the introduction, Angora rabbits are in practice fed balanced pelleted feeds (pellet diameter = 3–5 mm). In addition they must have permanent access to clean fresh water. Daily water intake is about 0.33 l animal⁻¹ day⁻¹ with a large variation between animals and season. Significant mortality can be observed if availability of water is not sufficient during a hot period. Dietary roughage, supplied once or twice a week as straw or hay or *ad libitum* as straw bedding, is not essential for health or wool production in the Angora rabbit (Rougeot *et al.*, 1980). However, when straw is fed, average daily intake falls from 19 g to 13 g, between the first and the third month following the harvest. Greater variations are observed from one rabbit to the other, e.g. straw intakes from 43 g day⁻¹ to only 3 g day⁻¹, without any apparent effects on wool production.

Feed restriction

Preliminary studies have shown that feed restriction decreases wool production by 14.7% (Rougeot and Thébault, 1977) or 9.2% (Schlolaut and Lange, 1983). However, in both of these studies, feed restriction was severe and no account was taken of variability in hair growth rate between harvests. More recently, Lebas and Thébault (1988) have shown that feed intake can be reduced by 61% in winter and 26% in summer with an adapted feed restriction (1200 g week⁻¹) during the first month following harvest without any adverse effects on wool production (Fig. 16.2 and Table 16.7).

However, the Angora rabbit seems unable to regulate daily intake as some does are able to consume more than 400 g day^{-1} or exceptionally 500 g day^{-1} during the first 2 weeks following harvest. This can cause nutritional disorders, such as enterotoxaemia, which occur when pellets are fed *ad libitum*.

A restricted feeding regime, as described overleaf, has been developed using the pattern (Fig. 16.1) of weekly hair production over the 3 months of hair growth between harvests (Rougeot and Thébault, 1984). This has now been adopted in commercial practice.

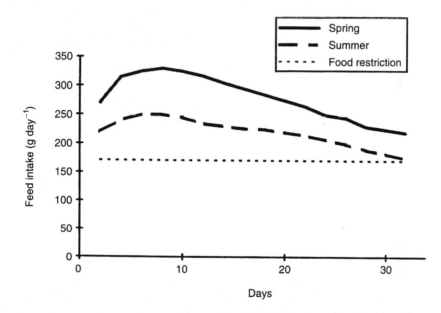

Fig. 16.2. Change in daily feed intake of Angora rabbits restricted-fed or fed *ad libitum* during the 5 weeks following defleecing, in spring and summer. From Lebas and Thébault (1988).

Table 16.7. Live weight (g) 5 weeks after the last defleecing and wool production at the next harvest of angora rabbits (mean ± SEM) with or without feed restriction during two different seasons (from Lebas and Thébault, 1988).

		Ad libitum	Feed restriction	Statistical probability
Live weight (g)	Spring	4457 ± 106	4244 ± 47	0.014
	Summer	4234 ± 73	4127 ± 44	0.08
Wool production (g)	Spring	258.3 ± 8.7	259.8 ± 6.6	NS
	Summer	235.9 ± 6.4	246.5 ± 8.7	NS

- First month: 1200 g animal^{-1} week^{-1};
- second month: 1100 g animal^{-1} week^{-1};
- third month: 1000 g animal^{-1} week^{-1}.

The weekly ration must be distributed equally over 6 days a week as Angora rabbits are not able to regulate feed intake themselves

One fasting day a week

A fasting day is essential when fibres are long or when hair losses are observed. Angora rabbits, like most mammals, lick their fleece when grooming. Hair is swallowed and it represents 0.3–0.4 g day^{-1} during the last month between harvests (Charlet-Lery *et al.*, 1985). As the rabbit is not able to vomit, long hair mixed with feed material is retained in the stomach. It rapidly forms a stomach hair ball (trichobezoard) which blocks the pylorus and prevents gastric emptying. The animal stops feeding and will die. Plitt-Hardy and Dolnick (1948) described this phenomenon and Rougeot and Thébault (1977) observed the death of 5 out of 11 females fed a pelleted diet *ad libitum*. Autopsy revealed the presence of a trichobezoard in the stomach of each animal. Feed restriction and 1 fasting day a week, when only straw or bulky forage is available, will facilitate voiding of ingested hair in hard faeces. On the day following the fast, hard faecal pellets connected to each other are often observed.

Conclusions

As Angora rabbits are housed in individual cages, it is very easy to control their feeding regime. When traditional raw feeds such as hay and cereals are used, no specific nutritional problems occur. Angora rabbits that have been selected for wool production cannot, however, achieve their genetic potential on such a regime. In addition, when hay is floor-fed inside the cage or even in a feeding rack, vegetable matter tends to 'contaminate' the fleece which drastically reduces its commercial value. For these reasons most commercial Angora rabbits are fed a specific pelleted balanced diet. Pelleted concentrates have the advantage that nutritional characteristics are precise and constant, feed storage is minimum, and labour costs for feeding are reduced. Some precautions are necessary when using a complete pelleted diet. Finally, water must be supplied *ad libitum* using an automatic watering trough. To avoid wastage and nutritional disorders, a restricted feeding regime adapted to variations in both feed requirements and hair growth should be adopted. Fasting once a week will avoid the formation of a stomach hair ball which, should it occur, is invariably fatal.

References

Charlet-Lery, G., Fiszlewicz, M., Morel, M.T., Rougeot, J. and Thébault, R.G. (1985) Variations annuelles de l'état nutritionel de la lapine angora durant les pousses saisonnières des poils. *Annales de Zootechnie* 34, 447–462.

Henics, Z., Tossenberger, J. and Gombos, S. (1989) [Digestion, and wool production of angora rabbits depending on the energy, protein and methionine content of feeds]. *1. Nyultenyéstési Tudományos Nap, Kaposvar (Hungary) 26 Május 1989*, pp. 80–99.

Henics, Z., Biróné Németh, E., Szendrö, Z and Tossenberger, J. (1990) [Effect of diet's sulphur amino acid content on wool production of males, female and castrated males Angora rabbits]. *2. Nyultenyéstési Tudományos Nap, Kaposvar (Hungary) 16 Május 1990*, pp. 68–74.

Kovácsné-Virányi, A. (1990) [Effect of vitamin A deficiency on the wool production of Angora rabbit]. *2. Nyultenyéstési Tudományos Nap, Kaposvar (Hungary) 16 Május 1990*, pp. 91–96.

Lebas, F. and Thébault, R.G. (1988) Influence of *ad libitum* feeding 5 weeks after plucking on wool production in Angora rabbits. In: *Proceedings of the 4th World Rabbit Congress, Budapest*, Vol. 3. Sandor Holdas, Hercegalom, pp. 249–253.

Lebas, F. and Thébault, R.G. (1990) Besoins alimentaires en acides aminés soufrés chez le lapin Angora. In: *5èmes Journées de la Recherche Cunicole, Paris*. Communication no. 49, ITAVI, Paris, France.

Liu, S.M., Zang, L., Chang, C., Pen, D.H., Xu, Z., Wang, Y.Z., Chen, Q., Cao, W.J. and Yuan, D.Z. (1992) The requirements of digestible energy, crude protein, and amino acids for Angora rabbits. *Journal of Applied Rabbit Research* 15, 1615–1622.

Plitt-Hardy, T.M. and Dolnick, E.H. (1948) Angora rabbit wool production. US Department of Agriculture Circular no. 785, USDA, Washington, DC, pp. 1–22.

Rougeot, J. and Thébault, R.G. (1977) Formation de trichobezoards chez le lapin Angora nourri *ad libitum* avec un aliment aggloméré. *Revue Médecine Vétérinaire* 153, 655–660.

Rougeot, J. and Thébault, R.G. (1984) *Le Lapin Angora, sa Toison, son Élevage*. Le Point Vétérinaire, Maisons Alfort, France, 182 pp.

Rougeot, J., Colin, M. and Thébault, R.G (1980) Définitions de conditions expérimentales pour l'étude des besoins nutritionnels du lapin: nature de la litière et présentation du lest alimentaire. *Annales de Zootechnie* 29, 1–11.

Schlolaut, W., 1985. *Abrégé de Production Cunicole*. Deutsche GTZ GmbH, Eschborn, Germany, pp. 215–216.

Schlolaut, W. and Lange, K. (1983) Untersuchungen über die Beeinflussung quantitiver Merkmale der Wolleistung beim Angorakanichen durch Geschlecht, Alter, Fütterungstechnik und Methioningehalt des Futters. *Züchtungskunde* 55, 69–84.

Thébault, R.G. and Allain, D (1995) Dietary requirements and feeding management of Angora rabbits. In: Laker, J.P. and Russel, A.J.F. (eds) *The Nutrition and Grazing Ecology of Speciality Fibre Producing Animals*. Occasional Publication No. 3, European Fine Fibre Network, pp. 71–84.

Tossenberger, J. and Henics, Z. (1988) Effect of energy, protein and sulfur amino acids on wool production of angora rabbits. In: *Proceedings of the 4th World Rabbit Congress, Budapest*, Vol. 3. Sandor Holdas, Hercegalom, pp. 274–280.

Vermorel, M., Vernet, J. and Thébault, R.G. (1988) Thermoregulation of angora rabbit after plucking. 2/ Heat loss reduction and rewarming of hypothermic rabbits. *Journal of Animal Physiology and Animal Nutrition* 60, 219–228.

17.   Pet Rabbit Feeding and Nutrition

Gilbertson and Page Ltd, PO Box 321, Welwyn Garden City,
Herts AL7 1LF, UK

Introduction

The rabbit (*Oryctolagus cuniculus*) was first domesticated around 2000 years ago in the Mediterranean, which is comparatively recently when compared with other species (Zeuner, 1963). The evidence for this timing is supported by the depiction of the rabbit on the coins of the period AD 120–130 and in Roman art.

The first breeding and controlled feeding of rabbits appears to have been in French monasteries in the period between the sixth and tenth centuries where the rabbits were kept in large leporaria. Subsequently rabbits became more commonplace such that by the 16th century, selection for tameness, size and variety was clearly evident. The rabbit does not feature in the art of the domestic animal until the appearance of a white rabbit in 1530 in the painting *Madonna with the Rabbit* by Titian.

Whilst the development of the domestic rabbit in mainland Europe was prominent, no records of the existence of the rabbit in Britain appear until after the Norman conquest. Yet by the 14th century the rabbit was commonplace, was captive by the 17th century but only truly domesticated by the end of the 18th century. It was not however until the mid 19th century that the pet and show rabbit began to appear (Sandford, 1986). As the rabbit is very adaptable and today is distributed over the five continents it is surprising that it was so late in becoming a domesticated species; although extensively farmed in Italy, France and Spain the rabbit has now become an increasingly popular and important pet across Europe, in particular in Germany, the UK and in the USA and Canada (Santomá *et al.*, 1989).

In the UK the numbers of pet rabbits have steadily risen in the last few years, with estimates in the range of 1.0–1.3 million in 1996 up to 1.3–1.6 million in 1998 and rising, making it the third most popular pet and present in almost 0.04 of the pet-owning households (PFMA, 1998). In the USA the pet rabbit population has similarly risen between 1991 and 1996 and is now estimated to be in the region of 5.7 million, making it the fifth most popular pet and present in 0.023 of households (American Veterinary Medical Association, 1997).

The pet rabbit is a small and pleasant animal to handle and, with a life expectancy of 6–8 years, can be kept as easily by the elderly and handicapped as by the child or fit adult. It has always been regarded as the 'classic' pet for children but has more recently begun to fulfil a similar position in the eyes of the pet-keeping public to the cosseted cat.

There are more than 50 breeds of rabbits now recognized, ranging in weight from 1 to over 10 kg which rabbit fanciers divide into two categories referred to as 'fur' and 'fancy', with the smaller breeds, for example Netherland Dwarf and Mini-lop, dominating as the popular pet. Much of the detailed information on the nutrient requirements and feeding of the rabbit are derived from studies and publications which are concerned with the rabbit as a research animal (NRC, 1977) or as farmed meat and fur producer (Lebas 1980, 1987; Schlolaut, 1982). There are, however, more recent studies which are specific to the pet animal (Hartmann *et al.*, 1994; Schwabe, 1995; Haffar, 1996) and its recognition as a significant pet in small animal veterinary practice has been established (Hartmann *et al.*, 1994; Malley, 1994, 1995, 1996). In assessing the nutrient requirements of the pet rabbit, reference to the basic principles of the former style of published information has therefore been made and appropriate adaptations of values for nutrient requirements have been determined using the more recent articles as a check for the feeding of the smaller rabbit, at maintenance, as a pet.

Consideration has also been given to the particular requirements that arise from the fact that the owner of the pet rabbit may wish to supplement the commercial diet with fresh material and the problems that may subsequently arise through malfeeding practices.

Feeding management

General considerations

It must be emphasized that keeping of rabbits will not be successful unless attention to detail is paid in terms of the housing, behaviour of the rabbit and the feed quality and balance of the diet, as well as the most suitable feeding practices for the rabbit. The rabbit is a social creature and requires companionship, ideally with another rabbit, and exercise on a daily basis.

The pet rabbit should be given adequate opportunity to feed in a quiet place. The provision of feed should be carried out on a regular basis, for example, in the early morning and evening and with appropriate recognition of the individual habits, for example, in the summer rabbits tend to eat more at night, when it is cooler. Whilst rabbits dislike dust in their food, it is important to ensure that the rabbit consumes all the components of a coarse or flake-mix proprietary product so that a balanced diet is consumed. Removal of any dusty or fouled portion of the diet is important but must not be such that the rabbit continually leaves the same component each day. It is sometimes suggested that the rabbit should have food withheld once each week to ensure that the digestive system is clear of fur balls and to encourage the consumption of essential roughage, usually in the form of hay.

Rabbits like variety in their diet, and mixtures are often considered more palatable than single component diets. However the importance of avoiding selective feeding and the consequential provision of an imbalanced diet and resultant health problems, which are discussed later in this chapter, must be emphasized. For simplicity, and with modern production processes, palatability can be maintained with a single pelleted complete compound diet to which the only additional provision required is hay and fresh drinking water. With the exception of long fibre it is not wise to dilute such a complete compound feed. Changing the diet to provide variety may also inadvertently result in obesity. This comes about as a result of the rabbit consuming the diet for pleasure rather than nutritional needs. Consequently the rabbit, although contented, becomes fat.

For home-made diets, variety often consciously or unconsciously offsets inadequacies in one or more of the foods. It is important to offer only clean and dust-free food. Where a cereal or bran which is prone to dustiness is included, this can be moistened with a little warm water to a crumbly texture to encourage intake. All foods must be free from mould and frost, and green foods must be fresh. Detailed considerations of each foodstuff is considered in subsequent sections of this chapter. The rabbit will selectively take concentrates if palatability of roughage is variable (Van Soest, 1982), however this behaviour will readily result in a scour from the consumption of too much protein relative to hay. A simple yet effective treatment for this is to remove feed and provide warm water and hay. A well-fed rabbit masticates its food extensively whereas, when the rabbit is hungry, this practice does not occur to any great extent. This forms a good practical guide to regulating food quantities.

Feeding guide

A guide to typical daily feed intakes for complete compound feeds for the adult at maintenance is around 112 g (30–35 g dry matter kg^{-1} body weight). For does in kindle, 10 days prior to kindling, 224 g day^{-1}, and from kindling to weaning at 21–28 days in the region of 350–750 g day^{-1} may be consumed. After weaning, a few days of small rations is practised followed by feeding to appetite. Growing young rabbits should be fed to appetite until 10 weeks of age and then at around 112 g day^{-1}.

Fresh foods can be fed to appetite in the case of the adult. The suggested best approach is to provide only as much as will be consumed by the next meal. If hay is supplied on a free-choice basis then a regular fresh supply should be ensured. Roots should be fed as lumps rather than slices or small pieces. It is worth remembering that an average adult rabbit can eat an amount of roots equivalent to the size of a tennis ball.

If feeding extensively on a home-grown diet, then it is suggested that one meal of grains and one of greens or roots is provided each day with dust-free hay and water offered on a free-choice basis. It should be noted that 'green foods' and roots do not affect bulk and thus influence digesta transit in the intestine, as does hay (long fibre).

Housing

Housing of the rabbit should be in neat hutches or pens which can be sited in a quiet part of the garden or yard, preferably in small wooden hutches with some protection from the weather in winter. If the hutch is within an outbuilding, then avoidance of draughts yet with adequate ventilation and light is important. It is recognized that the 'house rabbit' has become widely popular. Established in the US in the 1980s the loose-housed, house-trained indoor rabbit makes an affectionate and intelligent companion. Indeed there is now some evidence to suggest that, because of the social nature of rabbits, the behaviour patterns of the house rabbit are more natural than those kept in a hutch. In addition, such methods of keeping encourage exercise which improves digestive function and reduces skeletal problems (Anon., 1998). Even with the house rabbit, however, penning may be beneficial on certain occasions.

Pens can be made with either wire mesh or solid floors. A general guide to their dimensions is 0.1 m^2 0.5 kg^{-1} body weight plus an allowance for feeding utensils and bedding. Typically pens with mesh floors made of galvanized 12- or 14-gauge metal with a 19 × 19 mm mesh should be in the order of 900 × 600 × 500 mm high, whilst solid floor designed pens should have dimensions in the region of 1200 × 600 × 500 mm. The latter size is also suitable for breeding does rearing litters, whilst the former is adequate for single animals. The classical alternative is the outdoor triangular wooden hutch of approximately 750 × 900 mm with 1350 × 900 mm wire mesh run on to grass. All designs should be fastened with hasp and staple to prevent escape. Galvanized mesh should be avoided to minimize the risk of zinc toxicosis. If a nest box is required then a 500 × 250 mm box, with 250 mm high back and 120 mm front filled with soft meadow hay will suffice.

Feeding equipment

Feed containers should be heavy in construction as rabbits are prone to throwing them about. Earthenware or stainless steel, the latter hung on the door, seem to be the most effective. The best designs accommodate a turned-in lip to prevent the rabbit from scratching out the feed. Containers must not exceed 75 mm in depth as the rabbit will find access a problem. The inclusion of a fine wire sieve in the bottom to remove dust from feed pellets can be beneficial if feed replacement occurs only once daily.

Water is frequently provided via nipple or automatic drinkers in commercial situations and such arrangements are ideal. However, for the domestic situation, heavy earthenware pots or bottle drinkers are more practical. If a pot is provided within the cage or pen, precautions to prevent fouling by the rabbit are essential. The bottle drinker in a frame outside the pen and a small trough inside or a 6 mm diameter drinking tube overcomes such problems with ease.

The provision of free-choice roughage in the form of quality meadow hay is essential. The best method of supply is in a hay rack, which not only saves space but also keeps the hay fresh and free from fouling. In pens with a wire mesh roof, the hay can be put on top of the pen and the rabbits will happily pull it through.

Physiology/anatomical considerations

Within the 50 or more breeds of domestic rabbit there are four basic fur types: normal with fur length 30 mm in length; rex 12 mm; satin; and the one wool-producing angora which has 120 mm long, fine fur. Typical pet breeds weigh in the range of 1–5 kg, though larger breeds in excess of 10 kg as mature adults are occasionally selected (Sandford, 1986).

The rabbit is a true non-ruminant herbivore. It has different dentition from the rodent, with which it is sometimes mistakenly categorized, having two pairs of upper incisor teeth whereas rats and mice have only one. The upper and lower teeth meet and grind each other down with use. Herbivores by their nature consume high- fibre plant-based diets which present their own unique problems in terms of digestive efficiency.

Herbivores in general have developed in different ways to produce a digestive reservoir which permits an increase in the efficiency of utilization of their fibrous diets (Cheeke, 1988). The rabbit has a large stomach and well-developed caecum relative to other non-ruminant herbivores such as the horse. The rabbit has a very low stomach pH < 2.0 and substantial microbial digestion in the hindgut, and exhibits frequent feeding up to 30 times per day of 2–8 g intake over 4–6 min (Prud'hon *et al.*, 1975), relying on the bulk in the stomach to effect intestinal passage of digesta. This high voluntary feed intake (VFI), some four times higher *pro rata* than a 250 kg steer (Santomá *et al.*, 1989), is associated with a low gut retention time of 17.1 h in the rabbit compared with, for example, cattle at 68.8 h. The high VFI together with the reutilization of gut content by reingestion of caecal material, referred to as coprophagy or caecotrophy, supports its high nutrient requirement per unit of body weight and improves feed utilization for the rabbit.

Caecotrophy

At night the rabbit produces a soft faecal pellet which it takes directly from the anus. The normal faecal pellet is harder and is passed during both the day and night. Typically a 2.5–3.0 kg rabbit passes 150 faecal pellets per day. The soft faecal pellet results from the separation of digesta on the basis of particle size in the hind gut. Peristaltic action removes larger particles (> 0.5 mm) of predominantly lignocellulose through to the colon which is subsequently excreted as hard pellets. Antiperistalsis moves the smaller particles (0.1–0.2 mm) and soluble components into the caecum where fermentation occurs (Bjornhag, 1987). Periodically, often at the end of the night, the caecal contents are subsequently excreted as soft faecal pellets and are consumed directly from the anus. Once consumed, soft faeces remain in the stomach of the rabbit for 6–8 h and due to the mucosal cover remain protected from digestive attack. Microorganisms within the soft faecal pellets continue fermentation with the production of lactic acid. These pellets provide microbial protein which accounts for between 0.15 and 0.25 of total amino acid and between 0.09 and 0.15 of the digestible energy (DE) requirement of the rabbit (Lebas, 1989). In addition the pellets provide all of the B group and K vitamins and a certain amount of volatile fatty acids (VFA).

Digestive efficiency

Rabbits digest fibre poorly because of selective separation and rapid excretion. However, they require generous amounts of fibre to ensure intestinal motility and minimize disease. In crude fibre terms, a diet with less than 150 g kg^{-1} will almost always result in digestive upsets while diets with more than 200 g kg^{-1} crude fibre result in increased incidence of caecal impaction and mucoid enteritis (Fraser, 1991).

Fibre influences the digestibility of the diet and alters the analysis of caecal contents. A diet devoid of fibre results in a coefficient of apparent digestibility of organic matter of 0.90; this declines in a linear fashion to 0.40 when the diet contains 350 g crude fibre kg^{-1}. Increasing the crude fibre of the diet increases the crude fibre of the caecal contents and decreases the protein content (Carabaño *et al.*, 1988). The enzyme profile of the digestive system of the rabbit is similar to that of other non-ruminants and thus digestive efficiency of non-cell wall constituents is comparable.

Raw materials

General considerations

The secret of feeding, as with any pet animal, lies in the provision of a well-balanced diet. The range of feeds upon which the rabbit can survive is wide and varied. This means that the pet-rabbit owner can tailor the feeding from purchased compounds to entirely home-made diets from the garden to meet the budget and individual circumstances. The creation of an entirely home-grown complete diet does, however, demand a knowledge of both the requirements of the rabbit and the nutrient contents/risks and benefits of the plants in the garden. Conversely, the many commercially available compound feeds now offer a simple and convenient alternative package, with only the additional requirement of water required to provide complete and balanced nutrition. Such diets have steadily increased in their presence and popularity since their introduction in the 1950s. Some of the compound diets require the addition of hay in order to supply a complete diet. In general, the recommendation that hay should be supplied on a free-choice basis as a rule of good husbandry of the pet rabbit should be included in the feeding instructions.

Information on feed analyses is found in Aitken and King-Wilson (1962), Wiseman (1987) and Tobin (1996). A typical schedule for the feeding of 'classical' home-made diets based on cereals, bran and forages, fed fresh in the summer and dried in the winter, supplemented with beet and carrots, is provided in Table 17.1.

Raw material groups

The raw materials used to feed the rabbit can be grouped into succulents (greens and roots), roughages, concentrates (grains and proteins) and compounds. The latter is not strictly a raw material group, but general considerations will be covered.

Succulents

If feeding roots and greens, it should be remembered that their nutritional value will vary with season, age at harvest, soil type, weather and storage and that this will inevitably affect the nutrient supply to the rabbit and possibly the overall balance of the diet. There are many green foods suitable for the pet rabbit and a few general principles apply to them all:

- They must be fed fresh.
- If they have to be stored prior to feeding then they should not be left in a heap as they quickly ferment and this can be fatal to young rabbits.
- Introduce gradually. Any new food should be introduced in this way, not only to minimize digestive upsets but also to avoid excessive feed selection and therefore perceived palatability. A sudden change of diet may lead to inappetance and the animal may starve itself rather than eat the new diet. Force-feeding by starvation to achieve diet acceptance is not effective in the rabbit.
- Fresh cut grass is favoured, however lawn mowings should not be used as these heat almost before the rabbit can consume them.

Table 17.1. Succulent foods available to the pet rabbit feeder each month throughout the year (adapted from data in Sandford, 1973).

Jan	Feb	Mar	Apr	May	Jun	Jul	Aug	Sep	Oct	Nov	Dec
Carrots Cabbage and kale	Carrots Kale (thousand head)			Oats, clover and lucerne	Chicory Oats, clover and lucerne	Chicory Oats, clover and lucerne	Chicory Oats, clover and lucerne	Chicory	Cabbage and kale (marrow stem)	Carrots Cabbage and kale	Carrots Cabbage and kale
Swede	Swede						Green maize	Green maize	Swede	Swede	Swede
Mangolds Fodder and sugar beet	Mangolds Fodder and sugar beet	Mangolds Fodder and sugar beet	Mangolds					Kohlrabi	Kohlrabi Fodder and sugar beet	Kohlrabi Fodder and sugar beet	Kohlrabi Fodder and sugar beet
Cereals Hay	Cereals Hay	Cereals Hay	Cereals Hay	Cereals Hay	Cereals Hay	Cereals Hay	Cereals Hay	Cereals Hay	Cereals Hay	Cereals Hay	Cereals Hay

- Wilted greens are acceptable provided they are not yellow in colour or mouldy in any way.
- Plants to avoid include those with bulbous roots, also lobelia and lupins, potato leaves and tomato haulm.
- Rabbits will not in themselves reject poisonous plants. A further peculiarity of the rabbit is that it is unable to vomit and therefore cannot expel unwanted material or poisons if consumed.

Greens. Kale, chicory, kohl and carrot leaves as well as endive and spring greens are very acceptable to rabbits, though the leaves of the cabbage (brassicas) family should only be fed in small quantities to minimize the intake of the glucosinolate goitrogens which impair the uptake of iodine by the thyroid. Leafy vegetables are rich sources of minerals and vitamins.

Hedge or cow parsley (*Anthriscus sylvestris*), dandelion (*Taraxacum officinale*), coltsfoot (*Tussilago farfara*), sow thistle (*Sonchus*), plantain (*Plantago*) and knapweed (*Centaurea* sp. especially *C. nigra*) are similarly useful succulents. However if fed to excess, the dandelion may result in a condition known as 'red-water' which is a kidney complaint.

Lettuce should be fed in strict moderation as it is soporific and similar in action to opium (*Lactucarium*). Wild varieties are somewhat worse than cultivated in their effects. Rhubarb may be fed in small amounts; similarly clover and lucerne, all of which are prone to inducing bloat. Elder keeps flies away but is not recommended as a food.

Some plants are considered to act medicinally, for example shepherds purse (a white-flowered, hairy cornfield weed with a triangle of cordate pods), blackberry, raspberry or strawberry leaves, all of which are considered beneficial in cases of scouring in the rabbit.

The most useful and successful green feed is grass and the products of grass (hay and dried grass pellets). Traditionally young grass has been used to improve growth and general plane of nutrition. Similarly, lucerne (alfalfa), a modest protein source rich in fibre, calcium and the vitamins carotene and E are common materials in diets for rabbits.

Roots. Carrots, swede, turnip, parsnips and sugar beet (though this requires soaking before use) are all useful feeds. They are high in moisture (750–950 g kg^{-1}) and low in protein, starch and fibre and a poor source of vitamins. Most of the dry matter fraction is in the form of sugars. They should never be fed frozen or thawed out, mouldy or rotten.

The potato is somewhat higher in dry matter than the others and is primarily a starch source; cooking as a mash has the advantage of reducing the non-protein nitrogen alkaloid (solanidine) and substantially reduces the risk of gastrointestinal disturbances.

The Jerusalem artichoke provides a good food all year round, with the leaves and stems as a summer food and the roots in the winter. The plant is related to the sunflower and the roots are similar in nutrient content to potatoes. One of the advantages is that the root is impervious to frost and can therefore be left in the ground all year round. However, when harvested it must be washed prior to feeding.

Beetroot is considered to be a good appetite stimulant.

Roughages

Good hay, as explained in the previous section, is vital, with meadow hay before flowering being the best. True clover hay may be too coarse and of doubtful value, although a proportion of clover in the meadow is considered beneficial. Good hay smells sweet with no sense of mustiness and a faint tang of finest tobacco. Hay should be stored prior to use, as new-season hay (i.e. cut within the last 3–4 months) will often result in scouring in rabbits. Lucerne hay and pea haulm are also currently popular and favoured by rabbits. Hay from well-grown stinging nettles, carefully dried, makes a pleasant change and is protein-rich, though rabbits will not eat fresh nettles (Netherway, 1979).

Straw, although traditionally used for bedding, is consumed by rabbits and in the form of alkali-treated (sodium hydroxide) straw pellets is a useful ingredient in mixed feeds and compound diets.

Concentrates

Cereals. Oats are generally considered to be the best grain. The order of preference is oats, barley, maize and then wheat. Little or no information on rye, triticale and others has been found in the literature for pet rabbits. Whole plump oats are preferable to crushed oats as, with the latter, the rabbit will pick out the kernel and leave the husk. Rolling the oat so as to just crack the skin is an alternative preparation for use in flake mixes.

Wheat has a tendency to be pasty and is not generally fed alone. Wheat bran is a very popular and useful raw material and was traditionally fed in quantity to domestic rabbits in the early part of this century (Netherway, 1979). It was the only foodstuff of all rationed materials to be allowed to the rabbit breeder during post Second World War rationing in the UK (Sandford, 1986).

Maize although excellent may be relatively expensive compared with other cereals in certain countries.

Cereals in general are good energy sources with, typically, 80–120 g kg^{-1} protein. The primary deficiencies in the cereals are in lysine, vitamins A and D, and calcium. The phosphorus content, although frequently twice the level of calcium, is of limited availability, as approximately 0.5 is in the phytate form.

Proteins. Both beans and peas are useful, with peas being preferred. Only old season beans should be used as new season crops tend to heat when ground.

Both are rich in protein (200 g kg⁻¹) with a high lysine content which complements the cereals, although they are similar to cereals, being low in calcium and high in phosphorus.

Many legumes have anti-nutritional factors such as trypsin inhibitors, tannins and phytohaemagglutinins. Whilst heat treatment can reduce these factors, the inclusion of such materials in the diet should be limited. Beans of the genus *Phaseolus* (kidney, navy and butter beans) are toxic when fed raw and should not be used.

For commercial feeds both the Mediterranean carob and African locust bean are very palatable and popular favourites. Sunflower seeds are also popular and visually attractive in flake mixes. Oil cakes (peanut, sunflower, sesame and soya) usually in the decorticated form, while never being fed alone, can form an important part of a mixed feed. Linseed, which was traditionally used in a mash as a laxative and to achieve a 'bloom' on the coat, is now recognized as an important source of *n*-3 fatty acids as well as a protein source.

Animal proteins are considered unnecessary for the pet rabbit although they may be important in farmed situations.

Compounds

The design of a pelleted diet allows for a wide range of raw materials to be selected for their nutritional qualities rather than visual appeal. Further, the inclusion of supplements to balance the diet, together with vitamins and trace minerals, in a palatable form is easy. Diets do not necessarily need to be complicated in their raw material composition, as confirmed by Cheeke (1988) who used a combination of only five ingredients, relying heavily on lucerne, oats and a mineral vitamin supplement. There are, however, some general restrictions on the inclusion of certain raw materials that should be taken into account in the design of the complete compound feed (Table 17.2).

Pellet hardness and homogeneity of the mix are critical to diet manufacture. Currently only conventional pelleted diets and flake mixes are offered to rabbits although soft extrusions could become a possible future solution to the production of complete compound feeds which would be acceptable to the rabbit. Flaked mixes can be coated with oil, molasses or glucose syrup to aid palatability and reduce dust. The basic techniques for the processing of such diets are reviewed by Tobin (1996).

Compound feeds which are designed for *ad libitum* feeding should contain sufficient energy (>9.3 MJ DE kg⁻¹) such that the rabbit will be capable of regulating its intake based on energy consumption, whereas very high or very low (<9.0 MJ DE kg⁻¹) energy diets would require a twice daily, restricted feeding regime explained in the feeding instructions. Restricted feeding can also be used to reduce growth rates below the theoretical maximum genetic potential by up to one-third (Schultz *et al.*, 1988) which is desirable in the pet where growth rate is relatively unimportant.

Table 17.2. A guide to typical raw material constraints in compounds for pet rabbits (based on data from Santoma *et al.,* 1989; Maertens, 1992; D. Southey, personal communication).

Ingredient	Suggested maximum inclusion (g kg⁻¹)	Comments
Barley	300	
Wheat	250	All wheat products combined maximum 300
Oats	350	May be lower for processing reasons
Maize	250	
Wheatfeed	300	All wheat products 300
Bran	250	May be lower for processing reasons
Oatfeed	250	May be lower for processing reasons
Dried distillers grains	100	Some sources may contain high copper levels, thus restricting use to 50
Peas	100	50 more common
Field beans	50	Old crop to avoid heating on grinding
Sunflower extractions	1000	Commercially a 300 limit is more practical
Soya (dehulled)	200	Some consider this should have no upper limit
Full-fat soya	250	Typically no more than 125 is used
Citrus pulp	75	
Grass nuts	1000	No limit on grass inclusion
Lucerne	300	Ensiled lucerne may result in decreased food intake
Sugar beet	200	May consider lower value if molasses is included in the process
Molasses	50	Consider contribution from sugar beet
Wheat straw	150	Either plain or sodium hydroxide-treated Inclusions up to 250 have been used successfully
Maize gluten feed	200	Maize gluten 60 to no upper limit
Locust beans	1000	Carob similar inclusions
Grape seed	100	May depress feed intake at higher inclusion rates

Pelleted feeds are best produced to a finished size of 3–4 mm diameter by 10 mm in length. It is important not to exceed 5 mm in diameter as this has been shown to increase the wastage of the diet while a pellet which is much shorter than 10 mm is not well accepted by the rabbit (Lebas, 1987). It is possible to feed compounds as meals providing that the grist is large enough and the product essentially dust-free. However, this imposes certain constraints on the method of water supply which for the pet is more likely to become a problem (Lebas, 1987).

Water

Water is perhaps the most often neglected raw material (or, perhaps more correctly, nutrient). A fresh supply of clean, high-quality drinking water should always be available. The popular belief that rabbits which are fed fresh greens do not need water is not entirely without foundation (Schwabe, 1995) but should not be considered as a safe practice for the domestic pet rabbit. A loss of 10% body water results in death and in the summer rabbits in a hutch in direct sunlight may lose up to 28 g h^{-1} and 3.5 g h^{-1} in shade (Sandford, 1986). In winter it is often wise to provide warm water to encourage intake and prevent digestive upsets.

The various estimates provided in the literature for the water intake of the rabbit have been found to vary widely. Netherway (1979) reported that a large, 4–5 kg, rabbit on dry diet will drink 0.25 l 24 h^{-1} whereas Sandford (1986) suggested a range of 0.42–0.57 l 24 h^{-1} on a dry diet. A similar doe with a litter of seven or eight, at 3 weeks of age, will drink 0.5–0.625 l 24 h^{-1} whilst the litter on their own in a hutch will drink up to 1.5 l 24 h^{-1}.

Nutrient requirements

The suggested values presented below are based on a review of the literature, commercial diet design practice and general experience.

Protein

Whilst there are a number of excellent reviews on the protein requirements of the rabbit (NRC, 1977; Lebas, 1989) scant consideration is given to the effect of varying protein quality in the pet rabbit. The fact that commercial rabbits have successfully been reared on diets based on simple mixtures of plant proteins suggests that both protein quantity and quality are of little importance to the pet. There are, however, suggestions that an excess of dietary protein results in scours (NRC, 1966; Lebas, 1989).

The protein digestibility capacity of the rabbit is maximized early in life, around 4 weeks of age (Lebas *et al.*, 1971). In the adult rabbit, the ability to digest protein appears to be related to dietary protein source, with protein concentrates, cereals and forages having a coefficient of apparent digestibility of 0.80, 0.67 and 0.55, respectively. The fermentation within and contents of the caecum are critical to the supply of protein to the rabbit, as only 0.35 of the total protein digestion occurs in the small intestine (Gidenne, 1988).

Ammonia is the main end-product of nitrogen catabolism as well as the main nitrogenous source for the microbial population in the caecum. Ammonia content is between 6.0 and 8.5 mg 100 ml^{-1} caecal contents with normal diets (Carabaño *et al.*, 1988). If caecal ammonia concentrations are limiting for microbial growth, as in very low protein diets, then urea supplementation has not proved effective, as it is metabolized and absorbed before reaching the caecum, resulting in an increase in urinary nitrogen (Santomá et al., 1989). Consequently as caecotrophy provides approximately 0.18 of the total protein intake of the rabbit, although varying with dietary constituents, excessively low protein diets are to be avoided and some consideration of the protein quality of the rest of the diet is probably important even for the pet rabbit but more in terms of health and tissue repair than for production considerations.

The contribution of the soft faeces increases protein digestibility in the rabbit by a factor between 1.05 and 1.2 (Fraga and de Blas, 1977); this may explain the better utilization of protein from forages in rabbits than in other non-ruminant species.

In terms of specific amino acids, lysine and methionine are considered the most limiting (Santomá et al., 1989). Colin (1978) suggested that cystine can meet the total methionine plus cystine requirement. Given the typical raw material base of the pet rabbit diet, a shortfall in the sulphur amino acids is likely to be the first concern. NRC (1977) and Lebas (1987) suggest that the rabbit requires ten essential amino acids, whilst Cheeke (1987) suggested that glycine should also be considered essential.

It is concluded from the review of the literature, based on the adult maintenance recommendations, that a protein content for the domestic pet rabbit should be in the range 120–160 g kg^{-1}. The suggested minimum amino acid constraints on this should be for lysine 5.0–6.0 g kg^{-1} and methionine plus cystine 5.0–7.0 g kg^{-1}. Arginine should be in the region of 8.0–9.0 g kg^{-1} but this is unlikely to require constraining in typical formulations and some have argued that the rabbit is capable of some arginine synthesis (Santomá et al., 1989).

Fibre

Fibre is critical to the rabbit for health and well-being. Many chemical compounds are included in the broad definition of fibre. Their relative proportions in the diet will affect the way in which the rabbit responds to the diet. The nature of the fibre is important both in terms of its chemical (soluble vs. insoluble fibre) and physical (long, unground fibre vs. short, ground fibre) characteristics. Reviews of the nature and characterization of fibre have been published (Van Soest and McQueen, 1973; Sunvold and Fahey, 1994; Fahey, 1995).

Cell wall constituents (indigestible fibre) are important in the provision of bulk to the diet. This is reflected by the high crude fibre values on complete diets. Too little indigestible dietary fibre (IDF) will increase mortality as a result of a malfunctioning of the digestive system and proliferation of certain undesirable microflora and pathogenic bacteria (Gidenne, 1987), particularly in growing rabbits (Perez et al., 1994). Carabaño et al. (1988) demonstrated that this was the result of an increase in retention time with diets containing low levels of IDF by using a diet with <120 g crude fibre kg^{-1}. Increases in the caecal content and a reduction in the rate of caecal content turnover were observed.

The presence of adequate IDF in the diet of the rabbit maintains intestinal transit times whereas the presence of an appropriate amount of soluble fibre is important for satisfactory fermentation in the caecum. Gidenne *et al.* (1986) demonstrated that the production of a diet based on beet pulp, higher in soluble fibre, increased retention time in the rabbit when compared with an iso-crude fibre diet based on lucerne. Grape residue had the reverse effect (Santoma *et al.,* 1989). Thus, for the complete compound rabbit diet, it is important to consider the relative proportions of digestible fibre and starch to IDF or long fibre when using beet, sunflower hulls, rice bran, olive pulp, grape cake. It is recommended to include a minimum of conventional raw materials such as lucerne, straw, wheat bran and exclude the crude fibre contribution from beet and other digestible sources, or fix a minimum IDF nutrient constraint.

Because fibre influences transit time, and transit time is rapid in the rabbit, the energy supply from crude fibre is less than for other species. In conventional diets this may be 0.05 of the DE. Consequently fibre digestion is relatively poor in the rabbit, although coefficients of apparent digestibility in the region of 0.55–0.7 have been reported which may be explained by the extent of the lignification of the cell wall material in the diet (Lebas, 1989).

The degree of grinding of the fibre fraction of the diet, including the IDF fraction, is an important consideration in the design of the complete compound rabbit diet as it can exert similar physical effects on intestinal motility to those resulting from insufficient IDF (Pairet *et al.,* 1986; Bouyssou *et al.,* 1988). The finer the grinding the greater the digesta retention time and caecal content (Lebas and Laplace, 1977; Candau *et al.,* 1986) with consequent digestive upsets. Screen sizes of 1 mm induce such digestive upsets especially if the diet is marginal in IDF content (Pairet *et al.,* 1986; Auvergne *et al.,* 1987). However, diets ground on 2–7 mm screens did not produce such problems (Lebas and Franck, 1986; Lebas *et al.,* 1986). There is general agreement that screen sizes for complete compound feeds should be 2 mm.

The fine fibre material and soluble components of the diet that enter the caecum are fermented mainly to VFA between 34.5 and 351 μmol g^{-1} DM, predominantly acetic (0.73), butyric (0.17) and propionic (0.08). Energy is the limiting factor for the caecal microbial population. The VFA produced may contribute between 0.12 and 0.40 of the DE of the adult rabbit (Hoover and Heitman, 1972; Marty and Vernay, 1984). Thus, maintaining caecal fermentation and gut health is critical to diet design. To promote a beneficial microbial caecal population and to enhance the production of butyrate with its associated immune stimulation properties the inclusion of a fructose oligosaccharide is now becoming popular in pet rabbit formulations (P. Bruneau, personal communication). This has the added advantage that the shift in VFA proportions also enhances intestinal motility, thereby reducing the importance of a minimum inclusion of IDF.

As with all diet design, a balance of fibre type to other nutrients is important, and is perhaps more so in the case of the small pet rabbit. For example, with the Netherland dwarf it is possible that a high fibre diet with concomitantly low DE of <8.1 MJ kg^{-1} (as fed) may result in a situation where an insufficient intake would be achieved to provide sufficient energy for maintenance although for the larger breeds (>3.5 kg, e.g. Mini-lop and New Zealand white) this problem may not arise.

If the diet is designed so that the crude fibre levels are <100–120 g kg^{-1} then a decrease in intestinal transit time occurs which increases the caecal volume, resulting in a decrease in the carbohydrate supply for energy. This increases in relative terms the proportion of energy from protein sources. An increase in fibre content of the diet, at the other extreme, such that insufficient DE can be consumed also results in a concomitant rise in the proportion of energy derived from protein and has the effect of promoting proteolytic bacteria and ammonia production. This inevitably leads to digestive problems. If, however, the increase in fibre (150–160 g kg^{-1}) is associated with a reduction in the protein/DE ratio then digestive upsets will be avoided.

A similar increase in caecal content and caecal impaction to what occurs with an increased fibre protein to DE ratio can also happen with an excess mineral load in the diet, such as when a clay binder is used to extreme to enhance pellet quality (Grobner *et al.*, 1983).

De Blas *et al.* (1986) suggested that sufficient IDF would be provided with 100–110 g crude fibre kg^{-1} diet. However, to achieve a 90 g IDF kg^{-1} it is suggested that 130–140 g crude fibre kg^{-1} diet is necessary in many formulations. Exceeding a maximum crude fibre of 160 g kg^{-1} can lead to an increase in mortality in young rabbits (Lebas, 1989).

While ranges of crude fibre as wide as 140–250 g kg^{-1} fibre have been found, it is suggested for the complete diet that 130–160 g crude fibre kg^{-1} or 150–200 g acid detergent fibre kg^{-1}, with a total dietary fibre in the region of 300–400 g kg^{-1} and an IDF of 125–150 g kg^{-1} would be appropriate. Alternatively a minimum of IDF should be included by the addition of such traditional materials as hay, straw, wheat bran, etc.

Fat

It is generally considered that plant materials will meet the essential fatty acid (EFA) requirements of the rabbit which can be achieved with a diet containing 25 g fat kg^{-1}. Increasing the proportion of linolenic acid (C18:3) may improve the immune function and status of the rabbit (Kelley *et al.*, 1988). This can be achieved through the addition of stabilized flax. For the pet rabbit, where weight constraint is important, the major role of fat in the diet is for the supply of EFA and it is often in adequate supply within the normal raw materials used in the diet formulations. However, Cheeke (1974) reported a preference by rabbits for diets coated with 50 g corn oil kg^{-1} over one without fat addition. It is suggested therefore that, if added, the fat content should not exceed 50 g kg^{-1}.

Starch and energy

While starch even at high levels (>600 g barley kg^{-1} in the diet) is well digested, levels of available starch in excess of 150 g kg^{-1} at the caecum will lead to undesirable fermentation patterns. The sensitivity to high starch diets is controversial and appears to be much more apparent in the young weanling rabbit than in the adult. As discussed by Lebas (1989), if the relationship between starch and fibre is independent, a minimum fibre and maximum starch constraint is a sensible formulation consideration. There is evidence to suggest that the presentation of the starch in terms of flakes, ground or cereal type is immaterial to the total amount (Santoma *et al.*, 1985; Seroux, 1986).

While the DE content of diets with high levels of digestible fibre will be overestimated and those with added fat underestimated by many of the equations used to predict the DE of diets, for the conventional pet rabbit diet it is suggested that the equation of Maertens *et al.* (1988) provides a reasonable practical predictor of DE:

$$DE = -1801 + 7.10CP + 12.01EE + 5.59NFE,$$

where DE is in kcal kg^{-1}; CP is crude protein; EE is ether extract; NFE is nitrogen-free extract, each in g kg^{-1} as fed.

The literature provides some estimates of the DE of raw materials for rabbits, a selection of which have been summarized together with data for IDF in Table 17.3. However, it is interesting to note the similarity between these values and those generally used for pigs, even for forages.

In conclusion it is considered that the DE of the diet should be in the region of 9–10 MJ DE kg^{-1}. Based on typical values of coefficients of apparent digestibility for concentrates at 0.8, cereals and brans at 0.65–0.70 and forages between 0.45 and 0.65, an estimate of the energy concentration of the diet per unit of digestible protein (DP) can be estimated to be 98 kJ g DP^{-1}. Thus for a typical diet of 102 g DP kg^{-1} an energy intake of 10.0 MJ DE kg^{-1} would be considered suitable.

Vitamins and minerals

Vitamins

As indicated, the B-group vitamin requirements of the rabbit are supplied in sufficient quantity for pet rabbits from soft faeces (NRC, 1977; Harris *et al.*, 1983), though for complete compound feeds it is usual to supplement these to a limited extent together with up to 2 mg vitamin K kg^{-1}.

Table 17.3. A selection of values for the DE and IDF content of some raw materials. Data selected from Wiseman (1987), Santoma et al. (1989) and Maertens (1992).

Raw material	DE (MJ kg⁻¹)	IDF (g kg⁻¹)
Barley	12.50–12.8	38.0–41.0
Wheat	12.8–13.20	10.0
Oats[a]	11.2–11.7, 14.2	18.0–98.0
Maize	13.2–13.7	6.0
Beet pulp	10.9–11.25	0.0–51.0
Soybean meal	13.65–14.1	68.0
Sunflower meal	9.3–10.10	186.0
Wheat straw	2.70–5.4	395.0
Lucerne	7.6–7.9	140.0–260.0
Bran	9.2–10.0	68.0–74.0
Grass	7.2	141.0
Peas	11.7	44.0

[a]The low IDF and high DE values are for naked varieties.

While green foods when fresh may contain large amounts of carotene, the precursor of vitamin A, much of this will be lost upon drying, storage and/or processing. It is thus advisable to supplement complete diets with vitamin A in the range 5000–12,000 IU kg⁻¹ (Roche, 1998). There is also some evidence to indicate that breeding rabbits, at least, benefit from an additional 30 mg carotene kg⁻¹ diet even when vitamin A is in plentiful supply (Tobin, 1996).

Vitamin E is usually provided at 40–70 mg kg⁻¹ as rabbits are sensitive to its deficiency, developing muscular dystrophy, myocardial dysfunction and showing an increased incidence of coccidiosis.

Minerals

The rabbit appears to be unable to regulate the uptake of calcium (Ca) from the intestinal tract and thus the concentration of Ca in the complete diet must be close to the requirement. However, it is important to consider the extent to which low-Ca fresh foods may be fed as part of the diet, leading to a reduction in plasma Ca and a possible excess in phosphorus (P). Whilst the rabbit will tolerate wide Ca:P ratios, inverse ratios will soon lead to problems. Such imbalances of Ca to P have been implicated in dental problems in pet rabbits (Harcourt-Brown, 1996). However, Bucher (1994) indicated that the nature of the feed ingredients also exerts effects on incisor growth and attrition. The provision of too much Ca in the diet may lead to an increased incidence of urolithiasis. Deposition of Ca salts in the urine occurs due to the alkalinity of the urine, which is often enhanced by the high levels of potassium from grass or lucerne consumption contributing to the base excess. It is thus suggested that dietary Ca should be in the region of 5–10 g kg⁻¹ and to minimize the risk of the deposition of excess Ca in soft tissues the vitamin D content of the diet should be in the region of 800–1200 IU kg⁻¹. While the rabbit may be tolerant of a wide range of Ca:P ratio, it is however desirable to maintain the ratio between 1:1 and 2:1 in favour of Ca.

There is no reason to assume that other minerals and trace elements are different from those indicated in NRC (1977). However, it is suggested that 50 mg supplementary zinc kg⁻¹ should be considered to overcome the potential low bioavailability of the raw material zinc in the diet when large amounts of phytate are present. There may be a benefit from the addition of supplementary zinc in the form of a chelate or polysaccharide complex, as there are indications that these sources are not involved in such interactions (Lowe and Wiseman, 1997).

Suggested diet specifications

On the basis of the foregoing discussion and the author's experience the diet specification guide in Table 17.4 is suggested as being suitable for the pet rabbit.

Table 17.4. Suggested nutrient constraints for pet rabbit diets.

Nutrient	Minimum	Maximum	Nutrient	Typical range
Protein (g kg⁻¹)	120.0	160.0		
Crude fibre (g kg⁻¹)	130.0	200.0	Vitamin A (IU kg⁻¹)ᶜ	10,000–18,000
IDF (g kg⁻¹)	125.0	na	Vitamin D (IU kg⁻¹)	800–1200
Starch (g kg⁻¹)ᵃ	0.0	135.0	Vitamin E (mg kg⁻¹)	40–70
Fat (g kg⁻¹)	20.0	50.0	Vitamin K (mg kg⁻¹)	1–2
DE (MJ kg⁻¹)	9.0	10.0	Vitamin B₁ (mg kg⁻¹)	1–10
Lysine (g kg⁻¹)	5.0	na	Vitamin B₂ (mg kg⁻¹)	3–10
Methionine + cystine (g kg⁻¹)	5.0	na	Vitamin B₆ (mg kg⁻¹)	2–15
Ca (g kg⁻¹)ᵇ	8.0	10.0	Vitamin B₁₂ (mg kg⁻¹)	0.01–0.02
P (g kg⁻¹)ᵇ	5.0	8.0	Folic acid (mg kg⁻¹)	0.2–1.0
Mg (g kg⁻¹)	3.0	3.0	Pantothenic acid (mg kg⁻¹)	3.0–20.0
Zn (mg kg⁻¹)	50.0	100.0	Niacin (mg kg⁻¹)	30–60
K (g kg⁻¹)	6.0	7.0	Biotin (mg kg⁻¹)	0.05–0.20
NaCl (g kg⁻¹)	5.0	10.0	Choline (mg kg⁻¹)	300–1500

ᵃThe starch maximum only applies to diets for the very young rabbit. For the adult pet no constraint need be considered.
ᵇThe levels allow for the fact that the pet rabbit may be used for breeding. Adult maintenance can be satisfied with levels in the region of 4 g P kg⁻¹ and 6 g Ca kg⁻¹. It is also noted that >10 g kg⁻¹ may be unpalatable to the rabbit (NRC, 1977).
ᶜIt may be necessary to include higher levels of certain vitamins to allow for losses during manufacture. These will be specific to the raw materials and production processes used and must be taken into account when designing the supplementary additions.
na, not applicable.

Nutritional ailments

It is important when faced with the pet animal that is unwell to differentiate between symptoms and disease. Many so-called diseases, for example scouring, are not in themselves a disease but are a symptom(s) of a disease situation of different causal agent(s). As far as the complete or complementary diet manufacture is concerned, nutritional deficiencies and metabolic disorders may predispose to these conditions. It is worth bearing in mind that there is nothing so potent for producing ill health as improperly constituted food.

In many cases it is not the design of the food that is at fault but the way in which the diet is presented, fed or selected by the pet rabbit. Therefore the feeding instructions and guide as to how best to utilize food in the day-to-day feeding programme of the pet rabbit are as important in the design and manufacture of pet rabbit diets as are the nutrient and raw material specifications. Death from malnutrition is rare, but excess bulk in the diet of young rabbits can result in starvation. Such circumstances may arise when insufficient complete diet relative to hay is provided. Similarly selective feeding of the flake-mix type diets has been reported to lead to deficiencies and/or excess of certain nutrients.

It is beyond the scope of this chapter to discuss nutritional ailments in greater detail.

Conclusions

Diets for the pet rabbit can be produced from a wide selection of materials ranging from those grown in the garden or countryside to complete commercial diets. The creation of entirely home-produced diets must be treated with caution unless there is extensive knowledge of both the requirement of the rabbit and the nutrient content of the ingredients in place. The science of nutrition and diet formulation requires education and considerable experience and a haphazard approach is not usually successful. Furthermore, even the best designed and manufactured diets are only successful in the provision of appropriate nutrition to the pet rabbit if they are fed according to the instructions provided with them. The nutritionist is duty-bound to provide the best guidance possible in this area and the pet rabbit owner is wise to follow this advice.

References

Aitken, F.C. and King-Wilson, W. (1962) *Rabbit Feeding for Meat and Fur*, 2nd Edn. CAB International, Wallingford, UK.

American Veterinary Medical Association (1997) *US Pet Ownership and Demographic Source Book.* Schaumburg, Illinois.

Anon. (1998) *Hey! Look at Me, I'm a House Rabbit.* The British House Rabbit Association, Newcastle upon Tyne, UK.

Auvergne, A., Bouyssou, T., Pairet, M., Bouillier-Oudot, M., Ruckebusch, Y. and Candau, M. (1987) Nature de l'aliment, finesse de mouture et donnees anatomo-functionnelles de tube digestif proximal du lapin. *Reproduction, Nutrition and Development* 27, 755–768.

Bjornhag, G. (1987) *Deutsche Tierarztliche Wochenschriift* 94, 33–36.

Bouyssou, T., Candau, M. and Ruckebush, Y. (1988) *Reproduction, Nutrition and Development* 28, 181–182.

Bucher, L. (1994) Diet related influences on growth and attrition of incisors in dwarf rabbits. Doctoral Thesis, Freie Universität, Berlin.

Candau, M., Auvergne, A., Comes, F. and Bouillier-Oudet, M. (1986) *Annales de Zootechnie* 35, 373–386.

Carabaño, R., Fraga, M.J., Santoma, G. and de Blas, J.C. (1988) Effect of diet on composition of cecal contents and on excretion and composition of soft and hard faeces of rabbits. *Journal of Animal Science* 66, 901–910.

Cheeke, P.R. (1974) Feed performances of adult male dutch rabbits. *Laboratory Animal Science* 24, 601–604.

Cheeke, P.R. (1987) *Rabbit Feeding and Nutrition.* Academic Press, Orlando, Florida.

Cheeke, P.R. (1988) Rabbit nutrition: a quiet growth area with great potential. In: Lyons, T.P. (ed.) *Biotechnology in the Feed Industry, Proceedings of Alltech's Fourth Annual Symposium.* Alltech, Lexington, Kentucky, pp. 249–260.

Colin, M. (1978) Effets sur la croissance du lapin de la supplementation en l'lysine et en DL-methionine de régimes vegetaux simplifies. *Annales de Zootechnie* 24, 465.

Fahey, G.C. Jr (1995) Practical considerations in feeding dietary fibres to companion animals. In: Phillips, T. (ed.) *Pet Food Forum 95.* Watt Publishing, Mount Morris, Illinois.

Fraga, M.J. and de Blas, J.C. (1977) *Annales del Instituto Nacional de Investigaciones Agarias* 8, 43–47.

Fraser, C.M. (1991) *Merck Veterinary Manual*, 7th Edn. Merck and Co., Rahway, New Jersey.

Gidenne, T. (1987) Influence de la teneur en lilnines des auments sur la composition des digesta et la production de caecotropies chez le lapereau. *Annales de Zootechnie* 36, 85–90.

Gidenne, T. (1988) In: *Proceedings of the 4th World Rabbit Science Association Congress, Budapest*, Vol. 2. Sandor Holdas, Hercegalom, pp. 345–352.

Gidenne, T., Poncet, C. and Gómez, L (1986) In: *4èmes Journées de la Recherche Cunicole.* Communication No. 4, Paris.

Grobner, M.A., Robinson, K.L., Cheeke, P.R. and Patton, N.M. (1985) Utilisation of low and high energy diets by dwarf (Netherland Dwarf), intermediate (Mini Lop) and Giant (Flemish Giant) breeds of rabbits. *Journal of Applied Rabbit Research* 8, 12–18.

Haffar, A. (1996) Treating the pet rabbit. *Point Vétérinaire* 28, 347–353.

Harcourt-Brown, F.M. (1996) Calcium deficiency, diet and dental disease in pet rabbits. *Veterinary Records* 139, 567–571.

Harris, D.J., Cheeke, P.R. and Patton, N.M. (1983) *Journal of Applied Rabbit Research* 6, 15–1 7.

Hartmann, K., Fischer, S. and Kraft, W. (1994) Small pet animals as patients in veterinary practice. 1. Descent, physiology, husbandry, feeding. *Tierarztliche Praxis* 22, 585–591.

Hoover, W.H. and Heitmann, R.N. (1972) Effects of dietary fibre levels on weight gain, cecal volume and volatile fatty acid production in rabbits. *Journal of Nutrition* 102, 375–380.

Kelley, D.S., Welser, G.J., Serrato, C.M., Schmidt, P.C. and Branch, L.B. (1988) Effects of type of dietary fat on indices of immune status of rabbits. *Journal of Nutrition* 118, 1376–1384.

Lebas, F. (1980) Les recherches sur l'alimentation du lapin: évolution au cours des 20 dernières années et perspectives d'avenir. In: *Proceedings of the 2nd World Rabbit Congress*. Asociacíon Española de Cunicultura, Barcelona.

Lebas, F. (1987) Nutrition of rabbits. In: Wiseman, J. (ed.) *Feeding of Non-Ruminant Livestock*. Butterworths, London, pp. 63–69.

Lebas, F. (1989) Nutrient requirements of various categories of rabbits. In: Piva, G. and Wiseman, J. (eds) *Proceedings of the First International Feed Production Conference*. Piacenza, pp. 297–332.

Lebas, F. and Frank, T. (1986) Incidence du broyage sur la digestibilité de quatre aliments chez le lapin. *Reproduction, Nutrition and Development* 26, 235–236.

Lebas, F. and Laplace, J.P. (1977) Le transit digestif chez le lapin. VI. Influence de la granulation des aliments. *Annales de Zootechnie* 26, 83–91.

Lebas, F., Corring, T. and Courtot, D. (1971) Equipment enzymatique du pancreas exocrine citez le lapin mise en place et evolution de la naissance au sevrage: relation avec la composition du régime alimentaire. *Annales de Biologie Animale, Biochimie, Biophysique* 11, 399–413.

Lebas, F., Maitre, I., Seroux, M. and Frank, T. (1986) Influence du broyage des matieres premieres avant l'agglomeration de 2 aliments pour lapins, differant par leux taux de constituants membranaires. In: *4èmes Journées de la Recherche Cunicole en France*. Communication No. 9, Paris.

Lowe, J.A. and Wiseman, J. (1997) The effect of the source of dietary supplemental zinc on tissue copper concentrations in the rat. In: *Proceedings of the British Society of Animal Science*, p. 67.

Maertens, L. (1992) Rabbit nutrition and feeding: a review of some recent developments. *Journal of Applied Rabbit Research* 15, 889–913.

Malley, A.D. (1994) The pet rabbit in companion animal practice. 1. A clinician's approach to the pet rabbit. *Irish Veterinary Journal* 47, 9–1 5.

Malley, A.D. (1995) The pet rabbit in companion animal practice. 2. General clinical examination. *Irish Veterinary Journal* 48, 307–311.

Malley, A.D. (1996) The pet rabbit in companion animal practice. 4. Hematological and biochemical reference values. *Irish Veterinary Journal* 49, 354.

Marty, J. and Vernay, M. (1984) Absorption and metabolism of the volatile fatty acids in the hind gut of the rabbit. *British Journal of Nutrition* 51, 265–277.

Netherway, M.E.P. (1979) How to feed rabbits. In: *Home Rabbit Keeping*. E.P. Publishing, Wakefield, UK, pp. 23–32.

NRC (1966) *Nutrient Requirements of Rabbits*, 1st revised Edn. National Academy of Sciences, Washington, DC.

NRC (1977) *Nutrient Requirements of Rabbits*, 2nd revised Edn. National Academy of Sciences, Washington, DC.

Pairet, M., Bouyssou, T., Auvergne, A., Candau, M. and Ruckebusch, Y. (1986) Stimulation physicochimique d'origine alimentaire et motricite digestive chez le lapin. *Reproduction, Nutrition and Development* 26, 85–95.

Perez, J.M., Gidenne, T., Lebas, F., Caudron, I., Arveux, P., Bourdillon, A., Duperray, J. and Messager, B. (1994) Dietary lignin in growing rabbits. 2. Consequences on growth-performance and mortality. *Annales de Zootechnie* 43, 323–332.

PFMA (1998) *Profile of the UK Pet Food Manufacturers Association.* PFMA, London.

Prud'hon, M., Cherubin, M., Goussopoulos, J. and Charles, Y. (1975) Evolution au cours de la croissance des caracteristiques de la consommation d'aliment solide et liquide du lapin domestique nourri *ad libitum. Annales Zootechnie* 24, 289–298.

Roche (1998) *Roche Vitamin Supplementation Guidelines for Domestic Animals.* Roche Headquarters, Paramus, New Jersey.

Sandford, J.C. (1973) *The Domestic Rabbit,* 3rd Edn. Crosby, Lockwood and Staples, London, pp. 84–118.

Sandford, J.C. (1986) *The Domestic Rabbit,* 4th Edn. Collins, London, pp. 1–3, 51–61, 94–126.

Santoma, G., Carabaño, R., de Blas, J.C. and Fraga, M.J. (1985) *Annales del Instituto Nacional de Investigaciones Agrarias* 22, 75–82.

Santoma, G., de Blas, J.C., Carabaño, R. and Fraga, M.J. (1989) Nutrition of rabbits. In: Haresign, W. and Cole, D.J.A. (eds) *Recent Advances in Animal Nutrition.* Butterworths, London, pp. 109–138.

Schlolaut, W. (1982) *The Nutrition of the Rabbit.* Roche Information Service Animal Nutrition Department, Basel.

Schultz, W.H., Smith, W.C. and Mougham, P.J. (1988) Amino acid requirements of the growing meat rabbit. *Animal Production* 47, 303–311.

Schwabe, K. (1995) Feed and water intake of different species of pets (rabbits, guineapigs, chinchilla, hamsters) offered different water supply (drinkers vs. succulent feeds). Thesis, Tierarztliche Hochschule Hannover, Hannover.

Seroux, M. (1986) In: *4èmes Journées de la Recherche Cunicole.* Communication No. 10, Paris.

Sunvold, G.D. and Fahey, G.C., Jr (1994) The role of dietary fiber in the nutrition of dogs and cats. In: Phillips, T. (ed.) *Pet Food Forum 94.* Watt Publishing, Mount Morris, Illinois.

Tobin, G. (1996) Small pets – food types, nutrient requirements and nutritional disorders. In: Kelly, N. and Wills, J. (eds) *Manual of Companion Animal Nutrition and Feeding.* British Small Animal Veterinary Association, Cheltenham, UK, pp. 208–225.

Van Soest, P.J. (1982) *Nutritional Ecology of the Ruminant.* Q&B Books, Corvallis, Oregon.

Van Soest, P.J. and McQueen, R.W. (1973) The chemistry and estimation of fibre. *Proceedings of the Nutrition Society* 32, 123.

Wiseman, J. (1987) *Feeding of Non Ruminant Livestock.* Butterworths, London.

Zeuner, F.E. (1963) The small rodents. In: *A History of Domesticated Animals.* Hutchinson & Co., London, pp. 409–415.

Index

Note: page numbers in *italics* refer to figures and tables.